Edward Lee Greene

Manual of the Botany of the Region of San Francisco Bay

Edward Lee Greene

Manual of the Botany of the Region of San Francisco Bay

ISBN/EAN: 9783744662260

Printed in Europe, USA, Canada, Australia, Japan

Cover: Foto ©Andreas Hilbeck / pixelio.de

More available books at **www.hansebooks.com**

MANUAL

OF

THE BOTANY

OF THE

REGION OF SAN FRANCISCO BAY,

BEING A

Systematic Arrangement of the Higher Plants Growing
Spontaneously in the Counties of Marin, Sonoma,
Napa, Solano, Contra Costa, Alameda, Santa
Clara, San Mateo and San Francisco,
in the State of California.

BY

EDWARD LEE GREENE,

Professor of Botany in the University of California.

[Issued 2 Feb., 1894.]

SAN FRANCISCO:

CUBERY & COMPANY, BOOK AND JOB PRINTERS, 587 MISSION STREET,

1894.

PREFACE.

This volume has been prepared with reference to the needs of those who, whether as studying in our high schools, academies and colleges, or as private students and amateurs, desire to make some beginnings in the systematic botany of middle western California. In order that the volume should be small, it was necessary that the scope of it should be limited. So exceedingly varied is the flora of even limited areas in our State, that not even all the flowering plants of the counties touching San Francisco Bay could be briefly but sufficiently classified and described within the limit of three hundred and fifty octavo pages. We have therefore been obliged to conclude this Flora for beginners at the end of the ninetieth of our natural orders of flowering plants. These ninety embrace, however, all in our district which the novice in plant determination would be likely to take up, and a number of genera and species much greater than the beginner may master in one season's study, or in three. Thus no complaint will arise that the sedges and grasses, the pondweeds and the ferns have been omitted from this brief and more or less tentative treatise.

The old practices of preparing a digest of the genera under the larger natural orders, and of italicizing some of the salient marks of species as described, have been continued here; though with misgiving as to their real usefulness in general; it being too well known, by all teachers of the subject, that the pupil will rely on the "key" and on the italicized words unduly, and to his own misleading also, in some instances, instead of attending to all the terms of the full diagnosis. Still, the digests and keys are real aids to many a serious beginner in the work of plant classification.

In justice to the critical labor that has been bestowed on the plants of even this small area, by the author, it must be said that this Manual is one which the critical botanist will find indispensable, at least, until some worthier treatise shall take its place. On ground so new as this of the San Francisco Bay Region it still remains, and for years to come it will remain, that new convictions will be formed as to the limits of species and of genera; that every book or pamphlet of this kind the subject matter of which has been wrought out under the eye of a competent student, will present new specific and perhaps new generic propositions. This Manual will be found to contain not a few such.

Moreover: the present author, now longer engaged than any other American botanist in the very serious consideration of certain questions in botanical nomenclature, is more and more convinced that uniformity will never come but by closer conformity to the law of priority. He has therefore introduced into this volume not a few names of genera that are much older than those current in familiar books. There is therefore much that is new for the bibliographer and the nomenclator within these pages. This feature will not in the least affect the useful-ness of the Manual as a book for beginners; for to these it is as easy to call the California Horse-chestnut HIPPOCASTANUM CALIFORNICUM as *Æsculus Californica*. The inconvenience will be realized only by the experienced botanist, who is habituated to the use of other names. To these, however, the way is clear. No botanist will be obliged to adopt the nomenclature of the Manual of Bay-Region Botany. The author is convinced, however, that the day is coming, and at a fair rate of speed, when the employing generic names which Linnæus substituted for older ones of Micheli, Tournefort, Lobel or of Gesner, instead of such as have right of real priority, will no longer be thought of by those who name priority as a leading principle in plant nomenclature.

As to the completeness of the volume as an authentic list of the higher plants growing spontaneously within the limits specified in the title, it may be said that we included all the species which, at the time of writing, were known to us as occurring within this range. But vast areas within these counties have never yet been explored at all botanically; and the actual number of plant forms belonging to this aggregate of counties must be considerably greater than what these pages show. We invite all students, and others who may use the book as a field companion in this district, to make record of all additions to this list, and kindly report them to the author, that future editions of the Manual may be rendered more complete.

EDWARD L. GREENE.

UNIVERSITY OF CALIFORNIA, 24 *January, 1894.*

ANALYTIC KEY TO NATURAL ORDERS.

SUBCLASS I. EXOGENS.

Leaves netted-veined. Parts of the flower seldom in threes or sixes.
Cotyledons two. Wood of woody species showing concentric circles.

DIVISION I. CHORIPETALÆ HYPOGYNÆ.

Petals when present distinct, at least at base. Stamens hypogynous.

A. *Stamens more than 10, and more than twice as many as
the sepals or petals.*

	PAGE
Pistils distinct, simple, becoming achenes or follicles (in one a berry)..........RANUNCULACEÆ	2
Pistils compound, *i. e.*, the cells, placentæ or stigmas more than one.	
Petals outnumbering the sepals,	
Twice as many (4—6); sepals caducous...PAPAVERACEÆ	8
" " " (8—16); sepals persistent................Lewisia in PORTULACEÆ	58
Sepals and petals concave, intergrading, persistent. Plants aquatic........NYMPHÆEÆ	8
Sepals (or calyx-lobes) and petals 5 each;	
Calyx valvate in bud; stamens monadelphous........................MALVACEÆ	63
Calyx imbricate; stamens in indistinct bundles...........................HYPERICEÆ	62
Sepals very unequal; stamens neither united nor fascicled..............CISTOIDEÆ	28
Petals lacerate or palmatifid; ovary open before maturity.....................RESEDACEÆ	27

B. *Stamens 10 or fewer, not more than twice as many as the
petals or sepals.*

✱ *Pistils more than one, distinct.*

Sepals, petals and pistils equal in number; leaves fleshy.................CRASSULACEÆ	126
Pistils outnumbering petals or stamens ....RANUNCULACEÆ	2

DIVISION II. CHORIPETALÆ PERIGYNÆ.

Calyx more or less distinctly synsepalous, but petals distinct, at least
at base. Stamens perigynous.

A. *Ovary more or less completely superior.*

* *Stamens many, usually 20 or more.*

* * *Stamens 5 or 10 only.*

B. *Ovary mainly inferior.*

* *Stamens many, usually 20 or more.*

* * *Stamens few and definite.*

Division III. SYMPETALÆ PERIGYNÆ.

Sepals united below into a tube adherent to the ovary. United corolla, with adherent stamens, inserted on the calyx near its summit.

* *Stamens 1—10, distinct.*

DIVISION IV. SYMPETALÆ HYPOGYNÆ.

Corolla sympetalous, at least at base, the stamens attached to its tube or base, the whole inserted around the base of the (superior) ovary.

A. *Leaves opposite or whorled; corolla regular.*

* *Fruit indehiscent.*

Stamens alternate with the corolla-lobes;

B. *Leaves alternate; corollas regular.*

* *Stamens opposite the corolla-lobes; ovary 1-celled.*

* * *Stamens alternate with the corolla-lobes.*

C. *Corollas more or less distinctly bilabiate.*

* *Ovary not lobed; fruits capsular.*

* * *Ovary 4-lobed, the lobes maturing as nutlets.*

Division V. APETALÆ AMENTIFERÆ.

Flowers mostly apetalous and unisexual; the staminate usually in aments or catkins, as are also sometimes the pistillate.

* *Herbaceous plants; small green flowers in ament-like racemes.*

* * *Woody plants, mostly trees or large shrubs.*

Subclass II. ENDOGENS.

Leaves parallel-veined; flowers 3-merous; cotyledon 1 only.

MANUAL

BAY=REGION BOTANY.

SERIES I.

PHANEROGAMOUS or FLOWERING PLANTS.

Vegetables having stamens and pistils, and producing seeds, of which
the most essential part is a distinct embryo.

CLASS I. ANGIOSPERMÆ.

Seeds enclosed within a pericarp.   Cotyledons two or one.

SUBCLASS I. · DICOTYLEDONOUS or EXOGENOUS PLANTS.

Embryo with two cotyledons.   Leaves netted-veined.
Flowers having their parts usually in fives, fours or twos.

DIVISION I.   CHORIPETALÆ HYPOGYNÆ.

Corolla (often wanting) of petals which are distinct, at least at base.
Stamens hypogynous.

ORDER I. RANUNCULACEÆ.

Herbs *(Clematis* shrubby) with colorless juice. Leaves alternate (opposite in *Clematis;* the cauline whorled in *Anemone*), usually lobed or ternately divided. Sepals 3–6, deciduous. Stamens ∞, hypogynous; anthers adnate, opening lengthwise, by slits. Pistils usually ∞, distinct and simple, becoming achenes or follicles (in *Actæa* 1, becoming a berry).

Petals wanting;
      Sepals 4; achenes plumose-tailed,...........................CLEMATIS   1
         "   5 or more; achenes without tails,.....................ANEMONE   2
         "   green; flowers unisexual: achenes ribbed,............THALICTRUM   7
Flowers complete;
      Pistil 1; fruit berry-like,...............................ACTÆA   8
      Stamens few; achenes in a slender spike.................MYOSURUS   3
         "   many; achenes in heads............................RANUNCULUS   4
      Flowers irregular, one sepal spur-like.................DELPHINIUM   5
         "   regular, all 5 petals spur-like......................AQUILEGIA   6

1. CLEMATIS, *Diosc.* Half-woody, climbing by tortuous petioles of compound opposite leaves, in the axils of which are solitary or clustered flowers (ours white). Sepals 4, petaloid, valvate in bud. Pistils ∞; styles persistent, becoming feathery appendages of the large compressed and capitate-clustered achenes.

1. **C. lasiantha,** Nutt. Silky-pubescent; leaflets 3, ovate, coarsely toothed or 3-lobed or -parted: fl. large, *only one on each* bibracteate *peduncle;* sepals ¾ in. long.—Trailing over rocks and shrubs among the hills. April.

2. **C. ligusticifolia,** Nutt. Glabrous or nearly so, or the leaves silky-tomentose beneath: leaflets broadly ovate to lanceolate, usually 3-lobed: *fl. panicled in the axils;* sepals ½ in. long.—Often climbing 30 ft. upon small trees, in Alameda and Marin Counties. July.

2. ANEMONE, *Diosc.* Perennial herbs with radical leaves lobed or divided, and a cauline involucral whorl of 3. Flowers on erect peduncles. Sepals 5 or more, petaloid, imbricate. Achenes merely pointed.

1. **A. Grayi,** Behr. & Kell. Very slender, 6—14 in. high, from a horizontal rootstock: radical leaf remote from the stem, trifid, the segments serrate; the involucral not far below the flower, petiolate, 3-foliolate; leaflets all coarsely serrate, the lateral ones 2-lobed: sepals 5 or 6, oval, usually bluish outside: achenes 12—20, oblong, 2 lines long, pubescent, the fruiting pedicel coiled into a ring.—Coast Range, in moist shades. March—May.

3. MYOSURUS, *Lobel,* (MOUSETAIL). Small stemless glabrous annual, with narrow entire leaves, and slender 1-flowered scapes. Sepals

5, spurred at base. Petals 5; blade oblong, with a pit or gland at base; claw filiform. Stam. 5—15. Pistils ∞, crowded on a long slender receptacle, becoming thin-walled achenes.

1. **M. minimus, L.**—About San Francisco, and in the hills east of the Bay, commonly in very much reduced states; the spikes often less than an inch long, and very slender. March—May.

2. **M. alopecuroides,** Greene. Stouter and low: achenes with prominent spreading beak, in short thick spike.—Low plains near Antioch.

4. **RANUNCULUS,** *Pliny* (BUTTERCUP). Flowers solitary or scattered, regular, yellow or white. Sepals 5, commonly reflexed. Petals 5 or 10, with nectariferous scale or pit near the base within. The many pistils becoming beaked achenes disposed in rounded heads.

  * *Leaves undivided; achenes not strongly flattened.*

1. **R. Flammula,** L. var. **intermedius,** Hook. Stems slender, even to the filiform, rooting at the lower joints: leaves lanceolate, entire: fl. 2—5 lines broad: achenes few, with a very *stout straight but short beak.*— Small herb, found along the margins of lakes and pools.

2. **R. pusillus,** Poir.? Annual, slender, 2—10 in. high, glabrous except the villous-ciliate sheathing stipules: leaves round-ovate to lanceolate and linear, the radical ones coarsely toothed, ¼—½ in. long: stem simple and scapiform or with a few branches: fl. minute; sepals subscarious, not reflexed; achenes many in a small globose head, *delicately tuberculate, neither margined nor beaked.*—Moist places in Marin, Sonoma and Napa Counties.

  * * *Leaves ternately lobed, cleft or divided; fl. yellow; achenes flattened*

3. R. REPENS, L. Pubescent; stems rooting at the lower joints: leaves ternately parted, often subdivided: *sepals spreading:* petals 5; achenes 1¼ lines long, rather sharply margined, the beak nearly straight; 1½ lines long.—Frequent in lawns; scarcely naturalized.

4. **R. maximus,** Greene. Pilose or hirsute, the stout stems 2—5 ft. long, reclining but not rooting: leaves broad, ternate; *leaflets laciniately lobed:* petals 5- 8, oblong-obovate, *obtuse,* 7—10 lines long: head of achenes roundish or broadly ovoid; beak long, straight or slightly incurved.— In swampy places; not common. April—June.

5. **R. Bloomeri,** Wats. A foot or two in height, stout, usually glabrous, sometimes pilose: radical leaves sometimes undivided and round-cordate with coarse teeth or lobes; the later ones 3—9-foliate, the *leaflets with somewhat rounded teeth*: petals 5, *retuse,* ¾ in. long: achenes long-beaked, forming a subglobose head.—Common in wet ground. Feb.-May.

6. **R. Californicus,** Benth.   Erect or decumbent, 1—2½ ft. high, freely branching and many-flowered: *petals 10—15,* obovate-oblong, 4—5 lines long: achenes much flattened, 1½ lines long, the beak short, recurved; head dense, globose.   Var. **lætus,** Greene.   Strictly erect, stoutish and fistulous, hirsute, glaucescent below; herbage light green; leaflets much dissected.   Var. **canescens,** Greene.   Low and stout, the basal parts canescently long-villous: leaves less dissected: fl. large (fully 1 in. broad).   Var. **cuneatus,** Greene.   Slender, decumbent, the growing parts silky-pubescent: leaves cleft into 3 cuneate lobes or segments, these incisely toothed: fl. small: achenes very many, in a dense round-ovoid head.—Type abundant on open hills.   Var. *lætus,* in lowlands not far from salt marshes.   Var. *canescens,* in the Oakland Hills southward, towards Livermore.   Var. *cuneatus,* on the San Francisco peninsula southward.   Feb.—June.

7. **R. occidentalis,** Nutt., var. **Eiseni,** Gray.   Distinguished from the last by more slender habit, broad leaf-segments, small *petals, always 5,* and broader thinner achenes.   Var. **Rattani,** Gray.   Achenes smaller, hairy and papillose.—Higher hill country both north and south of the Bay, on dry open or sparsely wooded slopes.   April, May.

8. **R. canus,** Benth.   Stout and tall, with the habit of *R. Californicus,* but herbage more or less *silky-canescent;* leaves cut into narrow acute segments; *petals 5 only,* round-obovate.—Hills near Antioch.   April, May.

9. **R. hebecarpus,** Hook. & Arn.   Slender, erect, leafy, 5—15 in. high, pilose-pubescent: radical leaves rounded or reniform, deeply lobed or cleft, the segments 3-lobed: fl. minute, on filiform pedicels: *achenes few,* rounded, flattened, *papillose and short-hairy,* the beak very short, recurved.—Moist shades among the lower hills; not common.   April, May·

10. R. MURICATUS, L.   Stout and fistulous, yellow-green, glabrous: leaves round-reniform, slightly lobed: fl. small: achenes very large, with stout ensiform beak and coarsely *muricate-prickly sides.*—Rather common in wet soils, especially on the outskirts of San Francisco; naturalized from Europe; flowering throughout the year.

* * * *Aquatics; leaves mostly capillaceous-multifid and submersed; petals white, with naked nectariferous pit; achenes little flattened, transversely rugose.*

11. **R. aquatilis,** Dodoens.   Perennial, the emersed and floating leaves, when present, roundish, 3-lobed: *styles subulate; achenes slightly rugose,* usually hispidulous, 12—20 in a compact globose head.—Common in ponds; sometimes terrestrial on muddy shores.   May—Dec.

12. **R. Lobbii,** Gray.   Annual; floating leaves always present, deeply 3-lobed, the middle lobe usually elliptical and entire, the laterals larger,

oblong, obcordate at summit: *style filiform: achenes* few (4 –6), *rather sharply rugose.*—Common as the last, but of short duration. April, May.

5. **DELPHINIUM,** *Diosc.*, (LARKSPUR). Erect herbs. Leaves palmately cleft or divided. Flowers irregular, in terminal racemes. Sepals 5, colored and petaloid, the upper one produced backward into a long hollow spur, the others plane. Petals 2—4, two of them developed backwards and intruded into the spurred sepal. Pistils mostly 3, becoming ∞ -seeded follicles.

　　* *Flowers blue, varying to pink or flesh color (never scarlet).*

　　+ *Root a cluster thickish half-woody fibres.*

1. **D. Californicum,** Hook. & Arn. Stout, strict, 3—5 ft. high, pubescent: leaves ample, deeply 5-cleft, the segments variously lobed: raceme dense, 1—1½ ft. long: fl. smallish, dull greenish or whitish or with a purplish tinge, scarcely expanded, *externally rather densely velvety-pubescent:* follicles oblong, turgid, erect.—Plentiful on Mt. Diablo; also near Belmont; otherwise not common. April—June.

2. **D. hesperium,** Gray. Stoutish, 1½—2½ ft. high: canescent with a short and close pubescence: leaves much dissected, the lobes linear obtuse: raceme dense, elongated: fl. well expanded, deep blue (occasionally pink); spur stout, straight, about as long as the sepals: follicles erect, pubescent.— Common on dry slopes; late in flowering. June, July.

3. **D. variegatum,** Torr. & Gray, var. **apiculatum,** Greene. A foot high or more, coarsely and retrorsely pubescent: leaves few, 3-parted or -cleft into broad linear lobes: raceme *short, dense, cylindrical:* fl. large, dark blue: follicles pubescent. — Plains near Suisun and Antioch. March—May.

　　+ + *Roots more fleshy, often grumose or tuberiform.*

4. **D. Menziesii,** DC. Root a cluster of short roundish or compressed connected tubers: stem 1 ft. high or less, leafy below, but leaves few, long-petioled, palmately parted, pubescent or nearly glabrous: raceme loose; *fl. few and large,* on long ascending pedicels; spur short, stout, straight: follicles divergent.—Hills toward the sea, from San Mateo Co. northward. April –July.

5. **D. decorum,** Fisch. & Mey. Root-cluster short, grumose, the tuberiform branches producing many long fibres: stem solitary, slender, simple, seldom 1 ft. high: herbage pale green, pubescent or glabrate: leaves parted into 3—5 widely sundered segments, these broad cuneiform, 3-lobed in the radical ones, entire in the cauline: *fl. rather small, in an open raceme,* deep blue except the uppermost petals, these white: spur

straight: follicles glabrous, widely divergent.—Borders of thickets, and in open stony ground near hilltops. April.

* * *Scarlet-flowered species; roots not fleshy.*

6. **D. nudicaule,** Torr. & Gray. Leafy at base of stem only, 1—2 ft. high: raceme very lax, somewhat pyramidal: calyx 1 in. long or more, bright scarlet, not widely expanding, the spur straight: petals yellow: follicles glabrous, divergent at summit, often narrowed below to a short stipe.—Rocky slopes and summits of the higher hills. April, May.

6. AQUILEGIA, *Tragus* (COLUMBINE). Perennials. Leaves mostly radical and twice terrate; leaflets thin, their lobes rounded. Flowers large, nodding, solitary at the ends of the branches. Sepals 5, plane, colored. Petals 5, tubular, projecting behind the sepals. Pistils 5, becoming follicles.

1. **A. truncata,** Fisch. & Mey. Usually glabrous, 1—3 ft. high: fl. 1½—2 in. broad, scarlet tinged with yellow: sepals widely spreading or reflexed: petals with very short truncate limb.—Borders of moist shady thickets; common. April—June.

7. THALICTRUM, *Diosc.* (MEADOW RUE). Diœcious tall perennials with fibrous roots, hollow stems, bi- or triternately compound leaves and many panicled greenish apetalous flowers. Sepals 4—7, small, deciduous. Stamens ∞, with linear anthers on capillary filaments. Pistils several, becoming ribbed or veined achenes tipped with the persistent style.

1. **T. polycarpum,** Wats. Stout, 3—4 ft. high, glabrous, not glaucous, *aromatic-scented:* leaflets with *acute or acuminate lobes:* sepals lanceolate, not scarious: achenes very many in the head, broadly obovoid, short-stipitate, compressed, turgid, the sides marked with low more or less anostomosing veins.—Open places near streams, in the first Coast Range. May, June.

2. **T. hesperium,** Greene. Glabrous, except the growing parts, and lower face of leaves, which have a sparse glandular pubescence; herbage *ill-scented* (not aromatic): *lobes of leaflets rounded:* sepals 5, not scarious: achenes fewer, obliquely oval or semi-obovate, substipitate, the ribs or veins mostly distinct and parallel.—Oakland Hills and inner Coast Range generally.

8. ACTÆA, *Linn.* Perennial. Leaves few, ample, ternate and fern-like. Flowers small, white, in a single terminal short raceme. Sepals about 4, caducous. Petals 1 or more. Stamens ∞. Pistil 1. Berry-like pericarp with a false suture running down one side; seeds flattened and semiorbicular, packed in two vertical rows.

1. **A. arguta,** Nutt. Stem 2—3 ft. high; leaves 1 or more, with acute coarsely and incisely serrate leaflets: raceme 1—2 in. long, oblong, often

with one or more short branches at base: sepals obovate, concave: petal with rhombic-ovate acute limb and almost filiform claw: stamens 25–30; filaments filiform or slightly thickened under the roundish anthers: berries obliquely oval, as large as peas, the polished and shining surface cherry-red, or occasionally snow-white. — Wooded northward slopes under hazel bushes, etc. Feb.—April.

Order II. BERBERIDEÆ.

Shrubs or herbs with alternate or radical usually compound leaves. Sepals and Petals 3 or 6 each. Stamens 6 or 9, hypogynous; anthers opening by valves hinged at top. Pistil 1. Fruit a berry or a 1-celled capsule.

1. **BERBERIS,** *Brunfels* (OREGON GRAPE. BARBERRY). Ours low evergreen shrubs, with unequally pinnate coriaceous prickly leaves, and yellow inner bark and wood. Flowers yellow, in clustered terminal and axillary racemes. Sepals 6, subtended by 3 or more bractlets. Petals 6, opposite the sepals. Stamens 6. Berries (in ours) dark blue and glaucous.

1. **B..nervosa,** Pursh. Simple, the stem 1 ft. high or less, at summit bearing a crown of very large leaves, and many dry persistent chaffy bracts: leaves 1—2 ft. long; leaflets 11—17, ovate, acuminate, somewhat palmately nerved: racemes elongated: berries ovoid.—In deep woods near the coast.

2. **B. pinnata,** Lag. Branching 1—6 ft. high: leaflets 7—9 very prickly, the lowest pair near the base of the petiole: racemes profuse, clustered in the axils of all the leaves, as well as terminal: fr. ovoid.— Rocky hills; common. April, May.

2. **VANCOUVERIA,** *Morr. & Dcsne.* Perennial. Leaves all radical, 2—3-ternate. Scapes racemose or paniculate; the flowers small, nodding. Sepals 6, obovate, reflexed, subtended by 6—9 small bracts. Petals 6, deflexed, but with cucullate incurved tips. Stamens 6, erect, closely appressed to the pistil. Carpel 1; ovules 10 or fewer, in two rows along the suture. Capsule dehiscent by a dorsal valve.

1. **V. parviflora,** Greene. More or less villous with brownish hairs, 1 ft. high or more: leaves dark green, coriaceous, enduring through the year; leaflets 1 in. broad, petiolulate, subcordate, obtusely 3-lobed, emarginate: fl. small, 25—50 in a panicle, white, or with a lavender tinge: ovary glabrous.—Wooded hills at considerable elevations, both back of Oakland and in Marin Co. April, May.

ORDER III.   LAURINEÆ.

Represented by a single and monotypical genus.

1. UMBELLULARIA, *Nutt.* (CALIFORNIA LAUREL). An evergreen tree, with alternate coriaceous entire aromatic foliage, and perfect flowers in peduncled terminal and axillary small capitate umbels; these in bud covered by an involucre of about 4 broad caducous bracts. Perianth with no tube; segments 6, the 3 outer enfolding the others, all deciduous. Stamens 9; the outer series (6) spreading, the inner (3) erect and near the pistil; a circle of 6 stout stipitate glands intervening between the 2 series; anthers 4-celled, of valvate dehiscence, those the outer series introrse, of the inner extrorse. Fruit drupaceous.

1. **U. Californica,** (Arnott), Nutt. Tree 10–75 ft. high, the growing twigs and inflorescence minutely puberulent: leaves oblong-lanceolate, 2–4 in. long, short-petioled, bright green: peduncles ½–1 in. long; pedicels of the 5–10 greenish yellow flowers 1–5 lines: drupe dark purple, ovoid or subglobose, 1 in. long, the pulp and putamen thin. Common along streams and on northward slopes of hills. Jan.—May.

ORDER IV.   NYMPHÆEÆ.

Water-Lily Family; represented by one species.

1. NYMPHÆA, *Theophr.* (YELLOW POND-LILY). Aquatic: rootstock stout, creeping at the bottom of ponds and streams. Leaves large, leathery, cordate, entire, either floating or raised above the water. Sepals 6–12, imbricated, rounded and concave, yellow or reddish. Petals and stamens ∞, short, hypogynous; filaments short, anthers truncate, extrorse. Ovary, oblong or ovate 12–20-celled; the sessile broad flat stigma with as many radiating striæ. Seeds without aril.

1. **N. polysepala**(Engelm.), Greene. Sepals 9–12, all but the greenish and small outer ones of a rich brownish red: rays of stigma 15–21, the margin somewhat crenate.—Not rare in Marin Co.; also in some lakes or ponds in the outskirts of San Francisco.

ORDER V.   PAPAVERACEÆ.

Herbs (*Dendromecon* shrubby) with a colored or milky narcotic juice, commonly glaucous foliage, and mostly solitary showy 4-merous or 6-merous flowers. Sepals 1, 2 or 3, caducous. Petals 4–6, crumpled in the bud. Stamens 6–∞, usually hypogynous; anthers innate. Pistil compound and the ovary becoming a capsule, or the carpels nearly distinct, maturing as almost follicular pods. Seeds ∞; albumen fleshy or oily; embryo minute, straight.

1. **PAPAVER,** *Pliny* (Poppy). Sepals 2. Petals 4. Ovary with 4 or more intruded placentæ which partially divide the interior of the obovoid or subglobose capsule, this opening by short roundish or triangular apertures near the summit between the parietal ribs: stigma 4—8-lobed, sessile and the lobes radiating over the summit of the ovary and capsule, or raised on a short style and the lobes capitate-recurved. Seeds ∞, small, scrobiculate or reticulate.

1. **P. Californicum,** Gray. Sparsely pilose-pubescent, 1—2½ ft. high leafy below; leaves pinnately parted or divided into acutish toothed or 3-lobed or entire segments: peduncles elongated; corolla 2 in. broad; petals brick-red, with a green spot at the base bordered with rose-red: capsule ½ in. long or more, clavate-turbinate, 6—11-nerved; *stigmas sessile and radiating,* forming a flat cap to the pod; the short valvular openings somewhat quadrate: seeds coarsely and faintly reticulate.— Marin Co.

2. **P. Lemmoni,** Greene. Near the preceding, but larger, 1—3 ft. high: corolla 1—3 in. broad, apparently of a deeper red, the base of the petals green: capsule broader and merely obovate; stigmas 7—10, their lower half sessile and radiant upon the pod, the upper half coherent and *forming a conical apiculation.*—Marin Co. and southward.

3. **P. heterophyllum** (Benth.), Greene. Aspect and size of the two preceding, but the segments of the pinnately divided leaves singularly variable upon the same leaf, some linear, others in close juxtaposition oval; *stigmas capitate,* raised on a slender style: red *flowers large, nodding,* the *stamens declined.*—On wooded slopes.

4. **P. crassifolium** (Benth.). Smaller than the preceding, much more branching and floriferous, the foliage smaller and more fleshy: *fl. small, erect,* the stamens not declined.—In dry fields of the interior; never with *P. heterophyllum* or even near it.

2. **PLATYSTEMON,** *Benth.* Annual glaucescent glabrous or hirsute herbs with entire leaves; the cauline opposite or verticillate. Flowers rather small, white or cream-colored, on slender peduncles. Sepals 3, caducous. Petals 6. Stamens 6-∞; filaments filiform or flattened; anthers oblong to linear. Carpels 3—∞, in maturity variously more or less united, or quite distinct. Seeds smooth and shining.

*Carpels 6—25, torulose, jointed between the seeds.

1. **P. Californicus,** Benth. Branching from the base 6—12 in. high, sparingly hirsute; lowest leaves alternate; cauline opposite, all linear, entire, 2—4 in. long, sessile or clasping, obtuse: flower-buds ovoid; sepals hirsute: petals ½ in. long or more, pale yellow with a deep greenish-yellow spot at base, sometimes reddish tinged on the outside; stamens ∞;

filaments flattened and ligulate, carpels breaking transversely into 1-seeded indehiscent joints.—Throughout the western parts of the State. April—June.

* * *Carpels 3 only, partly united and forming a 3-lobed 1-celled ovoid capsule open at top.*

2. **P. linearis** (Benth.), Curran. Acaulescent, 3—12 in. high, sparsely hirsute: leaves narrowly linear, 1—3 in. long, acutish: peduncles scapiform, very slender: fl. ½—1 in broad: petals as in the last: stamens ∞ ; filaments filiform or flattened: capsule ovate-triquetrous, ½ in. long or more.--Gravelly hills. March—May.

* * * *Carpels 3 only, united, forming a slender elongated and twisted 1-celled capsule; stamens few and definite; stigmas linear.*

3. **P. Torreyi,** Greene. Erect, slender, dichotomous from the base, 3—8 in. high, glabrous: lowest leaves ovate-spatulate or oblanceolate; upper linear, acute, entire, ½—1 in. long: fl. ½—1 in. broad, white: stamens (usually 12) in two circles; filaments dilated upwards, those of the outer circle conspicuously shorter than those of the inner: capsule linear, ¾—1½ in. long. Feb.—May.

3. **DENDROMECON,** *Benth.* Shrubs with alternate coriaceous entire leaves, and solitary rather large yellow flowers. Sepals 2. Petals 4. Stamens ∞ ; filaments filiform, short; anthers linear. Ovary linear; style short; stigmas 2, short and erect. Capsule linear, many-nerved, 1-celled, 2-valved, the valves dehiscent somewhat elastically from base to apex.

1. **D. rigidum,** Benth. Shrub 2—8 ft. high, with many rigid ascending branches and slender branchlets; bark whitish: leaves ovate- to linearlanceolate, 1—3 in. long, very acute or mucronate, vertical, the very short petiole being twisted, the margin scabrous-denticulate: fl. 1—3 in. broad, the petals nearly rotate-spreading; capsules slightly arcuate, 1½—2½ in. long.—In clayey or gravelly soil on higher hills. March—June.

4. **ESCHSCHOLTZIA,** *Chamisso.* Nearly or quite glabrous glaucous flaccid herbs, with colorless bitter juice (that of the roots reddish) and ternately dissected leaves. Flowers solitary, yellow or orange-colored. Calyx an oblong or conical mitre-like organ deciduous from the more or less funnelform-dilated torus which bears the 4 petals. Stamens mostly ∞ ; filaments very short, attached to the base of the petals; anthers linear or oblong, usually longer than the filaments. Ovary linear; style very short; stigmas 2 or more, subulate-filiform. Capsule 10-nerved, 1-celled, ∞ -seeded, 2-valved, the valves elastically dehiscent from base to apex, forcibly ejecting the seeds; these spherical, reticulate or muricate.

* *Torus broadly rimmed; cotyledons bifid.*

+- *Perennials.*

1. **E. Californica,** Cham. Glabrous, glaucescent, the stems decumbent or at length procumbent, 1—2 ft. long, regularly dichotomous below, above bearing a flower opposite each leaf: leaves ternately dissected, the ultimate segments linear, obtuse: calyx oblong or ovoid, *abruptly short-pointed*; torus-rim broad: petals about ¾ in. long, light yellow with an orange spot at base: pods small for the size of the plant (2 in. long): seeds conspicuously reticulate.—Only along the seaboard in sandy soil.

2. **E. Douglasii** (H. & A.), Walp. Size of the last but less depressed, not obviously dichotomous: calyx *ovate-acuminate*: outer rim of torus narrow, not exceeding the erect inner one, in age deflexed: petals 1 in. long or more, yellow, shading into orange at base.—Plains of Solano and Contra Costa counties.

3. **E. crocea,** Benth. Stouter, erect or decumbent, the herbage of a deeper green and scarcely glaucescent: calyx very large, often 1 in. long or more, *long-conical*: outer rim of torus very broad, more or less undulate: petals, 1½—2 in. long, deep orange.—The most common species in our district, and very showy.

+- +- *Annuals.*

4. **E. compacta,** (Lindl.), Walp. Annual erect, 1—2 ft. high, glabrous, light green, more or less glaucescent: leaves finely dissected, the ultimate segments linear cuneiform, 3-toothed or cleft at the broad apex: calyx very *thin* and *partly diaphanous, slender-conical*; outer torus-rim broad, thin: petals ¾—1½ in. long, light yellow, shading into orange below the middle.—Dry plains of Contra Costa Co., and southward, but not in the typical form.

5. **E. ambigua,** Greene. Slender, branching from the base, decumbent, *glaucous and scabrous puberulent* throughout, 1 ft. high or less: leaves small, ternately dissected, the ultimate segments short, approximate in threes: calyx *ovate acuminate*, about 4 lines long or 5: torus small, but with ample rim: petals deep yellow, 1 in. long or less.—Near the summit of Mt. Diablo (not typical) and southward.

* * *Torus without rim; cotyledons entire; annuals.*

6. **E. hypecoides,** Benth. Scabrous or even hirsute-pubescent below,

glabrous above, glaucescent: branches many from the annual root, decumbent at base, 1 ft. high or less, slender, sparingly leafy: leaves small, segments rather few, linear-cuneiform: calyx oblong-conical, ½ *in. long, abruptly slender-pointed*: torus short-tubular, 1½ lines deep; outer margin a mere herbaceous ring, the inner erect, hyaline: petals 1 in. long: seeds with a faint irregular reticulation.—Vaca Mts. to Tamalpais.

7. **E. rhombipetala,** Greene. Glaucous and *tuberculate-scabrous* throughout; stemless or the stems stout, depressed, very leafy, the stout 4-angled peduncles little exceeding the subradical leaves: torus sub-cylindrical, with 2 minute approximate scarious margins: *petals* ½ in. long, *rhombic-ovate, fugacious*: capsules very large for the plant (3—4 in. long): seeds large, very distinctly and regularly favose-reticulate.— In grain fields along the eastern foothills of the Mt. Diablo Range.

ORDER VI. FUMARIACEÆ.

Glabrous often glaucous herbs, with watery juice, alternate pinnately or ternately divided or dissected leaves without stipules, and racemose flowers. Sepals 2, small, deciduous. Petals 4, in 2 dissimilar pairs; one or both of the outer ones saccate at base; inner pair cohering by the callous apex and enclosing the anthers and stigma. Stamens 6, hypogynous; filaments in 2 parcels placed opposite the outer petals, usually diadelphous; anther of the middle stamen in each parcel 2-celled, those of the lateral 1-celled. Fruit a several-seeded siliquose 2-valved, 1-celled capsule, or indehiscent.

1. **CAPNORCHIS,** *Boerhaave.* Perennials, with tuberiferous or granular or scaly subterranean stem or crown, fibrous rootlets, ternately or pinnately compound leaves, and racemose or paniculate flowers. Corolla flattened and cordate; the two outer petals larger, saccate or spurred at base.

1. **C. formosa** (Andr.), O. Ktze. Rootstock rather large, creeping, nearly naked: leaves and scapes 2 ft. high, the former twice or thrice pinnately compound, the final divisions incisely pinnatifid: fl. compound-racemose at summit of the naked scape: *corolla rose-purple, ovate-cordate*, with short spreading tips to the larger petals.—Common in the woods of Marin Co. April—June.

* 2. **C. chrysantha** (H. & A.), Planch. Very glaucous: leaves bipin-nate, the larger a foot long, the divisions cleft into few and narrow lobes: stem leafy, 2—5 ft. high, ending in a large racemose panicle of *yellow flowers: corolla linear-oblong*, only slightly cordate.—Santa Cruz Mts. and Mt. Diablo. May—July.

ORDER VII. **CRUCIFERÆ.**

Herbaceous or rarely suffruitescent plants with watery pungent juice, alternate exstipulate leaves, and usually racemose white or yellow or sometimes purple flowers. Sepals 4, imbricate, deciduous. Petals 4, often unguiculate, the laminæ spreading in the form of a cross (unequal, and differently arranged in many of our *Streptanthi*), hypogynous, deciduous. Stamens almost always tetradynamous, *i. e.*, 4 long, 2 short (except in some *Streptanthi*, where they are in 3 unequal pairs, and in *Athysanus* and *Heterodraba*, which have them all of equal length);[1] in some species reduced to 4 and even 2, hypogynous. Fruit usually a silique or silicle of 2 valves which separate from a central partition formed by the united placentæ. Seeds attached to the outer edge of the partition all around, usually forming a single row under the valves.

* Pods indehiscent.

Pods 1-seeded, not wing-margined.............................ATHYSANUS   3
  "   "   wing-margined.................................THYSANOCARPUS   17
  " several-seeded, flattened { short and rounded,.............HETERODRABA   2
                             { linear elongated...............TROPIDOCARPUM   21
  "   "   "   terete and pointed...................RHAPHANUS   19
  " breaking transversely into { 2 dissimilar 1-seeded joints..CAKILE   20
                          { several 1-seeded joints......RHAPHANUS   19
  " formed of 2 opposite 1-seeded nutlets......................CORONOPUS.   18

* * Pods dehiscent from below.

Pods obovoid, scarcely compressed..............................CAMELINA   4
  " orbicular, flattened parallel with the partition. ....ALYSSUM   1
  " flattened contrary to the partition { several-seeded.........BURSA   15
                             { 2-seeded..............LEPIDIUM   16
  " oblong, linear-oblong, or slender- { fl. white..............NASTURTIUM   9
        conical, turgid, slightly curved { fl. yellow.............RORIPA   10
        thin-walled, straight.....................................SISYMBRIUM   14
  " linear, 4-angled { leaves lyrate............................BARBAREA   11
                { leaves entire or runcinate...............ERYSIMUM   12
  "   "   terete { beaked...BRASSICA   13
               { beakless.......................................THELYPODIUM 7
  "   "   compressed; valves elastically dehiscent............CARDAMINE   8
  "   "   compressed; not elastic { claw of petals narrow......ARABIS   5
                          { " broad......STREPTANTHUS   6

* * * Pods dehiscent from the apex. TROPIDOCARPUM 21

1. **ALYSSUM**, *Diosc.* Low herbs with simple leaves and more or less of a stellate pubescence. Sepals equal. Petals white or yellowish. Pod orbicular; valves convex, nerveless. Seeds 1 or 2 in each cell.

1. A. **alyssoides** (L.), Gouan. Annual, branching from the base, decumbent, ½–1 ft. high, *canescent:* leaves linear-spatulate, ½–1 in. long: raceme rather slender, the white or pale yellow *petals little exceeding the sepals:* pod slightly emarginate, little exceeding the persistent sepals, 4-seeded.—Naturalized about San Francisco.

2. **A. marittimum** (L.), Lam. Perennial, ostensibly glabrous, but a *few appressed hairs* are revealed by a lens: broad white *petals twice the length of the* deciduous *sepals*: pod 2-seeded. —Naturalized more extensively than the last.

2. HETERODRABA, *Greene.* Slender diffuse annual, leafy only near the base, the elongated branches unilaterally racemose throughout. Leaves simple, toothed. Sepals equal. Petals without claw. Stamens 6 but equal, 3 on either side of the orbicular compressed ovary. Pod several-seeded, 2-celled by a very thin and filmy partition, indehiscent.

1. **H. unilateralis** (Jones), Greene. Pubescent with rigid short branching hairs: leaves cuneate-obovate, coarsely few-toothed above the middle, ½—1 in. long: branches horizontal and trailing or prostrate, ½—2 ft. long, in age rigid and wiry: pods on short rigid deflexed pedicels, 2 lines long, 1½ lines wide, stellate-pubescent and hispidulous, twisted when mature.—In fields among growing grain, along the eastern slopes of the Mt. Diablo Range. Feb.—May.

3. ATHYSANUS, *Greene.* Habit and character of the preceding, save that the very small orbicular and straight pods are 1-celled and 1-seeded.

1. **A. pusillus** (Hook.), Greene. Stems filiform, branching from the base, the branches mostly ascending, unilaterally racemose throughout: leaves few, ovate, sparingly toothed, ¼ in. long: fl. minute, often apetalous: pods lenticular, more or less uncinate-hispid, scarcely a line long, rather acute at base.—Common on hillsides. March, April.

4. CAMELINA, *Ruellius.* Erect herbs, sparingly branching, with clasping or sagittate leaves, and terminal loose racemes of small yellowish flowers. Sepals equal. Petals entire. Filaments without teeth. Silicle obovate or globose, beaked with a persistent style. Seeds several in each cell, oblong, marginless; cotyledons incumbent.

1. **C. sativa** (L.), Crantz. Pubescent, ½—2 ft. high: leaves lanceolate, sagittate at base, nearly entire: pods pyriform with acute base.—A weed in fields of grain in many countries; found at Berkeley by *Mr. Chesnut* in 1887.

5. ARABIS, *Linn.* Sepals erect, equal, or two of them slightly saccate at base. Petals white or purple with narrow claw and flat blade. Anthers short, straight, ovate or oblong, scarcely emarginate at base. Stigma entire or 2-lobed. Pod linear, compressed; valves nerveless or lightly 1-nerved. Seeds in one or two rows, flattened, often winged.

1. **A. glabra** (L.), Weinm. Stout biennial, usually simple, 2 to 5 ft.

high; lowest leaves and base of stem hirsute or hispidulous, the plant otherwise glabrous, *glaucous*: lower leaves spatulate, 2—4 in. long, sinuate-pinnatifid or toothed, ciliate at least on the petioles; cauline ovate or ovate-lanceolate, entire, clasping by a sagittate base: *petals dull white* or greenish white, *2—3 lines long*, little exceeding the sepals: *pods erect*, usually even appressed to the stem, 2—4 in. long, less than a line wide, straight, on pedicels 3—4 lines long: style short: seeds in 2 rows, narrowly winged or wingless.—Common. March—May.

2. **A. hirsuta** (L.), Scop. Seldom 2 ft. high, deep green, *not glaucous*, hirsute throughout with a *branched pubescence*: cauline leaves oblong-ovate or ovate lanceolate, sagittate or auricled at the clasping base, coarsely toothed or entire: fl. small, white: pods very slender, erect.—Near Tocaloma, *Bioletti*.

3. **A. blepharophylla**, Hook. & Arn. Stoutish, 4—12 in. high, deep green, glabrous or sparsely pilose-pubescent: lower leaves obovate to broadly spatulate, 1—2 in. long, entire or sinuate-toothed, *strongly ciliate;* cauline oblong, sessile: sepals usually purplish: *petals* ½ *in. long, rich red-purple:* pods 1¼ in. long, 1½ lines wide, suberect, beaked with a short stout style: seeds in one row, a line wide.—Rocky hills about San Francisco. Feb.—May.

4. **A. Breweri**, Wats. Low, tufted perennial, rather rigid, 2—10 in high, *canescent with dense stellate pubescence*: radical leaves spatulate, 1 in. long, short-petioled, entire; cauline ovate-oblong, sessile, not sagittate: petals rose-purple, 1—4 lines long: *pods spreading* or recurved, 1½—2½ in. long, scarcely a line wide.—Mt. Diablo Range, on rocks. Apr.—June.

5. **A. Ludoviciana** (Nutt.), C. A. Mey. Nearly glabrous, branched from the base and branches ascending, 6—10 in. long: *leaves all pinnately parted* into oblong or linear few-toothed or entire segments: *fl. small, white:* pods spreading on short pedicels, flat, rather broad-linear, 1 in. long: seeds orbicular, wing-margined.—Banks of the lower San Joaquin.

6. **STREPTANTHUS**, *Nutt.* Mostly annuals, the few branches loosely racemose throughout. Leaves pinnatifid or toothed, rarely entire, except the cauline, and these mostly sagittate and clasping. Calyx whitish or colored, open or closed, often irregular, 2 or all of sepals saccate at base. Petals with broad channelled claw and (in our species) a narrow usually more or less undulate limb. Stamens either tetradynamous or in 3 unequal pairs, the uppermost pair often with filaments united; anthers elongated, sagittate at base, curved in age. Pod from flat and thin to subterete; valves 1-nerved or rarely carinate. Seeds more or less flattened, margined or marginless.

Calyx not irregular.

1. **S. barbiger**, Greene. Slender, 1—1½ ft. high, pubscent or glabrous; cauline leaves *linear, entire*, scarcely auriculate: fl. subsessile,
3 lines long: sepals greenish, the rather acuminate tips becoming
whitish-petaloid and recurved, the whole calyx commonly bristly-hairy,
but often glabrous: *petals white: filaments dark purple*, the three
pairs very unequal, the uppermost connate almost to the summit, their
anthers much reduced and seemingly sterile: pods 1—2 in. long, narrowly linear, recurved.—In Napa Co., near St. Helena.

2. **S. suffrutescens**, Greene. Perennial, *suffrutescent*, the *stout leafy
trunk 6—8 in high;* flowering branches 1—2 ft. long: herbage glabrous,
glaucous: stem-leaves cuneate-obovate, coarsely serrate-toothed; floral
leaves round-cordate or more elongated: sepals purplish-green, their
tips not reflexed: one pair of filaments connate; all the anthers equal
and fertile.—Hood's Peak, Sonoma Co., *Bioletti.*

++ ++ *Calyx irregular, three sepals more or less connivent behind the upper
petals, the fourth separated from these and somewhat deflexed;
1 pair of filaments connate.*

3. **S. niger**, Greene. Branching loosely from near the base and above,
1—3 ft. high, *glabrous, glaucous:* leaves linear, 2--3 in. long, the lowest
with narrow pinnate gland-tipped lobes or teeth, the upper entire, auriculate-clasping: *racemes loose, flexuous:* pedicels ascending, 1 in. long:
calyx 3 lines long, of a very *dark metallic shining purple; sepals ovate-cymbiform*, the 3 upper slightly separated from the lowest, and connivent at
apex: blade of petals very slender, white: upper pair of filaments
connate almost throughout, their anthers small and sterile: pod 2 in.
long, erect or ascending, nearly straight: seed narrowly winged.—Hills
at Tiburon.

4. **S. albidus**, Greene. Stouter than the last, equally glabrous and
glaucous, even the cauline leaves with callous-tipped prominent teeth,
the base sagittate-clasping: *racemes not flexuous:* pedicels short:
sepals 3 —4 lines long, *white*, with purple base: petals ½ in. long, the
lamina ample, crisped, white, with purple veins: upper pair of filaments
united to the tip, their *anthers small but polliniferous.*—On hillsides not
far from San Jose.

5. **S. Mildredæ**, Greene. Slender, 1—1½ ft. high, more or less pilose-
hispid: lower leaves coarsely and sinuately toothed; cauline linear-
lanceolate, entire, clasping: racemes somewhat flexuous, not secund:
fl. small, *very dark metallic-purple:* petals with small, slenderly attenuate white-margined purple blade: upper pair of filaments almost wholly
united, their *anthers mere rudiments* closely approximate, the other 4
stamens much shorter and little unequal: pods 3 in. long, slender,

arcuate-spreading on the short pedicels: seeds oval, the upper half narrowly margined.—Mt. Hamilton.

6. **S. glandulosus,** Hook. Pubescence and sinuately toothed foliage of the last, but larger, 1—2½ft. high: racemes more or less inclined to be secund: *fl. very large, bright red-purple:* sepals ½ in. long, ovate-cymbiform, carinate, 3 strongly connivent at tip, the fourth hanging loosely apart from the others: petals well-exserted, white-margined: upper pair of filaments connate above the middle, thence rather widely divergent, their anthers smaller than the others, but not greatly reduced, apparently sterile: pods 3 in. long, a line wide, *arcuate-recurved:* seed narrowly winged.—On clayey hillsides and banks, in the Oakland Hills, and southward.

7. **S. Biolettii,** Greene. Habit and pubescence of the last, but smaller and more slender, the *leaves glaucous beneath:* racemes not secund: fl. 4—5 lines long: *sepals very dark purple:* upper pair of filaments much the longest, united two-thirds their length, thence divergent, their *anthers half the size of those of the shorter stamens,* apparently polliniferous: pods slender, suberect, hispid.—Hood's Peak, Sonoma Co.

8. **S. pulchellus,** Greene. Low and much branched, ¼—1 ft. high, pilose-hispid: leaves linear-oblong or oblong-lanceolate, the cauline sessile by a broad, partly clasping base, and with a *few coarse and very salient teeth:* racemes rather dense, subsecund: *calyx deep lilac-purple,* the sepals subequal, broadest at base, sharply carinate, the *keel with some bristly hairs:* upper pair of filaments united almost throughout, their subsagittate anthers little reduced, the other 4 stamens in very unequal pairs: pods very narrow, hispidulous, spreading and slightly incurved. —Southern flanks of Mt. Tamalpais.

9. **S. hispidus,** Gray. *Stiff-hirsute* or *hispid throughout,* only 3—6 in. high, branching: lowest leaves obovate- or cuneate-oblong, coarsely and somewhat incisely toothed, the teeth obtuse; cauline narrower, scarcely clasping: raceme short, loose, the fl. at length recurved: *sepals red-purple* with white petaloid tips, half as long as the similarly colored petals: *pods hispid,* 1½—2 in. long, 1 line wide, *straight,* ascending: seeds winged.—Mt. Diablo.

10. **S. secundus,** Greene. Slender, sparingly branched above, 1—2 ft. high: the long pinnately toothed or lobed lower leaves hispid-strigose; cauline lanceolate, sagittate, entire or toothed, and, with the branches, pedicels and pods, sparsely hispidulous with spreading short hairs: racemes rather dense, wholly secund: fl. flesh-color, 4 lines long: *sepals sharply carinate, the keel hispid-ciliolate,* the short tips greenish, the remote lower one distinctly, the opposite uppermost one obscurely *unguiculate:*

petals with ample purple-veined crisped limb: upper pair of filaments connate to near their tips, the free parts scarcely divergent, the anthers reduced in size, but polliniferous: pods 2 in. long, very slender, falcate-recurved, the valves carinate-veined: seeds small, wingless.—Northern base of Mt. Tamalpais, a rare and peculiar species.

7. **THELYPODIUM,** *Endl.* Coarser than *Streptanthus,* often biennial; the racemes often shorter and condensed. Calyx green, whitish or purplish; sepals equal at base. Petals with narrow claw and flat linear to obovate limb, exserted, white, yellowish or rose-color. Stamens tetradynamous; filaments never connate; anthers sagittate at base, curved. Pod usually long, linear, terete or slightly compressed, sessile or short stipitate. Seeds in 1 row, oblong, somewhat compressed, not winged.

1. **T. procerum** (Brew.), Greene. Annual, stoutish, branched from near the base, 3–7 ft. high, glabrous except at base: lower leaves petiolate, coarsely pinnatifid; upper lanceolate, sessile, acuminate: racemes long and lax: fl. greenish or yellowish white, 4—5 lines long, on ascending pedicels half as long: *pod very slender, terete, 3—5 in. long,* less than a line wide, acuminate, erect or somewhat spreading.— In fields at the eastern base of Mt. Diablo.

2. **T. flavescens** (Torr.), Wats. Sparsely pilose-hispid below: lower leaves elongated, petiolate, sinuately toothed; upper entire, sessile, not auricled: raceme long and lax: fl. yellowish or rarely purplish, 4—5 lines long: *sepals narrow and with the pedicels hispidulous:* petals long-exserted, with linear and *narrow claw;* the blade dilated: pod 1½ in. long, nearly terete, sparsely hispidulous, erect or somewhat spreading.— Fields of the lower Sacramento.

3. **T. Hookeri,** Greene. Size, habit and whole aspect of the pre-ceding, but lower leaves often pinnatifid, though as often sinuate-toothed; inflorescence the same, also size and color of fl. but *sepals* broader, less spreading, *glabrous:* petals with a *rather broad claw* and relatively *narrow blade:* pods 2 in. long, slender, terete, erect.—In the Mt. Diablo Range.

4. **T. rigidum,** Greene. Stoutish and very rigid, 1—3 ft. high, with few wide-spread branches: hispidulous below, glabrous above, deep green, not glaucous: lower leaves somewhat lyrately pinnatifid; upper oblong-lanceolate and laciniate-toothed: fl. yellowish, small, rather crowded and subsessile, the fruiting raceme long and loose: *pods* 1¼ in. long, *nearly sessile,* ascending or somewhat spreading or curved, *rigid, sharply tipped* with a short style.—Eastern base of the Mt. Diablo Range, from near Antioch southward, chiefly on clayey hillsides.

5. **T. lasiophyllum** (H. & A.), Greene. Glabrous, or more or less hirsute below, ½—6 ft. high, usually stoutish, rather rigidly erect, simple, or sparingly branching above the middle: leaves 2—4 in. long, pinnatifid with divaricate toothed segments, or the upper only sinuate-toothed: petals white or yellowish 1¼—2½ lines long: *pods slender, nearly terete,* 1—2 in. long, short-pedicellate, straight or somewhat curved, ascending or strongly deflexed.—Common and variable. The sandhill form at San Francisco is small, early flowering, and has suberect pods. In the Coast Range the plant is often a yard high or more, late in flowering, with pods strongly deflexed.

8. **CARDAMINE,** *Diosc.* Annuals or perennials of woods or moist places; rootstock often tuberous. Stems mostly simple, often very sparingly leafy. Flowers white or purplish, in short racemes. Sepals equal. Petals unguiculate. Silique elongated, linear, compressed, beaked or pointed, the valves plane, almost nerveless, more or less elastically dehiscent. Seeds compressed, not margined.

* *Without fleshy or tuberous rootstocks; leaves all pinnate.*

1. **C. oligosperma,** Nutt. Annual, erect, slender, ½—1 ft. high, nearly or quite glabrous: *leaflets small, in 3—5 pairs,* roundish 1—6 lines long, often obtusely 3—5-lobed, petiolulate: petals white, 1—1½ lines long, twice the length of the calyx: pods few, ½—¾ in. long, ½ line wide short-beaked, not becoming dry, the mature valves, while yet green-herbaceous, separating elastically and falling in a close coil; cells about 8-seeded.—Common on shady banks along streams. March, April.

* * *Stems from elongated or rounded and tuberous perennial rootstocks.*

2. **C. integrifolia** (Nutt.), Greene. Rather robust, 1 ft. high, glabrous, somewhat fleshy: *radical leaves 1—5 foliolate,* the leaflets usually rounded and more or less cordate and nearly or quite entire, 1—2½ in. broad; upper deeply lobed, or pinnately 3—5 foliolate, the *segments linear or linear-oblong, entire:* corolla large, white, nodding, the petals only campanulately spreading: pod conspicuously beaked.—Common in wet meadows, in open ground. Jan.—May.

3. **C. Californica** (Nutt.), Greene. Near the last, but slender, tall, less fleshy; the leaves, both radical and cauline, with *broad and ample repandly and mucronulately denticulate leaflets:* fl. pale rose-color.— Very common in rich woods, or shady banks. March—May.

4. **C. cardiophylla,** Greene. Stoutish, 1 ft. high or less: radical leaves undivided, *round-reniform to broadly cordate,* slightly and somewhat angularly 5-lobed and mucronately denticulate, 1 in. wide or more; cauline nearly as large, *broadly cordate, acute,* mucronate-denticulate,

tapering from within the broad sinus to a petiole ¾ in. long: fl. white: pods slender-beaked.—In Weldon Cañon of the Vaca Mountains, Solano Co. March, April. *Jepson.*

9. NASTURTIUM, *Bauhin.* Perennials with lyrately compound or simple and pinnatifid or undivided leaves. Flowers white. Sepals erect. Petals unguiculate. Pods short, turgid, little compressed, nerveless. Seeds small, rounded, somewhat flattened, impressed punctate.

1. N. OFFICINALE, R. Br. (WATERCRESS). Stems rooting at the decumbent base, the branches ½—5 ft. long, stoutish, hollow; *roots all fibrous; leaves pinnate,* leaflets rounded or elongated, the terminal one largest: pods ½ in. long or more, acute at each end, equalling the pedicels: style short, thick.—Plentiful in sluggish streamlets, and cool springy places.

2. N. ARMORACIA, Fries (HORSERADISH). Erect from a stout *perpendicular perennial root,* 2—4 ft. high: earlier radical leaves pinnatifid; later ones very large, *not cleft, crenate:* pods (seldom formed) ellipsoid or subglobular: style very short.—Escaped from cultivation.

10. RORIPA, *Gesner.* (FALSE CRESS). Annuals or biennials, commonly referred to *Nasturtium,* but the flowers yellow, the sepals greenish-yellow, ascending or spreading, the petals only short-unguiculate and ascending.

1. R. palustris (Leyss.). Erect, stoutish, 1—3 ft. high, branching above, glabrous: leaves oblong-lanceolate, coarsely and irregularly toothed or pinnatifid, 2—6 in. long: fl. 1 line long: *pods linear-oblong, 3—5 lines long,* on slender pedicels.—Margins of ponds; Oakland Hills; and a rank form in marshes of Sonoma Co.

2. R. lyrata (Nutt.). Branching from the base, the branches decumbent or ascending, seldom a foot long, glabrous or sparsely hispidulous: leaves quite regularly pinnatifid into divaricate linear- or oblong-lanceolate entire segments: *pods ½—¾ in. long, linear, more or less curved;* pedicels half as long: seeds in 2 rows.—Along streams in Marin Co.

3. R. dictyota (Greene). Stout, erect, 2—4 ft. high, hirsute-pubescent: racemes rather dense: *pods ovate-lanceolate;* valves firm in texture, with strong *tortuous midvein* and anastomosing veinlets; partition thick, favose-reticulate. —Marshes of the lower Sacramento.

11. BARBAREA, *Dodonæus* (WINTER CRESS). Erect branching glabrous biennials or perennials of rather low stature, with angular stems and more or less distinctly lyrate or pinnatifid leaves. Flowers rather small, bright yellow. Sepals equal at base, erect. Pods linear, either somewhat flattened, or more distinctly quadrangular, pointed; valves more or less carinate. Seeds in 1 row, oblong, turgid, marginless.

1. **B. vulgaris**, R. Br. Stoutish, 1—3 ft. high; herbage bright green and glossy: leaves mostly radical, the very lowest sometimes simple, oftener with 1 or more pairs of relatively small lobes below a very large terminal one; cauline either simple and toothed, or pinnately parted: fl. 2—3 lines long: pods 1—2 in., erect, ascending, or even arcuate-spreading.—Common in moist shady places. March—June.

12. ERYSIMUM, *Diosc.* Biennials or perennials; ours stout, simple or with few branches. Leaves narrow, entire or runcinately toothed, not clasping. Flowers large, yellowish or orange. Sepals erect, one pair strongly gibbous at base. Petals with low claw and flat blade. Anthers sagittate. Pod 4-angled or flattened, and the valves merely nerved. Seeds in 1 row, not margined, oblique.

1. **E. asperum** (Nutt.), DC. Canescent with short straight closely appressed hairs: stems solitary, rarely with a few branches above, 1—3 ft. high, angular: leaves narrowly spatulate or oblanceolate, entire or runcinate-toothed, 1—3 in. long: fl. large, fragrant: sepals narrow, 4—6 lines long: petals from light yellow to deep orange, 8—12 lines long: *pods slender*, spreading, *quadrangular*, commonly 3—4 in. long, 1 line wide, beaked with a stout style.—Common in the mountains almost everywhere. April—July.

2. **E. capitatum** (Dougl.), Greene. Stout and low, ½—1½ ft. high, sparingly pubescent with appressed bifid or 2-parted hairs; leaves narrow, entire, or sinuately or angularly toothed or lobed: fl. large, cream-color or yellowish, in a depressed terminal corymb, scarcely fragrant: *pods nearly flat*, with a strong midvein, 1½ lines wide, the whole 1½—2½ in. long, abruptly and stoutly short-pointed: seeds flattened.—Among the sandy or rocky hills of the seaboard only. Feb.— May.

13. BRASSICA, *Pliny.* Large annuals or biennials, with erect often widely branching stems, lyrate or pinnatifid lower leaves, and yellow flowers. Sepals equal at base. Petals unguiculate; limb obovate. Pods linear or oblong, terete or quadrangular, with a stout 1-seeded or seedless beak; valves 1 –5-nerved. Seeds in 1 row, globose; cotyledons incumbent.

* *Sepals erect, enfolding the claws of the petals.*—BRASSICA proper.

1. **B.** CAMPESTRIS, L. Glabrous and glaucous, 2 –3 ft. high: lower leaves somewhat rough-hairy, lyrate with large terminal lobe; cauline oblong or lanceolate with a broad auriculate-clasping base: fl. 3—4 lines long: *pods nearly terete, 2 in. long or more,* on ascending or spreading pedicels, the stout beak 8 –10 lines long.—Abundant in fields, flowering in the late winter and early spring months; commonly but erroneously called Mustard. Jan.—April.

* Sepals spreading, releasing the claws of the petals.

2. **B. NIGRA** (L.), Koch. (BLACK MUSTARD). Not glaucous but dark green, roughish with scattered stiff hairs, stout, 3—12 ft. high: leaves all petiolate; the lower lyrate, with a very large and lobed terminal lobe; the uppermost lobed or toothed or entire: petals 3—4 lines long, twice the length of the yellowish sepals: *pods closely appressed* to the rachis of the raceme, 4-angled, ½—¾ in. long, sharply beaked with the long style.—Common as the preceding, but taking more exclusive possession of fence corners and rich waste lands. June, July.

3. **B. SINAPISTRUM**, Boiss. Annual, 2—5 ft. high, the herbage light green, rough with spreading hairs: lower leaves usually with a large coarsely toothed terminal lobe and smaller ones of angular outline on the rachis: fl. 4—6 lines long: pods 1—1½ in. long, ascending, nearly cylindrical, with a stout somewhat *2-edged beak a third as long as the* prominently nerved *valves*, often containing a seed, the seeds under each valve 3—8.—Common by waysides in the vicinity of Berkeley and Oakland; flowering later than *B. campestris*, but earlier than *B. nigra*.

14. SISYMBRIUM, *Diosc.* Erect and rather slender annuals. Leaves not clasping, lyrate-pinnatifid, or (in our species) finely dissected. Flowers small, yellow. Sepals scarcely gibbous at base. Petals unguiculate. Anthers mostly linear-oblong, sagittate. Pods linear or oblong-linear, terete or nearly so, obtuse or short pointed; valves slightly 1—3-nerved. Seeds usually numerous, small, oblong and teretish; cotyledons incumbent.

* Seeds in 2 rows; leaves finely dissected.

1. **S. multifidum** (Pursh), MacM. Simple or with few branches, ½—2½ ft. high, canescent with short branching hairs: leaves 1—2-pinnate, the segments more or less deeply toothed or pinnatifid: petals 1 line long or less, about equalling the sepals: *pods oblong to linear,* or *subclavate,* ¼—½ in. long, on slender *spreading pedicels* of equal or greater length, acute at each end, and beaked with a very short style.— Plains near Livermore.

* * Seeds in 1 row; leaves pinnatifid or entire.

2. **S. OFFICINALE** (L.), Scop. (HEDGE MUSTARD). Rigid, erect, sparingly and divaricately branching above, somewhat hirsute; lowest leaves depressed and rosulate, lyrately and somewhat runcinately pinnatifid, 3 —6 in. long: *pods terete,* ½ in long, *tapering* from base to summit, *nearly sessile,* closely appressed to the rachis in a long slender raceme.—By waysides and in waste grounds.

3. **S. ACUTANGULUM**, DC. Hirsute with scattered simple hairs, 1—2 ft. high, with ascending branches: leaves 2—6 in. long, runcinate-pin-

natifid: *pods terete, 1—1½ in.* long, less than a line wide, erect or ascending on very short pedicels.—Not as common as the last.

15. BURSA, *Siegesb.* Slender nearly glabrous annuals, with simple or pinnate leaves, and small white flowers. Pods oblong or obcordate, more or less obcompressed, ∞-seeded; valves carinate, 1-nerved. Seeds not winged; cotyledons incumbent.

1. **B. pastoris,** Wigg. Usually hirsute at base, otherwise glabrous, erect, ½—2 ft. high, the stems racemose almost from the base, simple or with few branches: radical leaves usually in a depressed rosulate tuft, runcinate pinnatifid, or oblanceolate with coarse teeth; cauline sagittate, entire or toothed: *pods cuneate-triangular, retuse,* 1—2 lines long and broad, on rather long spreading pedicels.—Cosmopolitan weed, flourishing with us at all seasons.

2. **B. divaricata** (Nutt.), O. Ktze. Slender, often diffusely branching and decumbent or procumbent, 3—8 in. high: lowest leaves sinuate-pinnatifid; cauline entire or nearly so: petals minute, barely equalling the sepals: *pod oblong or ovoid, little flattened,* 2 lines long or less, obtuse, the valves rather thin; Pedicels slender, longer than the pods—Borders of salt marshes. March—May.

16. LEPIDIUM, *Diosc.* Low herbs with pinnatifid or toothed leaves, and small white or apetalous and greenish flowers. Stamens only 4, or even 2. Pod orbicular or ovate, strongly obcompressed, emarginately 2-winged at summit; valves acutely carinate; cells 1-seeded. Seeds not winged; cotyledons usually incumbent, rarely accumbent.

* *Annuals; pedicels flattened.*

+ *Pods reticulated.*

1. **L. latipes,** Hook. Branches stout and depressed, far surpassed by the leaves; these several inches long, irregularly and coarsely pinnatifid, the segments linear, entire or lobed; pubescence scanty on the leaves, more dense on the branches, hispidulous: racemes short, dense; pedicels 1—2 lines long: sepals very unequal: petals broadly spatulate, ciliate, greenish, exceeding the sepals: pod broadly oval, 2 lines broad, sparingly pubescent, strongly reticulate, the *broad acute wings nearly as long as the body* of the pod.—In saline soils at Martinez, Alameda, etc. March —May.

2. **L. dictyotum,** Gray. Habit and pubescence of the preceding, but much smaller, the branches at length ascending: leaves narrowly linear, entire or with a few narrow divaricate linear lobes: petals little exceeding the sepals or wanting: *pods rounded,* 1½ lines broad, emarginate,

with short acute wings, finely reticulated and pubescent, exceeding the thick erect pedicels.—Livermore Valley; also along the borders of marshes at Alameda.

3. **L. oxycarpum,** Torr. & Gray. Very *slender,* the elongated and racemose branches decumbent or assurgent, nearly glabrous: leaves linear, with a few linear segments or entire: sepals caducous: petals 0: stamens 2: pods on slender *deflexed pedicels,* glabrous, rounded, 1½ lines broad, the terminal *wings tooth-like, short, acute, divergent.*—Borders of salt marshes at Vallejo, *Greene;* also in subsaline soils east of Wild Cat Creek in the Berkeley Hills, and near Alameda.

4. **L. Oreganum,** Howell. Erect, simple or with a few ascending branches, 3—6 in. high, ostensibly glabrous (more or less hispidulous under a lens): leaves linear, with a few linear segments or entire; sepals and petals less fugacious: stamens 4: *pods round-ovoid,* 2 lines broad, the terminal *teeth more or less prominent* and divergent, the body somewhat hispidulous or glabrate.—Plentiful in subsaline soil in the Livermore Valley. March—May.

+ + *Pods faintly or not at all reticulate.*

5. **L. nitidum,** Nutt. Erect and with few ascending branches, or more diffusely branching from the base, ½—1½ ft. high, rather slender, almost glabrous, or the branches distinctly hirsutulous; these racemose almost throughout: lower leaves loosely pinnatifid, segments linear; cauline often entire: stamens 2 or 4: *pods rounded,* glabrous and shining, often of a dark purple, or with purple dots, 1½ lines broad, with *a small abrupt sinus* between the *short* terminal *teeth.* Var. **insigne,** Greene. Stout, mostly simple, 4—8 in. high; fruiting raceme shorter and denser: *pods twice as large,* round-obovoid.—Very common, especially towards the seaboard. The variety is of the Mt. Diablo Range. Jan.—June.

6. **L. Menziesii,** DC. Low and diffuse, herbage light green, hispid-puberulent or glabrate; branches 3—6 in. long; *racemes* numerous, rather *narrow* and *dense:* leaves of oblong outline, pinnatifid, the segments usually 3-cleft or -toothed: petals 0: pods rounded, 1—1½ lines broad, glabrous, or around the margin more or less hispidulous, faintly reticulate: *teeth* at the summit very *short and obtuse;* pedicels short, ascending or spreading, often very little flattened.—Common, especially by waysides and in hard clayey soil; late flowering, *i. e.,* April—June.

* * *Stouter and taller; pedicels terete.*

7. **L. intermedium,** Gray. Erect, branching above the middle, ½—1½ ft. high, puberulent or glabrous: lower leaves 1—2 in. long, toothed or pinnatifid: upper entire or only sparingly toothed, oblanceolate or

linear: petals 0: *pods glabrous, rounded,* 1—1½ lines broad *very shortly winged,* the *obtuse teeth* slightly divergent; pedicels 2 lines long.—Only occasionally met with in western California.

8. **L. Draba,** L. Biennial or perennial, erect, a foot high or taller, the several *stems corymbosely branched* at summit; herbage canescently pubescent: lower leaves oblong-obovate, 1—3 in. long, sparingly serrate or entire; cauline narrower, sagittate and clasping: petals white, conspicuous: *pods cordate,* not winged, turgid, acutish, tipped with a slender short style.—In old fields at Berkeley.

17. THYSANOCARPUS, *Hook.* (Lace-Pod). Erect, slender sparingly branched annuals, with minute, white or rose-colored flowers, in slender elongated racemes. Petals cuneate-obovate, or linear-oblong. Stamens 6, tetradynamous, or sometimes 4 only. Pistil a compressed rounded uniovulate ovary, short slender style, and small obtuse stigma, becoming a plano-convex or concavo-convex samara; the hard substance of the body of the fruit branching into several (12 to 16) radiating lines with diaphanous spaces or even complete rounded perforations between them, the whole forming a crenate wing. Seed solitary.

1. **T. curvipes,** Hook. A foot high or more; *radical leaves* pinnatifid, with short obtuse lobes or subentire, *hirsute;* cauline oblong- or linear-lanceolate, entire, sagittate-clasping: fr. obovate, seldom 2 lines wide, strongly concavo-convex, glabrous or slightly tomentose, the marginal rays broad, dilated above, rather crowded, with narrow diaphanous spots (rarely a few perforations) between them. Var. (1) **involutus,** Greene. Taller and more strict: fr. elliptical, only a line wide; rays nearly obsolete, the purplish subscarious margin closely involute all around; style (rather prominent in fl.) deciduous. Var. (2) **pulchellus,** Greene. Radical leaves merely toothed: pods densely tomentose; the wing rather broader. Common in middle California. April—June.

2. **T. elegans,** Fisch. & Mey. Rather stouter, with fewer racemose branches: lower leaves ascending, repand-toothed: fr. 3—4 lines broad, of more rounded outline, nearly plane, the body densely tomentose, the *rays separated by regularly ovoid perforations* and joined together beyond them into a very distinct diaphanous nearly entire margin.—Common on low hills of the interior. March—May.

3. **T. laciniatus,** Nutt. Glabrous throughout and *glaucous:* leaves linear, entire, or with a few incised or opposite and divaricate narrow segments: fr. from elliptical with narrow margin, to almost orbicular with broad evenly crenate border, scarcely plano-convex, 1½ - 2½ lines broad, imperforate, with irregular *deep sinuses between the rays,* or rarely with a few perforations, glabrous and very distinctly reticulate-venulose. —Lake Merced; Mt. Diablo, etc.

4. **T. radians,** Benth. Glabrous, 1 ft. high: lower leaves runcinate-pinnatifid; cauline ovate-lanceolate, auriculate-clasping: silicle *round-obovate, almost plane*, 4—5 lines wide, tomentose, the *rays narrowly linear*, ending abruptly near the edge of the broad diaphanous margin. —Sonoma Co. to Solano.

18. **CORONOPUS,** *Ruellius.* Diffuse prostrate heavy-scented annuals, with pinnatifid leaves, and the general aspect of some species of *Lepidium.* Flowers minute, greenish. Stamens often 2 only. Pods small, short, didymous, 2-celled; cells indehiscent, subglobose, when ripe separating from the persistent linear axis, strongly rugose, 1-seeded.

1. C. DIDYMUS (L.) Smith. Stems diffuse, ½—1½ ft. long; the heavy-scented somewhat aromatic herbage more or less hirsute; leaves with small narrow segments: pod a line broad or more, *emarginate at base and at summit*, strongly reticulate.—Plentiful on bluffs overhanging the sea at Point Lobos; occasional at Berkeley, etc.

2. C. RUELLII, Gærtn. Pods cristate-muricate, *not emarginate at summit*, but tipped with a stout style.—San Francisco.

19. **RAPHANUS,** *Pliny* (RADISH). Coarse annuals, with large somewhat fleshy lyrate lower leaves, and loose racemes of purple or yellowish large flowers. Sepals erect, the two outer gibbous at base. Petals entire or emarginate, unguiculate. Pod indehiscent, elongated, somewhat moniliform or at least constricted between the seeds, long-beaked. Cotyledons enfolding the radicle.

1. R. SATIVUS, L. More or less hispid with scattered stiff hairs: fl. 8—10 lines long: petals purplish, with veinlets of darker color, rarely white or yellowish: *pod thick*, fleshy when young, *spongy* in maturity, 1—2½ in. long, 2—5-seeded.—One of the prevalent and troublesome weeds in Californian fields everywhere.

2. R. RAPHANISTRUM, L. Petals yellow: *pods moniliform*, long-beaked, *breaking transversely* into 1-seeded joints.—San Francisco.

20. **CAKILE,** *Serapio* (SEA ROCKET). Glabrous very succulent sea-side annuals, with simple leaves and short racemes of smallish purple flowers. Sepals suberect, the two outer gibbous at base. Petals entire, unguiculate. Pod of 2 unequal joints, each 1-seeded, the upper and larger joint deciduous from the other. Seed in the upper cell erect; in the lower pendulous; cotyledons usually accumbent.

1. C. edentula (Bigel.), Hook. A foot high or more, the stout stem and few ascending branches somewhat flexuous: leaves obovate, sinuately toothed: lower joint of silicle oblong, 3—4 lines long; upper twice as large, ovate, compressed and emarginate at apex.—Common along sandy beaches.

21. TROPIDOCARPUM, Hook. Annuals, with light green pubescent herbage, pinnatifid leaves, and loose leafy-bracted racemes of middle-sized yellow flowers. Sepals concave, spreading, equal at base. Petals spatulate-obovate. Stamens tetradynamous; anthers rounded. Silique sessile, elongated, more or less obcompressed, flat or inflated, without partition, indehiscent or the valves (2—4 !) opening from above.

1. **T. gracile, Hook.** Erect, very slender, usually only a few inches high, nearly glabrous: leaves linear, with opposite pairs of linear segments, the floral similar but reduced: stamens very unequal, all exceeding the short pistil: silique linear, 2 in. long, glabrous, flat, indehiscent: seeds in 2 rows. Var. **scabriusculum**, Greene. Much larger, with many decumbent branches, and roughish-pubescent throughout, even to the pods.—Foothills of Mt. Diablo Range and on the plains.

2. **T. capparideum**, Greene. Usually erect, less than a foot high, simple, or with few ascending branches, the stem stoutish but hollow: pods ½—¾ in. long, linear-oblong, inflated, 2 lines wide, slightly obcompressed (the cross section transversely elliptical), conspicuously 6-nerved; valves 4, 2 deciduous and 2 persistent, the dehiscence beginning at the apex: seeds in 4 rows, i. e., 1 row along either margin of each of the 2 persistent valves.—Abundant in alkaline soil about Byron Springs.

ORDER VIII. RESEDACEÆ.

Herbs with alternate exstipulate leaves, and terminal racemes or spikes of small flowers. Sepals 4—6, often somewhat united at base, unequal, herbaceous, persistent, open in the bud. Torus bearing a rounded and glandular hypogynous disk which is produced posteriorly between the petals and the stamens. Petals 4—6, open in the bud, the lamina often lacerate or palmately parted. Stamens 3—20, inserted on the disk; anthers oval, fixed by the middle, introrse. Ovary 1-celled, 3—4-lobed, of 3—4 carpels at apex distinct and divergent; stigmas sessile, minute. Fruit membranous, 1-celled, open before maturity. Seeds reniform.

1. RESEDA, *Pliny* (MIGNONETTE. DYER'S WEED). Characters of the genus almost those of the order. Three Old World species, fugitives from the flower gardens, are here and there spontaneous with us.

1. R. ALBA, L. A tall stout sparingly branching perennial, with long spikes of whitish flowers: leaves deeply pinnate: sepals 5 or 6: petals as many, all equal, 3-cleft.

2. R. ODORATA, L. Annual; leaves oblanceolate or spatulate, often undulate: spike or raceme short in fl., elongated in fr.: fl. greenish, the large anthers dull red: petals parted into about 6 spatulate-linear segments.

ORDER IX. DATISCEÆ.

With us represented by a species of

DATISCA, *Linn.* Stout glabrous diœcious perennials. Leaves laciniate-pinnatifid; the segments coarsely toothed. Flowers axillary, subsessile, fascicled. Calyx of sterile fl. very short, with 4—9 unequal lobes.
Stamens 10—25; filaments short. Calyx of pistillate fl. with ovoid tube
somewhat 3-angled, 3-toothed, the stamens when present 3, alternate
with the teeth. Styles 3, bifid, opposite the teeth, the linear lobes stigmatic on the inner side. Capsule oblong, coriaceous, 1-celled, opening
at apex between the styles. Seeds ∞, small, in several rows on the 3
parietal placentæ; embryo cylindrical, in the axis of small albumen.

D. glomerata (Presl.), Brew. & Wats. Erect, 3—6 ft. high, simple or
sparingly branching: leaves of ovate or lanceolate outline, acuminate, 6
in. long; the floral shorter: fl. 4—7 in each axil of the long leafy raceme,
the fertile mostly perfect: anthers subsessile, 2 lines long, yellow: styles
exceeding the ovary: capsule oblong-ovate, 3—4 lines long, slightly narrowed toward the truncate triangular 3-toothed summit.—Along mountain streams.

ORDER X. CISTOIDEÆ.

In Asia an extensive family, of which we have one species.

HELIANTHEMUM, *Valerius Cordus.* Low, branching, suffrutescent.
Leaves alternate, simple, entire. Flowers perfect, regular. Sepals
mostly 5, unequal, persistent. Petals 5, yellow, fugacious. Stamens ∞,
hypogynous; filaments filiform; anthers short. Style 1, short, deciduous. Capsule ovoid, 1-celled, few- or many-seeded; the seeds borne on
the middle of the valve.

H. scoparium, Nutt. Plant a bushy tuft 1 ft. high, glabrate, or
stellate-pubescent: leaves narrowly linear, ⅓—1 in. long: fl. on slender
pedicels, solitary or cymose at the ends of the branches: sepals 3 lines
long, acuminate, the 2 outer linear and much shorter: petals 4 lines:
stamens about 20: capsule equalling the calyx.—Common on dry hills.

ORDER XI. VIOLARIEÆ.

Represented by a fair number of species of the principal genus of the
order.

VIOLA, *Pliny* (VIOLET). Low perennial herbs, with alternate leaves
of involute vernation, foliaceous persistent stipules, and 1-flowered
axillary peduncles. Flowers 5-merous. Sepals unequal, more or less
auricled at base, persistent. Petals unequal, the lower one often spurred

at base. Stamens hypogynous, the adnate anthers connivent over the
pistil, broad, often coherent, the connectives of the two lower often bearing spurs which project into the spur of the petal. Ovary 1-celled, with
3 parietal placentæ; style clavate; stigma 1-sided. Capsule 3-valved; the
valves bearing the seeds along the middle.

* *Acaulescent.*

1. **V. odorata**, Renealm. Rootstock stout, branching, stoloniferous;
leaves round-cordate, obtuse, crenate, more or less villous or glabrate, on
petioles 3—10 in. long: peduncles shorter than the leaves: fl. large,
violet, fragrant.—Occasionally spontaneous, as an escape from gardens.

* * *Stems short or elongated, leafy.*

← *Leaves undivided; flowers not yellow.*

2. **V. canina**, L., var. **adunca**, Gray. Scarcely stoloniferous, mostly
tufted and low, 2 –6 in. high, glabrous or puberulent: leaves ovate or
ovate-oblong, with subcordate or almost truncate base, obtuse, or rarely
acutish, obscurely crenate, ½—1½ in. long; stipules narrowly lanceolate,
lacerate-toothed: fl. rather large, *violet, turning to red-purple;* lateral
petals bearded; spur variable, much shorter than the petals, or quite as
long, usually straight and obtuse. Common on grassy hilltops along the
seaboard. Feb.—May.

3. **V. ocellata**, Torr. & Gray. Erect or ascending, ½—1 ft. high,
nearly glabrous, or pubescent: leaves cordate or cordate-ovate,
acutish, crenate, 1—2 in. long; stipules scarious, entire or slightly
lacerate: petals 5—7 lines long, the upper ones *white within, deep purple
without,* the others pale yellow-veined with purple, the lateral ones with
a purple spot near the base, and slightly bearded.—Woods of the
Coast Range. Apr.—June.

← ← *Leaves undivided; fl. yellow within, often brown-purple without.*

4. **V. glabella**, Nutt. Stems slender from a creeping rootstock,
nearly or quite leafless below, 5—12 in. high; minutely pubescent or
glabrous: radical leaves on long, the cauline on short petioles, *reniform-
cordate* to cordate, *acute,* crenately toothed or crenulate, 1—4 in. broad;
stipules usually small and scarious, entire or serrulate: fl. bright-yellow,
½ in. long; petals more or less purple-nerved, the lateral ones bearded:
capsule obovate-oblong, 4—5 lines long, abruptly beaked.—In wet
shades among the higher hills.

5. **V. pedunculata**, Torr. & Gray. Stems 2—6 in. long, prostrate or
assurgent; almost glabrous or puberulent: leaves *rhombic-cordate,*
usually *almost truncate at the broad base,* obtuse, coarsely crenate, ½—1½
in. long; stipules foliaceous, narrowly lanceolate, entire or incised:

peduncles erect, greatly exceeding the leaves, 4—8 in. high: fl. 1 in. broad or more, golden-yellow, the upper petals dark-brown on the outside, the others purple-veined within; lateral petals bearded: capsule oblong-ovate, 4—6 lines long, glabrous.—On low hills, in open ground.

6. **V. sarmentosa,** Dougl. Stems prostrate, more or less creeping, slender, sparsely leafy; slightly pubescent: leaves *rather thick and persistent*, reniform, round-cordate or ovate, ½—1½ in. broad, finely crenate, deep green above, often rusty beneath, *usually punctate* with dark dots: peduncles slender, elongated: fl. light-yellow, not large.—In woods of the Coast Range.

7. **V. purpurea,** Kell. Stems clustered, from a branching perpendicular root, 2—6 in. high: pubescence very scant but under a lens hispidulous, somewhat retrorse or at least spreading: herbage rather succulent, in early stages purple, except the upper surface of the leaves: leaves from broadly *ovate to lanceolate, tapering to the petiole*, entire or more or less coarsely and often somewhat crenately toothed: peduncles little exceeding the leaves: petals 3—5 lines long, light yellow within, dark purple externally: capsule almost globular, 3 lines long, pubescent. —Mt. Diablo Range.

+ + + *Leaves divided or lobed; fl. yellow.*

8. **V. lobata,** Benth. Stoutish, erect, ½—1 ft. high, from an erect rootstock, leafy at the summit; puberulent or nearly glabrous; leaves of reniform or cordate outline, 2—4 in. broad, the cauline short-petioled, all *palmately cleft into 5—9 narrowly oblong lobes*, the central lobe largest or longest; some of the radical leaves less lobed or only coarsely toothed: petals 6 lines long, yellow, the upper brownish externally, the lateral slightly bearded: capsule 5—6 lines long, acute. Var. **integrifolia,** Wats. Leaves deltoid, acuminate, evenly crenate-serrate, not at all lobed.—Inner Coast Range.

Order XII. CARYOPHYLLEÆ.

Herbs or suffrutescent plants with inert watery juice, mostly opposite leaves and swollen nodes. Inflorescence usually dichotomous. Flowers mostly 5-merous, complete and regular. Sepals united or distinct, imbricate in bud, persistent. Petals imbricate or convolute, often bifid, sometimes wanting. Stamens usually 10, occasionally 5, distinct, hypogynous around a ring-like disk, or perigynous by cohesion of disk with calyx-tube. Styles 2—5, mostly distinct and with decurrent stigmas. Fruit a capsule opening by valves or teeth.

* *Sepals joined into a tubular calyx.*

Calyx prominently 5-angled........................ .... VACCARIA 1
  " 10-ribbed: capsule coriaceous, 5-toothed .... AGROSTEMMA 2
  " many-striate; capsule 6-toothed ....SILENE 3

* * *Sepals distinct.*

← *Stipules none.*

Petals bifid {	capsule tubular, 10-toothed.............................CERASTIUM	4
	" 6-valved .. ..ALSINE	5
Petals entire or only emarginate {	capsule 3-valved; valves cleft......ARENARIA	6
	" 4—5 valved; valves entire.ALSINELLA	7

← ← *Stipules present, scarious or setaceous.*

Petals conspicuous {	capsule 5-valved..SPERGULA	8
	" 3-valved.................................TISSA	9
Petals minute or 0 {	leaves flat, not pungent........................POLYCARPUM	10
	" rigid, pungent.........................LŒFLINGIA	11

1. **VACCARIA,** *Dodoens.* A glabrous glaucous annual. Calyx synsepalous, pyramidal, with 5 prominent angles. Petals 5, unguiculate, not appendaged. Stamens 10. Styles 2. Capsule ovate, 1-celled, but with rudimentary partitions at base, 4-toothed at apex.

1. V. VULGARIS, Host. Erect, 1—2 ft. high, simple below, cymose-paniculate above: leaves cordate-ovate, acute, entire, sessile: petals red; blade obcordate; claw linear: styles short: seeds dark-colored.—An Old World weed of grain-fields.

2. **AGROSTEMMA,** *L.* (CORN-COCKLE). Tall annual, sparingly branched above; pubescent, not viscous. Calyx synsepalous, tubular, coriaceous, 10-ribbed, 5-toothed. Petals 5, unguiculate. Capsule coriaceous, 1-celled, 5-toothed.

1. A. GITHAGO, L. Erect, 2—4 ft. high, soft-hirsute: leaves linear-lanceolate, connate at base: fl. solitary on long upright peduncles: calyx 1½ in. long, the linear teeth as long as the tube, deciduous from the mature fruit: petals purple, not equalling the calyx-teeth; limb broad, obtuse, entire; claw unappendaged.—A weed of the grain-fields, not yet common in California.

3. **SILENE,** *Lobel,* (CATCHFLY). More or less viscid herbs. Calyx synsepalous, membranaceous, striate, 5-toothed. Petals (usually vespertine), commonly with cleft limb and appendaged claw. Stamens 10. Styles 3. Pod dehiscent at summit by 3 or 6 teeth.

* *Annuals.*

1. **S. antirrhina,** Linn. Erect, slender, *glabrous*, glandless except a viscid belt of an inch, more or less, in the middle of each internode of the branches: leaves lanceolate, acute, 1 in. long: pedicels erect: mature calyx oval, 3 lines long, the teeth short: petals red, the *blade emarginate, a line long;* crown inconspicuous: seeds minutely papillose.—In sandy soil, both along the seaboard and in the interior. March, April.

2. S. GALLICA, Linn. Slender, 1 ft. high, sparingly branched or nearly simple, *hirsute:* leaves spatulate, 1—1½ in. long: fl. racemose on very short pedicels, rose-color: *petals with obovate entire blade* and small appendages.—The commonest weed of fields and waysides; the small flowers usually forming a one-sided spike or raceme; the petals not withering so early in the day as in other species. Mar.—June.

3. S. RACEMOSA, Otto. Stoutish, rather roughly pubescent, 1½—2 ft. high, dichotomously racemose from near the base: leaves lanceolate: *fl. white, fragrant,* ½ in. broad, unilateral: *blade of petal cuneate-obovate, deeply bifid.*—Occasional in fields about Berkeley; flowers pure white, very fragrant, strictly vespertine, about twice as large as those of *S. Gallica* and quite showy.

4. **S. multinervis,** Wats. Erect, 1 ft. high, pubescent throughout, viscid-glandular above : leaves oblong-linear, acute : inflorescence cymose: calyx ovate, 5 lines long, *conspicuously 20—25 nerved:* petals small, purplish, unappendaged, not exceeding the subulate spreading calyx-teeth.—Mt. Tamalpais and southward.

* * *Perennials.*

5. S. INFLATA, Smith. Rather slender, 2 ft. high, glabrous: leaves obovate and oblanceolate to lanceolate-acuminate: cyme dichotomous and loose: calyx inflated, ovoid, with deltoid teeth, the nerves fine and numerous: petals large, white, bifid: capsule round-ovoid.—Naturalized about Vallejo, *Mr. Towle.*

6. **S. Californica,** Durand. Puberulent and more or less glandular, 4 in. to 4 ft. high, simple or sparingly branched above: leaves ovate to oblanceolate, 1½—4 in. long, acute or acuminate: fl. large, on short pedicels: calyx 7—10 lines long: *petals scarlet,* deeply parted, the segments bifid, their lobes 2—3-toothed or entire, often with a linear lateral tooth; appendages oblong-lanceolate: capsule ½ in. long, ovate, short-stipitate. —Coast hills.

7. **S. verecunda,** Wats. Pubescent and viscid throughout: stems ½—1½ ft. high, erect or decumbent: leaves oblanceolate, acute, 1—2 in. long: fl. few, erect, on stoutish pedicels ½—1 in. long: calyx oblong-cylindric, ½ in. long; teeth triangular, acutish: *petals* ¾ in. *long, rose-color;* limb bifid to the middle; lobes linear, the inner entire, outer commonly with a tooth near the base; appendages notched at apex; claw narrowly auricled: capsule oblong-ovate: seeds strongly tubercled on the back.—San Francisco peninsula from near the Presidio and the Mission Hills to Point San Pedro; also on Mt. Diablo.

4. **CERASTIUM,** *Dillen.* (MOUSE-EAR CHICKWEED). Soft-pubescent and slightly clammy low herbs, with white flowers in leafy- or scarious-bracted dichotomous cymes. Sepals 5, neither carinate nor 3-nerved.

Petals 5, bifid or emarginate. Stamens 10. Styles 5, rarely 4 or 3. Capsule cylindric, often incurved, thin and translucent, 1-celled, ∞ -seeded, dehiscent at apex by about 10 teeth. Seeds roundish-reniform.

1. **C. viscosum**, L. Annual, *soft pubescent* and somewhat *clammy*, the branches erect or ascending from a decumbent base, ½—1 ft. high: leaves ovate, obovate, or oblong-spatulate, ½—1 in. long: cymes in early state rather dense: pedicels even in fruit only 2 lines long; the calyx as long, the sepals acute: petals shorter than the calyx: capsule nearly straight, much longer than the calyx. Common weed in early spring; corolla expanding only in sunshine.—Native of Europe. Feb.—May.

3. **C. arvense**, L. Perennial, cespitose, *downy with reflexed hairs*, the inflorescence somewhat viscid: branches 4—8 in. high: leaves linear-lanceolate, 4—10 lines long, acutish: cyme contracted, bearing about 3 flowers (sometimes 5; as often 1 only), the branches ascending, often little exceeding the pedicel of the first flower; sepals ovate-oblong, obtuse, scarious-margined, 1½—2 lines long; the obcordate petals twice as long: capsule little exceeding the calyx.—San Francisco, and in Marin County. March—June.

4. **C. pilosum**, Ledeb. Perennial, erect, stout, more or less densely *pilose*, the inflorescence glandular-viscid: leaves oblong-lanceolate, ½—1 in. long, 1—6 lines broad, acute, almost sheathing at base: fl. few, large, in a terminal leafless cyme: sepals 3—4 lines long, obtuse; petals longer: capsule 6—10 lines long, the slender *teeth* at length *circinate-revolute*.—A Siberian and Alaskan species, said to have been found on Point Reyes.

5. **ALSINE**, *Diosc.* (CHICKWEED). Low herbs with mostly quadrangular stems, no stipules, and small axillary and solitary, or terminal and cymose white flowers. Flowers as in *Cerastium*, but styles usually 3 only, sometimes 2 or 4. Capsule globose or oblong, cleft below the middle into twice as many valves as there are styles.

1. **A. media**, Camerarius (1558). Weak, procumbent, rooting at the lower joints; stems marked by a pubescent line: *leaves ovate*, ¼—¾ in. long, *on slender petioles*, or the upper sessile: floral bracts foliaceous; pedicels slender, deflexed in fruit: calyx pubescent: stamens 3—10: capsule oblong-ovate, 2-3 lines long, equalling or exceeding the calyx.—A very common weed of shady places. Dec.—June.

2. **A. nitens** (Nutt). Stems almost capillary, diffuse, sparingly leafy, 3—6 in. high, the whole plant very glabrous and shining, or with a slight pubescence below: *leaves lanceolate*, ¼—½ in. long, acute, the lower short-petiolate: fl. erect, on short pedicels, in *a very lax bractless cyme*: sepals 3-nerved, narrow, acuminate, 2 lines long: petals deeply bifid, only half as long, sometimes 0; capsule oblong, shorter than the calyx.—Very common, yet so delicate and inconspicuous as to be easily overlooked.

3. **A. littoralis** (Torr.) Pubescent, ascending, stoutish, 1 ft. high: *leaves 1 in. long, ovate, acute,* rounded at base, rather thick: fl. in a terminal compound cyme: sepals lanceolate, acute, 3 lines long, obscurely 3-nerved, shorter than the 2-parted petals: capsule included within the calyx.—Point Reyes and Point Lobos.

6. **ARENARIA,** *Chabræus* (SANDWORT). Mostly low tufted herbs with sessile often rigid leaves and no stipules; flowers white, cymose-panicled. Sepals 5 or 4. Petals as many, entire, emarginate, or 0. Styles 3, opposite as many sepals. Capsule globose or ovoid, dehiscent into as many entire, 2-cleft, or 2-parted valves as there are styles. Seed reniform-globose, or laterally compressed.

* Low annuals; cymes foliaceous-bracted; valves of capsule 3, entire.

4. **A. Douglasii** (Fenzl), Torr. & Gray. Sparsely pubescent with spreading gland-tipped hairs, or glabrous, slender, branching, 3–12 in. high: *leaves filiform,* ¼—1½ in. long, ascending or spreading, slightly connate at base: fl. large, on long slender pedicels: sepals oblong-ovate, acute, 1½ lines long, 1—3 nerved: petals obovate, 2 lines long or more: capsule globose, equalling the calyx: seeds large, smooth, compressed and acutely angled.—On stony hill-tops and sandy or gravelly plains. March—May.

5. **A. Californica,** Brewer. Glabrous, very slender, 2—3 in. high: *leaves lanceolate, obtusish,* 1—2 lines long: fl. small, on slender pedicels: sepals oblong-ovate, acute, 3-nerved. 1—1½ lines long; petals spatulate, 2 lines: capsule oblong, as long as the calyx: seeds small, sharply muriculate.—Napa and Livermore valleys. April, May.

6. **A. palustris** (Kell.), Wats. Glabrous, flaccid, decumbent, leafy throughout, ½—2 ft. high: *leaves linear-lanceolate, acute,* ½—1 in. long: fl. few, large, long-pedicelled: sepals elliptic, obtuse or acutish, nerveless, herbaceous, but with a narrow scarious margin, 1½—2 lines long: petals oblong, twice longer: capsule oblong, shorter than the calyx: seeds numerous.—Dr. Kellogg, writing of this from San Francisco almost thirty years ago, says: "A plant very abundant in swamps in this vicinity, known to us for the last ten years." Probably now extinct.

* * Perennials; valves of capsule bifid.

7. **A. macrophylla,** Hook. Stems low, ascending from *running rootstocks,* mostly simple, leafy, puberulent above: *leaves in 3 or 4 pairs, lanceolate,* acute at each end, 1—2 in. long, thin and flaccid: fl. few, on slender pedicels: sepals ovate-oblong, acuminate, 1½—2½ lines long, 1-nerved, longer than the obtuse petals: capsule ovoid, nearly equalling the calyx: seeds few, large, smooth.—Shady northward slopes of Mt. Diablo, Mt. Hamilton, etc.

7. ALSINELLA, *Dillen.* (PEARLWORT). Diminutive herbs with subulate or filiform exstipulate leaves, and minute long-pedicelled often apetalous flowers. Sepals 4 or 5, commonly rotate-spreading in fruit. Petals when present as many, entire or emarginate. Styles 4 or 5. Capsule 1-celled, ∞-seeded, dehiscent to the base into as many entire valves as there are styles: the valves alternate with the sepals.

1. **A. occidentalis** (Wats.), Greene. Annual, glabrous or nearly so, almost capillary, decumbent at base or ascending, 1—6 in. high: leaves in pairs (none fascicled), slightly connate, acute, ¼—½ in. long: fl. 5-merous, on long pedicels, these erect in fruit: sepals 1 line long: petals nearly as long: stamens 10: capsule exceeding the calyx.—Very common. March—May.

2. **A. crassicaulis** (Wats.), Greene. Perennial, stoutish and succulent, decumbent: leaves broadly linear, acute, 2—6 lines long, scarious and connate at base: pedicels 4—8 lines long; fl. erect or nodding, large, the sepals more than a line long; petals smaller: styles very short: capsule ovate, scarcely exserted from the closed fruiting calyx.—A little known species found at Dillon's Beach, Marin Co.

8. SPERGULA, *Dodoens* (CORN-SPURREY). Herbs with linear and apparently whorled leaves; the opposite pair (subtended by a pair of scarious stipular scales) being augmented by several crowded and spreading fascicled ones of nearly their own size which along with them seem to form a verticil. Flowers perfectly symmetrical (stamens 10 or 5); the 5 styles alternate with the sepals, the 5 valves of the capsule opposite the sepals. Petals entire. Seeds acutely margined or winged.

1. S. ARVENSIS, L. Glabrous or pilose-pubescent and slightly clammy, 1—2 ft. high, simple or with many decumbent basal branches: leaves almost filiform, 1—2 in. long: cyme terminal, ample, dichotomous, the long pedicels nodding after flowering, but erect in flower and again when the capsule is mature: sepals oblong or ovate, 2—3 lines long, the white petals rather long, unfolding only in sunshine: capsule ovoid: seeds acutely margined.—In fields and by waysides everywhere. Jan.—Sept.

9. TISSA, *Adanson* (SAND SPURREY). More or less succulent herbs of maritime districts or subsaline plains inland. Leaves linear or subulate, with scarious stipules. Flowers arranged dichotomously or unilaterally. Sepals 5. Petals 5, entire (sometimes fewer than 5 or even 0). Stamens 2—10. Styles 3, rarely 5. Capsule 3-valved. Seeds winged or wingless. Embryo annular.

  * *Perennials with fusiform fleshy roots.*

  ← *Internodes not short (about 1 in.); fascicled leaves few.*

1. **T. macrotheca** (Hornem.), Britt. Stems ascending, stoutish, terete, often 1 ft. high; whole *herbage deep green* and rather densely

viscid-pubescent: leaves semiterete, linear-subulate, acute, often longer than the internodes (1—2 in.); stipules ovate-triangular, 2 lines long: pedicels ½ in. long or more, subtended by leafy bracts often nearly as long: sepals ¼ in. long, with narrow scarious margins: petals as long, lilac: capsule ovoid, about equalling the calyx: seeds triquetrous-obovate, *smooth, dark-brown,* with a very narrow or sometimes obsolete scarious wing.—Maritime only, and common in sandy soil along the borders of salt marshes. April—Dec.

2. **T. leucantha,** Greene. Habit of the preceding, but *glabrous* except a glandular pubescence on the more ample and loosely dichotomous inflorescence; branches more or less distinctly quadrangular: leaves linear, acute, little exceeding the internodes (1 in. or. more); stipules deltoid-ovate, acuminate, 2—3 lines long: pedicels 1 in. long or more, at length abruptly deflexed, subtended by reduced and linear-subulate bracts: sepals 2—3 lines long, with broad scarious margins: corolla ½ in. broad or more, *pure white: filaments broadly subulate* and almost petaloid: apex of capsule exserted, distinctly triquetrous: seed brown, smooth, of round-obovate outline and with a broad scarious wing.—Confined to clayey subsaline or alkaline plains of the interior; plentiful on the eastern side of the Livermore Valley. March—May.

+ + *Internodes short; axillary leaf-fascicles conspicuous.*

3. **T. pallida,** Greene. Prostrate, *diffusely branching and densely cespitose,* the geniculate stems stoutish below, often naked and appearing suffrutescent; *herbage pale,* densely pubescent and *very viscid:* primary leaves oblong-linear, very acute, ½ in. long or more; those of the fascicles shorter and relatively broader; stipules ovate-acuminate, often 4—5 lines long: fl. either scattered singly on short branchlets, or in reduced terminal cymes: pedicels ½ in. long: calyx ¼ in. long: *petals lilac:* capsule equalling the calyx: seeds obliquely orbicular, light brown, very smooth, broadly margined.—On high and dry clayey bluffs overhanging the ocean in San Francisco Co., also in Marin. April—July.

4. **T. Clevelandi,** Greene. Prostrate, slender, very diffuse, *forming deep green mats* ½—1½ ft. broad; herbage pubescent but only slightly viscid: leaves narrowly linear, the fascicled ones subulate, all equalling or exceeding the internodes: fl. in terminal cymes only, small (⅛ in. broad), *pure white.*—Only on high and dry sandy ground, back from the sea-bluffs. April—June.

* * *Annuals; flowers usually lilac or lavender-color.*

5. T. RUBRA (L.), Britt. Stems slender, terete, prostrate, a few inches long, glabrous below, pubescent and more or less glandular above:

leaves narrowly linear or subulate, acute or mucronate, $\frac{1}{4}$—$\frac{1}{2}$ in. long; stipules lanceolate, acuminate, 1—2 lines long: pedicels slender, 2—3 lines long: sepals oblong, obtuse, scarious-margined: petals reddish, about equalling the sepals: capsule ovate, obtuse, not exserted: seeds brownish, tuberculate, wingless, triquetrous-obovate, with a marginal elevation.—Roadsides; frequent, and sometimes perennial. April—Oct.

6. **T. marina** (Wahlb.), Britt. Root thickish, not much branched, sometimes perennial: stems ascending, 3—8 in. high, somewhat compressed or angular, glabrous or somewhat glandular-pubescent: *leaves semiterete, narrowly linear*, acute, light green, glabrate, seldom exceeding the internodes; stipules broadly ovate, abruptly acuminate : cymes scarcely leafy: pedicels about twice as long as the capsules: sepals acute or acuminate, with a broad or narrow scarious margin: petals broadly ovate, obtuse, scarcely equalling the sepals, whitish or pale rose-color: capsule ovate, obtuse, nearly twice the length of the calyx: *seeds orbicular*, with an elevated margin, reddish-brown, *smooth*, winged or wingless.— Common and variable, occurring mostly near the sea.

7. **T. salina** (Presl.), Britt. Roots slender and tufted, simple or much branched: stems 6 in. high, much branched, usually ascending, rarely divaricate and prostrate: herbage glabrous or pubescent: *leaves flat, linear-filiform*, obtuse or acutish, glabrous, light or livid green, seldom longer than the internodes; stipules broadly ovate, short-acuminate, not shining: pedicels leafy-bracted, or the upper bractless, none of them longer than the capsules: sepals oblong or oblong-ovate, obtuse, scarious-margined: capsule acute, much longer than the calyx: *seeds roundobovate, tuberculate or muriculate*, the marginal elevation distinct; hyaline wing narrow or wanting. Var. **sordida**, Greene. Stems ascending; herbage very viscid and hairy; fl. in unilateral leafless racemes: seeds nearly black, sharply muriculate, wingless. Var. **Sanfordi**, Greene. Stems erect repeatedly dichotomous: herbage scarcely viscid and only slightly pubescent: inflorescence partly dichotomous, only the ultimate branchlets unilaterally racemose: seeds dark brown, nearly smooth, wingless.—Common on the seaboard. The first variety is very abundant in low rich soil above the salt marshes on the "Island," near Alameda. The second belongs to the plains in the interior. March—May.

9. **T. tenuis**, Greene. Slender, prostrate, very diffuse, the whole plant 1 ft. broad, glabrous, or the inflorescence sparsely glandular-pubescent: leaves linear-filiform, 1 in. long, equalling the internodes; stipules broader than long, acute, but small and inconspicuous: fl. very numerous, crowded and often subsessile on the countless dichotomous-cymose branchlets, apetalous: stamens 2: *capsule triquetrous*, acute, more than *twice the length of the* oblong obtuse scarious-margined *sepals*: seeds reddish-brown, obliquely obovate, compressed, *smooth*, margined, wingless.—Alameda.

10. POLYCARPON, *Lœfling.* Low annuals, diffusely dichotomous, with flat leaves, small scarious stipules and minute cymose flowers. Sepals 5, carinate-concave. Petals 5, minute, hyaline. Stamens 3—5. Ovary 1-celled; style short, 3-cleft. Capsule 3-valved, several-seeded.

1. P. TETRAPHYLLUM, Linn. Branched from the base and depressed, the branches 3—7 in. long: leaves ½ in. long, obovate, short-petiolate, the lower pairs with a second rather smaller pair at right angles to them so as to appear whorled: stipules triangular-lanceolate: cymes many-flowered, dense, the small flowers short-pedicelled: sepals green or purplish, scarious-margined: petals scarious, shorter than the sepals: capsule about equalling the calyx: seeds yellowish brown, semiorbicular, compressed, with a sharp muriculation disposed in irregular striæ.—Naturalized in Napa and Solano counties. *Bioletti, Jepson.*

11. LŒFLINGIA, *Linn.* Low much branched rather rigid and pungent-leaved annuals; the leaves with adnate and connate setaceous stipules. Flowers small, sessile in the axils of the leaves and branches. Sepals 5, rigid, carinate. Petals minute or 0. Stamens 3—5. Ovary 1-celled; style very short or 0. Capsule 3-valved, several-seeded.

1. L. squarrosa, Nutt. The numerous prostrate or erect-spreading branches 2—6 in. long; herbage glandular-pubescent: leaves and sepals subulate-setaceous, rigid and recurved, the leaves 2—3 lines long, the sepals somewhat shorter: capsule elongated, triquetrous, at length exserted, ∞-seeded.—Plains of the lower San Joaquin, and Sacramento.

ORDER XIII. FRANKENIACEÆ.

An order embracing scarcely more than the genus

FRANCA, *Micheli.* Herbs or undershrubs with opposite entire small exstipulate leaves usually sessile and even united at base by a slight membranous continuation of the blade. Fl. small, solitary and sessile in the axils of the very numerous branches and branchlets, usually 5-merous and complete. Calyx tubular, furrowed; the lobes valvate and induplicate in bud. Petals hypogynous, narrowed to a claw which bears an appendage on its inner face. Style cleft into 2—4 filiform divisions; ovary 1-celled. Capsule invested by the persistent calyx; the few seeds attached to the margins of the 2—4 valves.

1. F. grandifolia (Esch.). Somewhat woody at base, erect, much branched and slender, ½—1 ft. high, glabrous or soft-pubescent, very leafy: leaves obovate to narrowly oblanceolate, revolute, ¼—½ in. long, of a dull green: calyx linear, ¼ in. long, strongly furrowed, the lobes short, acute: petals small, red, the blade 1 line long or more, erose at summit, the appendage of the claw bifid: stamens 4—7: style 3-cleft: capsule shorter than the calyx, linear, angular: seeds numerous.—A homely plant of the salt marshes along the seaboard.

Order XIV. ILLECEBREÆ.

Differing from the Caryophylleæ only in having a 1-seeded and utricular fruit.

1. PARONYCHIA, Clusius. Herbs with opposite entire leaves and a pair of scarious stipules at each node; flowers (in ours) clustered in the axils. Sepals 5, imbricate, somewhat cucullate under the apex and aristate or mucronate at the very tip. Stamens 5 or fewer, inserted on the base of the sepals, these often slightly united. Petals represented by 5 small setiform organs alternating with the stamens. Ovary 1-celled, 1-ovuled; ovule attached by a slender basal funiculus, ascending or subpendulous. Utricle enclosed in the persistent calyx, at length bursting longitudinally. Seed smooth. Embryo annular.

1. **P. Chilensis, DC.** Perennial, diffuse, cespitose, the tough and pliable short-jointed stems suffrutescent: leaves oblong-linear, 1½—3 lines long, membranaceous, pungent at tip, minutely appressed pubescent; stipules thin-hyaline, ovate-lanceolate, 1—2 lines long: fl. few in the axils, very shortly pedicelled: calyx scarcely ¾ line long, purplish; sepals spinulose-tipped and only slightly cucullate: seed reddish-brown. Frequent on grassy hillsides and summits at the Presidio; evidently indigenous; otherwise known only as South American.

2. HERNIARIA, Dodoens. Differing from *Paronychia* in habit, and in that the sepals are united at base.

1. **H. cinerea, DC.** Annual, slender, parted from the base into a few ascending branches, these with many short distichous branchlets; herbage canescent with setulose straight or uncinate-tipped hairs: leaves oblong-lanceolate, acute, sessile, 1—2 lines long; stipules hyaline, minute, broadly ovate: fl. crowded, sessile, minute, the calyx ½ line long: seed black, smooth and lustrous.—Eastern base of Mt Diablo; native of S. Europe. (*Paronychia pusilla* of Fl. Fr.) April.

3. PENTACÆNA, Bartling. Perennials of cespitose habit, with alternate subulate rigid and pungent leaves, silvery-hyaline stipules, and sessile flowers clustered in the axils. Sepals 5, united at base, very unequal, cucullate, the 3 outer large and with a stout divergent terminal spine, the 2 inner much smaller and with but a short awn. Petals minute, scale-like. Stamens 3—5: staminodia 0. Style very short, bifid. Utricle enclosed in the rigid persistent calyx. Embryo curved.

1. **P. ramosissima (Weinm.), Hook. & Arn.** Stems prostrate, forming mats 5 in. to 2 ft. broad, woolly-pubescent: leaves 3—5 lines long, squarrose when old; stipules lanceolate, shorter than the leaves, 1-nerved: calyx-tube nearly a line long, the divergent outer lobes 2 lines: utricle apiculate.—On sandy plains and dry gravelly hilltops toward the sea throughout our district.

Order XV. POLYGONEÆ.

Herbs, or rarely shrubs, with alternate or whorled leaves of revolute vernation; stipules when present cohering around the stem and forming a sheath. Inflorescence various, but commonly racemose and terminal. Calyx of 4—9 nearly or quite distinct sepals, often colored and petaloid, persistent. Stamens as many as the sepals, or fewer, perigynous. Styles 2—4, distinct or somewhat connate, opposite the angles of the lenticular or triquetrous 1-ovuled ovary. Fruit a compressed or triquetrous achene. Seed erect; embryo straight, in the midst of a farinaceous albumen, or curved around it.

Leaves alternate, stipulate:
Sepals 4—6, equal, appressed to the achene..........................Polygonum 1
" " the outer smaller...Rumex 2
Perianth tubular below, 6-lobed above...............................Vibo 8

Leaves often verticillate, exstipulate:
Involucre tubular or campanulate, with 4—8 teeth.................} achenes triquetrous.....Eriogonum 4
" lenticular......Oxytheca 5
" " " " with 3—6 cuspidate teeth.........Chorizanthe 6
Involucre 2-lobed, 1-flowered...................................Pterostegia 7

1. **POLYGONUM,** *Columna.* Herbs or undershrubs with alternate entire leaves and sheathing stipules. Flowers small, in axillary fascicles or terminal spikes or racemes. Perianth of 5 or 6 nearly distinct often colored and petaloid sepals. Petals 0. Stamens 4--9, commonly in 2 sets or circles. Styles 2 or 3, distinct, or connate below, often very short; stigmas capitate. Fruit a triangular or lenticular achene, usually closely invested by the persistent perianth.

* *Leaves jointed upon a short petiole adnate to the 2-lobed or lacerate sheath; flowers axillary to leaves or bracts; filaments of the 3 inner stamens broad at base; achenes triquetrous.*

← *Glabrous and suffrutescent; sheaths conspicuous; sepals colored.*

1. **P. Paronychia,** Ch. & Schl. Stems stoutish, ascending or prostrate, 1—2 ft. long, leafy above, below clothed with the scarious sheaths, these ½ in. long, brownish and 5-nerved below, lacerate above: leaves sub-coriaceous, 1 in. long, linear-lanceolate, revolute: fl. densely crowded at the ends of the branches, the spikes more or less leafy-bracted: perianth white or rose-color veined with green or brown, ¼ in. long: achene 2 lines long, smooth and shining.—In sandy soil near the sea; flowering almost all the year through.

2. **P. Bolanderi,** Brew. Stems slender, tufted and strictly erect, ½—2 ft. high: sheaths much shorter than the nodes, herbaceous below, scarious and lacerate above, persistent: leaves narrowly linear or subulate, acute or cuspidate, ¼—½ in. long, not revolute: fl. solitary or few

in the axils of short leafy branchlets, each involucrate with a sheath-like scarious bract on the joint of the short pedicel: sepals 1½ lines long, rose-color or white, slightly spreading.—Plentiful near the Soda Springs above Napa; also in Sonoma Co. Aug.—Oct.

+ + Annuals, with striate stems and less conspicuous sheaths; branches leafy to the summit, floriferous throughout.

3. P. AVIOULARE, L. Stoutish, much branched, prostrate, the branches 1—3 ft. long: herbage *glabrous, bluish-green:* leaves oblong or lanceolate, acutish, ½—2½ in. long: fl. on very short pedicels: sepals 1 line long, green, with white or rose-colored margin: *achene broadly ovate,* 1 line long or less, *dull black and minutely granular.*—A prevalent weed in summer fields and vineyards. April—Oct.

4. P. coarctatum, Dougl. Erect, freely branching, the herbage more or less *scabro-puberulent throughout:* leaves firm in texture, acute: fl. spicate-crowded and on erect pedicels: sepals rose-color or white with only a broad midvein of green: achenes very *minutely punctate toward the apex.*—Petrified Forest, Sonoma Co. July—Sept.

* * Leaves not jointed with the petiole, striately 3-nerved; sheaths 2-lobed or fimbriate; stamens 8, the inner 3 scarcely dilated.

5. P. Californicum, Meisn. Erect, slender, 3—6 in. high, panicled-spicate, the stem and branches glabrous, dark brown: leaves rigid, linear or filiform, ½—1¼ in. long, pungently acute: spikes very slender, elongated, the subulate bracts 1—2 lines long; sheaths 1 line long, deeply lacerate-fringed, nearly equalling the pale rose-colored flowers: achene narrow, slightly exposed; styles slightly divergent.—Valleys and dry hills of the interior, near Napa, etc. · July—Sept.

* * * Leaves not jointed, more ample, pinnately veined; sheaths cylindrical, oblique or truncate; fl. in dense spikes or loose cymelets; stamens 4—8, all the filaments filiform; styles deciduous, often only 2 and the achene lenticular.

+ Weedy annuals of fields and gardens.

6. P. nodosum, Pers. Stoutish, erect or ascending, 1—4 ft. high, freely branching, glabrous except the rough glandular peduncles, and scabrous leaf-margins and veins beneath; stem often purple-dotted throughout: leaves lanceolate, 2—5 in. long, acuminate, short-petioled; *sheaths naked in age,* glandular-ciliolate when young: spikes linear, usually drooping, 1 in. long or more: fl. white or pale rose, 1 line long: stamens 6: styles 2: *achene lenticular, ovate.*—Very common in cultivated lands, preferring moist places. July—Oct.

7. **P. Persicaria, L.** Much like the last but the *sheaths* and bracts *conspicuously ciliate:* leaves less acuminate, subsessile: spikes shorter and erect: fl. rose-color: achenes often triquetrous.—Not common.

+ + *Perennials, either aquatic or of wet places.*

8. **P. acre, HBK.** Decumbent, rooting at the lower joints, 2—5 ft. high; herbage *light green, pellucid-punctate and acrid,* glabrous or a little scabrous: leaves lanceolate, acuminate, short-petioled; *sheaths bristly-ciliate:* spikes narrow and lax, 1—3 in. long, erect: *sepals greenish* and glandular-dotted, 1 line long: stamens 8: achene commonly triquetrous. —Very common in marshy places, along streamlets, etc. June—Nov.

9. **P. Hartwrightii,** Gray. Stems stout and simple, rooting at the decumbent base, above equably leafy to the summit; herbage more or less *strigose-hirsute:* leaves broadly lanceolate, acute, 2—7 in. long, on very short petioles; stipules with an abruptly spreading foliaceous bor-der: fl. rose-red, in a dense ovate or oblong terminal spike: stamens 5: style 2-cleft: achene lenticular.—In low ground; not common.

10. **P. Muhlenbergii,** Wats. Stoutish, erect, 2—3 ft. high, leafy throughout; *scabrous* with short appressed or glandular hairs, with more or less of a softer pubescence: *leaves broadly lanceolate,* narrowly acumi-nate, 4—7 in. long, on petioles of nearly 1 in.; sheaths with no spreading margin: *spikes* 1 or 2, elongated and narrow, *1—3 in. long:* fl. and fr. as in the last.—Shore of the lakelet in front of the U. S. Marine Hospital, San Francisco; rare in California.

11. **P. amphibium, L.** Aquatic and with floating leaves, or geniculate and rooting in the mud along the shores of ponds and lakes; herbage *glabrous* or nearly so: *leaves elliptical or oblong,* obtuse or acutish, very smooth and shining above, 2—5 in. long, on petioles half as long: *spike* mostly solitary, *dense, ovate or oblong,* 1—1½ in. long: fl. rose-color: fr. lenticular.—Common in mountain lakes.

* * * * *Twining or climbing annuals with broad leaves, and flowers in loose axillary panicles or racemes; achenes triquetrous.*

12. P. CONVOLVULUS, L. Twining or trailing, 1—3 ft. high, minutely scabrous: leaves 1—2 in. long, hastate-cordate, acuminate: fl. in axillary interrupted racemes: fruiting perianth 1½—2 lines long, equalling the somewhat opaque granulate-striate achene.—A weed in cultivated lands; native of Europe, not yet prevalent in California, but already met with near Berkeley. July—Sept.

2. **RUMEX,** *Pliny* (DOCK. SORREL). Coarse perennials (rarely annual or biennial), with leafy stems, and cylindrical obliquely truncate scarious stipules; the small green or reddish perfect or unisexual flowers fascicled or verticillate, forming panicled racemes. Perianth of 6 nearly

or quite distinct sepals; the outer herbaceous, spreading or reflexed; inner larger, in some becoming greatly enlarged in fruit, appressed to the 3-angled achene. Stamens 6; filaments very short. Styles 3 (or 2); stigmas tufted. Embryo lateral, slender, slightly curved.

* *Fl. perfect or polygamous; valves accrescent, often with a grain-like protuberance on the back; leaves elongated, never hastate, pinnately many-veined; herbage scarcely acidulous.*

+— *Valves small (2 lines, more or less), one or more of them grain-bearing.*

++ *Valves with slender awned teeth; herbage pubescent or scabrous.*

1. R. OBTUSIFOLIUS, L. Tall (3—5 ft.), slender, somewhat scabrous: radical leaves oblong, obtuse, cordate or truncate at base, long-petioled, the blade 6—15 in. long: fl. in loose whorls, on long pedicels, these jointed below the middle: *valves ovate-deltoid,* 2—3 lines long, *with 1—3 setaceous teeth* on each side, usually only one valve grain-bearing.— Naturalized, but rather sparingly, and in low lands only.

2. R. PULCHER, L. Erect, 2—3 ft. high, with rigid branches divaricately and widely spreading: leaves scabrous beneath, the radical oblong or lanceolate (sometimes panduriform), acute, at base cordate or obtuse: fl. on short stout rigid pedicels: *valves ovate,* 2—3 lines long, with *4—6 rigidly awned teeth* on each side.—Very abundant by way-sides.

3. R. polygonoides, L. Low (about 1 ft.), erect, stout, from an annual or biennial root: herbage minutely pubescent and of a pale or yellowish green: leaves linear-lanceolate, the margin somewhat crisped or undulate, short-petioled, the blade 1—4 in. long: inflorescence compact, the verticils dense: valves 1 line long, *ovate-lanceolate,* all grainbearing, and with *2 or 3 long-awned teeth* on each side.—Common in brackish marshes, and on lake shores.

++ ++ *Valves entire or only denticulate; herbage glabrous.*

4. R. CONGLOMERATUS, Murr. Stoutish, 3—4 ft. high, leafy-paniculate above: radical leaves ovate or lanceolate, cordate, slightly undulate: pedicels short, stout and geniculate in fruit, jointed near the base: valves small, all grain-bearing, *ovate-lanceolate,* acute.—In wet places.

5. R. CRISPUS, L. Size and habit of the last, but panicles less leafy and more condensed: leaves long-petioled, truncate at base, *strongly undulate or crisped:* pedicels 2—4 lines long, rather slender: valves all grain-bearing, *ovate or cordate,* strongly reticulate.—In waste lands.

6. R. Berlandieri, Meisn. Stout, erect, 2—4 ft. high: leaves narrowly lanceolate, very undulate, more or less acuminate, narrowed below to an abruptly cuneate or almost truncate base, 6 in. long, short-petioled:

pedicels 1—2 lines long, jointed below the middle: valves *ovate-lanceolate*, 1½ lines long, finely reticulate, all grain-bearing.—Said to have been found at San Francisco.

7. **R. salicifolius,** Weinm. Stems clustered, ascending, 1—3 ft. high: lowest leaves oblong, upper linear-lanceolate, 3—6 in. long, acuminate, narrowed to a short petiole, not undulate, pale green: panicle open, somewhat leafy, the flowers crowded: pedicels slender, 1—3 lines long: valves *ovate-rhomboid or broadly deltoid*, 1½—2 lines long, entire or denticulate, one or two of them with large whitish grains.

+ + *Valves ¼—½ in. long, not grain-bearing; herbage glabrous.*

8. **R. occidentalis,** Wats. Erect, 3—6 ft. high, sparingly branched: leaves oblong-lanceolate, usually narrowing upward from the truncate or somewhat cordate base, not decurrent upon the petiole, 1 ft. long or more, scarcely undulate, usually acute: panicle narrow, elongated, nearly leafless: pedicels slender, ¼—½ in. long, obscurely jointed near the base: valves *broadly cordate*, with a very shallow sinus, becoming about ¼ in. broad, often denticulate near the base: achenes 1½ lines long. — Frequent in marshy places.

* * *Glabrous perennials with reddish usually diœcious flowers; valves not grain-bearing; leaves mostly either broad and rounded, or hastate, sparingly veined; herbage tender and acid.*

9. R. Acetosella, L. Stems erect from running rootstocks, slender, 6—18 in. high: leaves oblong- to linear-lanceolate, or oblanceolate, 1—3 in. long, usually hastate, the lobes often toothed: panicle naked, long and narrow; fl. diœcious, small, red, in loose fascicles; pedicels short, jointed at top: achene small, ovate-triquetrous, ⅔ line long.—Very common.

3. **VIBO,** *Medic.* Annual herbs with alternate leaves, and axillary solitary or clustered unisexual flowers. Staminate perianth 5—6 parted; segments equal, spreading. Stamens 4—6; filaments filiform. Fertile perianth with urceolate tube and 6 unequal lobes in 2 series, the whole accrescent in fruit and indurated; the outer lobes spreading and spinescent, the inner plane, erect-connivent. Fruit a triquetrous achene enclosed in the tube of the perianth but free from it. Seed subterete.

1. V. australis (Steinh.). Glabrous; the stout and rigid prostrate branches 1—2 ft. long: leaves triangular-ovate, entire, 2 in. long; at base abruptly narrowed to a long petiole: staminate fl. often clustered at the end of a peduncle; the pistillate sessile: fructiferous perianth ⅓—½ in. long, thick and almost woody; outer lobes broadly subulate and thorn-like, the inner broadly ovate, mucronate.—Native of S. Africa and Australia; adventive on our sea-beaches.

4. ERIOGONUM, *Michaux*. Annual, perennial or suffrutescent plants with radical or alternate or verticillate exstipulate leaves and a greatly diversified inflorescence of involucrate, mostly small and dense primary flower-clusters. Involucre campanulate, turbinate or oblong, 4—8-toothed or -lobed without awns; the pedicels few or many, more or less exserted, subtended by scarious and narrow or quite setaceous bractlets. Perianth 6-cleft or -parted, colored, enclosing the achene. Stamens 9, upon the base of the perianth. Styles 3; stigmas capitate. Achene 3-angled, rarely 3-winged.

* *Plants with scape-like peduncles, from a more or less woody and leafy base; involucres umbellate or capitate, not virgately disposed.*

← *Perianth narrowed to a slender stipe-like base.*

1. **E. stellatum,** Benth. More or less tomentose, the stems diffuse and leafy: leaves ovate-spatulate to oblanceolate: peduncles ½—1 ft. high, bearing an *umbel of 2—4 usually elongated and cymosely divided rays;* the nodes all leafy-bracted: fl. yellow: stipitate *base of perianth elongated.*—On Mt. Diablo and Mt. Hamilton.

2. **E. compositum,** Dougl. More or less white- or yellowish-tomentose, the leaves densely so beneath; these oblong-ovate, cordate at base, acute or acutish, 1—3 in. long on rather long petioles: *peduncles stout, naked,* ½—1½ ft. high, nearly glabrous: umbel of 6—10 long *rays, each bearing a short several-rayed umbellule,* subtended by whorls of linear-oblanceolate leaflets: fl. 2—4 lines long, cream-colored or yellow, the *stipe-like base relatively short.*—Napa and Sonoma counties, at middle elevations of the Coast Range.

← ← *Perianth abruptly contracted at base.*

3. **E. latifolium,** Smith. *Stout, tomentose throughout,* the short caudex sparingly branched and leafy: leaves oblong or oval, obtuse or acute, 1—2 in. long, rounded or cordate, or rarely cuneate at base, commonly undulate, often glabrate above, 1—2 in. long, the stoutish petiole often short and margined: peduncles stout, 6—20 in. high: bracts triangular: *involucres* very-many-flowered, crowded *in 1—3 large terminal heads,* or the peduncles more than once forked above and the heads smaller: bractlets densely villous-plumose: fl. white, the sepals broadly obovate. In rocky or sandy places near the sea.

4. **E. nudum,** Dougl. Much taller and more slender than the last, the ovate or oblong *leaves* (½—2 in. long) densely tomentose beneath, *glabrate above:* peduncle and loose panicle 1—2 ft. high, glabrous and glaucescent, or somewhat floccose-tomentose: involucres 2—3 lines long, nearly or quite glabrous, 3—6 in each cluster: fl. glabrous or villous, 1—1½ lines long, white.—Clayey banks and dry hills; a more inland species than the last. July—Oct.

* * *Annuals, leafy near the base only; panicles dichotomous; involucres sessile, not virgately disposed.*

5. **E. truncatum,** Torr. & Gray. Slender, *floccose-tomentose throughout,* 1 ft. high: leaves mostly rosulate near the base of the stem, sometimes a whorl subtending the lowest node; blade oblanceolate, 1 in. long, attenuate to a slender petiole, the margin undulate: inflorescence very lax, in a kind of umbel of 4—6 elongated and di- or trichotomous rays: involucres few, oblong-turbinate, 2 lines long: fl. rose-color, 1 line.— Seemingly local, but plentiful at the eastern base of Mt. Diablo.

6. **E. Nortoni,** Greene. Near *E. truncatum,* but smaller, more branching, the stem branches and involucres glabrous, reddish in age: *leaves* small and thick, broadly obovate or somewhat obcordate, cuneately but abruptly narrowed to a long slender petiole, *glabrous above,* white-tomentose beneath: involucres terminal upon the dichotomous branches and sessile in the forks, solitary, turbinate, 1½ lines long, the 5—7 angles conspicuous, teeth short and blunt: fl. very numerous, rose-color, ¾ line long: sepals equal, obovate, nearly truncate.—Marin Co.

* * * *Annuals; the diffusely di- or trichotomous panicle leafy at the nodes; involucres pedicellate.*

7. **E. angulosum,** Benth. Grayish-tomentose, 6—18 in. high, loosely and widely branching from near the base, the *branches 4—6 angled:* lowest leaves ovate or rounded, cuneate or somewhat cordate at base, obtuse, often undulate, ½—1 in. long, on rather short petioles; upper oblong or lanceolate, subsessile: pedicels of the involucres ¼—1¼ in. long, filiform: involucre hemispherical, 1—2 lines broad, many-flowered, smooth or glandular: bractlets mostly dilated and rather firm: fl. rose-color, purplish, or even greenish-white, ½ line long, not quite glabrous; outer sepals *ovate, concave,* the inner *lanceolate, plane,* somewhat longer.— Plains of the interior of Solano Co.

* * * * *Involucres sessile, scattered up and down the virgate branches.*

← *Annuals; leaves mostly basal and rosulate.*

++ *Involucres tubular, 2 lines long.*

8. **E. virgatum,** Benth. Usually white-tomentose throughout: leaves oblong, 1 in. long on slender petioles: peduncle simple, or with only a few erect virgate branches, 1—2½ ft. high, the involucres remote, the 5 teeth very short: perianth 1 line long, buff or yellow; outer sepals broadly obovate, cuneately narrowed at base, the inner about as long, spatulate-oblong: achene with a *minutely puberulent* rather slender *beak.* —Banks of Putah Creek, Solano Co.

9. **E. dasyanthemum,** Torr. & Gray. Usually hoary-tomentose, sometimes nearly glabrate: leaves oval or rounded, 5—10 lines long,

abruptly narrowed to a slender petiole: peduncle 1 ft. high, mostly rather loosely but widely branching, the branches often more or less cymose-dichotomous: involucres rather remote, not always solitary, narrowly tubular, very shortly toothed, tomentose except the prominent ribs, these glabrous: fl. scarcely exserted, erect, not numerous, 1 line long, white or rose-color; more or less densely *villous on the outside.* Var. **Jepsonii,** Greene. Panicle ample, as broad as high, the dichotomous branches widely spreading: involucres campanulate-tubular, very-many-flowered, the pedicels exserted and recurved; fl. rose-red.—The type, along Putah Creek; the variety, from Gate's Cañon, near Vacaville. Sept., Oct.

++ ++ *Involucres* 1—1½ *lines long, usually turbinate.*

10. **E. vimineum,** Dougl. Seldom at all tomentose except on the lower face of the ovate or orbicular slender-stalked leaves: peduncle 1 ft. high, branched from near the base, the branches slender and virgate: involucres very narrow and rather prismatic, the *teeth very short:* fl. few, rose-color or white, exserted; outer sepals obovate, the inner obovate-oblong and only half as broad.—Sonoma Co.

11. **E. gracile,** Benth. Slender, 1—2 ft. high, usually white-woolly throughout: leaves rosulate or scattered, ovate, oblong or oblanceolate, tomentose on both faces: panicle of few or many usually rather strict and virgate very slender branches: involucres many-flowered, turbinate, the 5 *teeth stout, prominent, acutish:* fl. white, rose-color or yellowish, ¾ line long; outer sepals obovate, inner oblong.—Mt. Diablo Range.

+— +— *Suffrutescent; stems leafy up to the base of the peduncles.*

12. **E. trachygonum,** Torr. Woody stems erect, rather slender, 6—10 in. high, densely clothed with the living and dead leaves; these narrowly oblanceolate, 1 in. long, narrowed to a slender petiole, the dense tomentum persistent on both faces: panicle short-peduncled, 3—5 in. high, twice or thrice dichotomous; lower involucres scattered, upper more condensed, sessile, campanulate-tubular, prominently but obtusely angled, glabrous except the woolly and obtusely toothed orifice: sepals white with a green midrib, the inner longer and somewhat narrower than the outer: ovary pubescent on the angles.—Abundant in dry gravel beds along Putah Creek, and on Mt. Hamilton.

13. **E. saxatile,** Wats. Caudex stoutish, sparingly branched, very leafy, ½—1 ft. high: leaves obovate, obtuse, 6—8 lines broad, 1 in. long, cuneate at base, short-petioled, densely tomentose on both sides: branches of the inflorescence short, spreading: bracts subfoliaceous, triangular or oblong and acute: involucres 1½—2 lines long; teeth acute: fl. rose-colored, 2 lines long, the sepals all spatulate-oblong and carinate, about equal, the inner appressed to the achene.—Near the summit of Mt. Hamilton.

5. OXYTHECA, *Nutt.* Slender annuals, glandular-pubescent (not tomentose), with a rosulate basal tuft of leaves and a repeatedly dichotomous paniculate inflorescence. Bracts of the flowering branches foliaceous, more or less connate. Involucres small, few-flowered, more or less distinctly pedicellate, the lobes awn-tipped or unarmed. Perianth 6-parted, usually glandular-pubescent on the outside, the segments alike. Stamens 6. Achene commonly lenticular.

1. **O. inermis,** Wats. Slender and low, 3—6 in. high, rather diffuse: leaves broadly oblanceolate, 1 in. long, glabrous except the scabrous-ciliate margins: bracts linear-oblong, acute, awnless: involucres short-pedicelled, 4-cleft almost to the base, the oblong-lanceolate lobes 1 line long, acute but awnless: fl. rose-color, ½ line long: achenes obtusely triangular.—Supposed to have been found originally on Mt. Diablo, but better known from beyond our limits.

2. **O. hirtiflora.** (Gray), Greene. Glandular-puberulent and viscid, 6 in. high, erect, cymose-paniculate above: *leaves* 1 in. long, *oblong-spatulate*, with scabrous-ciliate margins and a broad red midvein: bracts hispidulous, oblong, ¼ in. long or less, acutish: involucres awnless, ½ line long, on slender erect or nodding pedicels 1—3 lines long: fl. 3—5, very hirsute, rose-red, ½ line long.—Summits of Mt. Diablo and Mt. Hamilton.

6. CHORIZANTHE, *R. Br.* Dichotomous annuals with few and mostly basal leaves; the branches with ternate bracts at the nodes. Involucres 1—3 flowered, sessile, more or less tubular, coriaceous or chartaceous, often corrugated or reticulate, 3—6-angled or -ribbed, with as many cuspidate or rigidly awned teeth or segments. Flowers rarely exserted, 6-parted or -cleft; bractlets minute or obsolete. Stamens 9 (rarely 6 or 3). Achenes triangular.

* *Villous or hirsute; involucres usually clustered, 6-angled and sulcate, the teeth cuspidate; bractlets obsolete: perianth 6-cleft, the stamens inserted at or near its base.*

← *Erect or erect-spreading; involucres mostly in dense cymose clusters.*

++ *Margins of involucral lobes scarious.*

1. **C. membranacea,** Benth. Floccose-tomentose, erect, sparingly branched, with long internodes and leafy nodes, ½—2 ft. high: leaves linear, acute, 1 -2 in. long: bracts similar to the leaves but cuspidate: heads sessile, solitary or few upon the branches: involucres tomentose, 2—2½ lines long, the limb at length dilated and with uncinate teeth: tube contracted in the middle: perianth villous, becoming 1½ lines long, 6-parted, the segments oblong or spatulate: achene broadly triangular and rostrate-attenuate.—In the Coast and Mt. Diablo ranges.

2. **C. robusta,** Parry. Stout, erect, 6—18 in. high, dichotomously branched, the main stem below with several whorls of oblanceolate petiolate leaves; herbage hirsute, the inflorescence and growing parts almost canescently so: capitate cymes sessile and solitary in the lower forks, several and peduncled along the upper branches: bracts linear, with acerose tips: involucres oblong-campanulate, sharply angled; segments unequal, the scarious margin very narrow, purplish, the uncinate teeth not widely spreading: perianth short-pedicellate; lobes nearly equal, erose-denticulate and mucronulate.—In dry sandy soil at Alameda. June—Sept.

++ ++ *Lobes of involucre without scarious margins.*

3. **C. valida,** Wats. Stout, 6—18 in. high, branching above, villous: lower leaves oblanceolate, 1 in. long, on long petioles: involucres in dense heads 2—3 lines long, the lobes nearly equal, slightly spreading, the awns straight: perianth subsessile, narrowly tubular, 2½ lines long, villous or glabrous, cleft one-third of the length, the lobes oblong, very unequal, the shorter ones erose: filaments adnate to the middle or even higher.—In Sonoma Co., near Petaluma, etc.

+— +— *Of diffuse habit; involucres scattered, or in loose clusters.*

++ *Lobes of involucre with narrow scarious margins.*

4. **C. pungens,** Benth. Branches prostrate, 6—12 in. long, hirsute-pubescent: leaves spatulate or oblanceolate, 1 in. long, mostly opposite; bracts similar but narrower, acerose at apex: involucres crowded on short lateral branchlets, 1½—2 lines long, unequally toothed, usually margined; teeth strongly uncinate: perianth obconic, subsessile, shortly cleft: segments equal, oblong, entire: filaments more or less adnate to the lower part of the tube.—Sandy hills about San Francisco.

++ ++ *Lobes of involucre without scarious margins.*

5. **C. cuspidata,** Wats. Habit of *C. pungens,* leafy-bracted: leaves narrowly oblanceolate, 1 in. long; floral bracts acerose: involucres loosely cymose-clustered, 1 line long, 6-toothed, without scarious margins, the alternate teeth shorter, all armed with hooked awns: perianth subsessile, pinkish; lobes nearly equal, oblong, acutish, the strong nerve excurrent as a short cusp.—This was regarded by Dr. Parry as only a form of *C. pungens;* but by the description, it should be very distinct. Sandy hills at San Francisco.

6. **C. Clevelandi,** Parry. Prostrate or assurgent, the rather few branches 2—3 in. long, villous-pubescent: leaves mostly radical, broadly oblanceolate, narrowed to a rather long and slender petiole: involucres soft-pubescent, the triquetrous tube contracted above; segments very unequal, 3 as long as the tube, the other 3 scarcely half as long, all unci-

nate: perianth shortly cleft; outer segments broadly ovate, erose, retuse or emarginate, the inner narrow and lacerate: stamens 3; anthers orbicular.—Wooded hills of Sonoma and Lake counties. June—Sept.

7. PTEROSTEGIA, *Fisch. & Mey.* Our species diffusely dichotomous slender and flaccid (or in age somewhat wiry) annual with opposite, petiolate, exstipulate 2-lobed leaves, and small foliaceous bracts. Involucres each of a single bract shorter than the solitary sessile flower, rounded and 2-lobed, in age larger, reticulated, loosely enfolding the achene, and gibbously 2-saccate on the back. Perianth 5- or 6-parted; segments equal, oblong-lanceolate. Stamens as many or fewer, inserted at the base of the segments. Achene triquetrous.

1. **P. drymarioides,** Fisch. & Mey. Glabrous or hirsute-pubescent: leaves obovate, obcordate or reniform-bifid, often with the lobes again 2-lobed, the lowest petiolate, the upper sessile, ¼—¾ in. long: fl. minute: fructiferous involucre 1½ lines long, closely enfolding the minute light brown achene.—Common on rocky hills, and sandy banks. June.

Order XVI. NYCTAGINEÆ.

Herbs or suffrutescent plants (ours mostly coarse and fleshy seaside herbs) with tumid joints, opposite exstipulate entire leaves and showy perfect flowers in axillary pedunculate and involucrate clusters. Involucre calyx-like, closely subtending the flower-cluster. Perianth corolla-like, campanulate, salverform or tubular, the persistent base indurated and constricted over the 1-celled 1-seeded free ovary. Stamens few, hypogynous; filaments slender; anthers small and rounded. Pistil 1, simple. Seed erect; embryo encircling a copious mealy albumen.

1. **ABRONIA,** *Juss.* Viscid-pubescent rather succulent herbs with opposite and somewhat unequal leaves. Flowers in umbel-like heads on rather long axillary peduncles. Involucre of 5—15 distinct or slightly united somewhat scarious bracts enfolding the base of the heads. Perianth salverform; limb of 5 or 4 emarginate or obcordate lobes. Stamens mostly 5, adnate to the tube and not exserted. Stigma linear-clavate. Fruit coriaceous. 3—5-winged, enclosing a smooth cylindric achene. Embryo with but one cotyledon.

* *Wings thin but solid; body of fruit rigid or ligneous.*

1. **A. umbellata,** Lam. Perennial, prostrate, rather slender, *viscid-puberulent*, the stems 1—3 ft. long: leaves almost glabrous, ovate to narrowly oblong, 1—1½ in. long, narrowed to a slender petiole, obtuse, the margin often somewhat sinuate: peduncles 2—6 in. long: bracts of the involucre narrowly lanceolate, ¼ in. long; head 10—15 flowered:

perianth rose-purple, 6—8 lines long: fr. 4—5 lines long, nearly glabrous, the body oblong, attenuate at each end; wings thin, nearly as long, broadest above, narrowing toward the base: achene 1½ lines long.— Sandy places along the seaboard. June—Oct.

* * *Wings thicker, the central cavity of the fruit extending through them.*

2. **A. latifolia**, Esch. Perennial, *stout and succulent*, very viscid, the stems prostrate, 1—2 ft. long: leaves broadly ovate or reniform, ½—1½ in. long, obtuse: peduncles usually exceeding the leaves: bracts 5, rounded to ovate or oblong, 2—4 lines long: *perianth* 5—6 lines long, *yellow:* fr. 4—6 lines long, coriaceous, acute at each end; wings usually narrow.—Plentiful along the seashore. May—Dec.

Order XVII. AMARANTOIDEÆ.

Herbs with simple exstipulate leaves, and small inconspicuous (mostly greenish) axillary solitary or clustered perfect or unisexual flowers. Calyx of 3—5 hypogynous more or less scarious persistent sepals, occasionally with a pair of bractlets at base, generally enveloped by dry and almost chaffy bracts. Corolla 0. Stamens usually 5 or more, distinct or monadelphous. Stigmas 2 or 3, sessile on an undivided style. Fruit utricular, sometimes circumscissile, or bursting irregularly. Seed small, compressed, vertical. Embryo curved.

1. **AMARANTUS,** *Dodoens.* Annual weeds; leaves alternate, usually broad, veiny, and tipped with a short sharp mucro. Flowers green or purplish, in axillary spiked clusters or spikelets, the staminate usually mingled with the pistillate in the same cluster. Sepals distinct or united at base, seldom less than 3 or more than 5, more or less scarious, erect, or the tips spreading. Stamens as many as the sepals, distinct. Stigmas linear. Utricle ovate, 2—3 beaked, circumscissile or indehiscent often deciduous with the perianth.

* *Stems erect; sepals 5 or 3.*

1. A. RETROFLEXUS, L. Stout, 1—4 ft. high, paniculately branched above; herbage dull green, roughish and more or less pubescent: leaves ovate or rhombic-ovate, 1—4 in. long, on slender petioles: fl. green, in erect or somewhat spreading nearly cylindrical spikes: bracts lanceolate-subulate, scarious except the green carinate midrib, attenuate to a rigid awn, 1½—3 lines long: *sepals 5*, narrowly oblong, mostly acute or even mucronate, *exceeding the utricle:* seed ½ line broad, black and shining, with a rather obtuse margin.—Gardens and waste lands.

2. A. ÁLBUS, L. Erect, ½—2 ft. high, rigidly and widely branched from the base; herbage of a light green, glabrous or nearly so: leaves

oblong-spatulate to obovate, ½—1½ in. long including the slender petiole, obtuse or retuse, often crisped: spikelets axillary, 4—5-flowered: bracts subulate, rigid, pungently awned, 1—2½ lines long, the lateral ones reduced or wanting: *sepals 3*, oblong-lanceolate, subulate-mucronate, *shorter than the somewhat rugose utricle:* seed ⅓ line broad, black and shining, very sharply margined.

* * *Stems prostrate; sepals 3 or 1.*

+— *Spikelets small, axillary.*

3. A. BLITOÏDES, Wats. Somewhat succulent, weak and prostrate, the branches often 1—2 ft. long, whitish, the foliage of a rather deep shining green, glabrous or nearly so: spikelets few-flowered and contracted: bracts ovate-oblong, shortly acuminate, about equal, 1—1½ lines long, little longer than the oblong obtuse and mucronulate or acute sepals: *utricle smooth*, little surpassing the 3 sepals: seed ¾ line broad, abruptly but rather obtusely margined.

4. A. Californicus (Moq.), Wats. Stems stoutish and rather fleshy, branched from the base, prostrate, the branches 1—1½ ft. long, with many short lateral branchlets: *leaves* obovate or oblong, 1 in. long or less, including the short petiole, obtuse or acutish, *with white veins and margins:* fl. green or purplish, in many small dense axillary clusters: bract more or less scarious, little exceeding the utricle: sepals of staminate fl. ¾ line long; of the fertile (1 only) shorter: *utricle slightly rugose*, tardily circumscissile: seed ½ line broad, obscurely margined.—In low moist soils.

+— +— *Spikelets mainly in terminal clusters.*

5. A. DEFLEXUS, Willd. Stems slender, prostrate, 1 ft. long: leaves rather small, rhombic-lanceolate, obtuse, subemarginate, somewhat puberulent, dull-green: spikelets obtuse, some glomerate in the axils of the leaves, others forming a dense terminal thyrsoid cluster: sepals 3: utricle surpassing the calyx, marked with 3 obscure lines: seed black and shining, obtusely margined.—Native of southern Europe; common in some gardens about the Bay.

ORDER XVIII.　SALSOLACEÆ.

Herbs or shrubs, often succulent, glabrous, pubescent, mealy or scurfy, sometimes leafless. Flowers clustered, apetalous. Perianth of a solitary bract-like sepal, or of 2 which are distinct and valvate or more or less united, or of five distinct or united at base and calyx-like, never scarious. Stamens as many as the sepals and opposite to them, or fewer; anthers 2-celled. Ovary 1-celled, 1-ovuled, becoming an utricle or achene enclosed in the persistent perianth. Embryo annular or spiral; albumen mealy or wanting.

Stems leafy; leaves not terete;
  Fertile perianth 5-cleft or divided;
    Ovary superior..CHENOPODIUM 1
    " partly inferior......................................BETA 2
  Fertile perianth of 2 bracts often more or less united..........ATRIPLEX 3
Stems leafy; leaves terete...SUÆDA 5
Stems leafless, terete...SALICORNIA 4

1. CHENOPODIUM, *Tabernæmontanus* (GOOSEFOOT. PIG WEED).

Herbs with alternate petiolate mostly angular foliage. Flowers small, greenish, sessile, clustered in axillary or terminal spikes or cymes, perfect, or pistillate only, bractless. Perianth herbaceous, 3—5-parted; lobes imbricate, often carinate or crested, persistent and more or less covering the fruit, remaining green and herbaceous or becoming colored and fleshy. Stamens 5 or fewer. Styles, 2, 3 or 4, slender. Pericarp membranous, closely investing the lenticular horizontal or vertical seed. Embryo annular or curved around a copious albumen.

  * *Annual, more or less mealy, not pubescent; seed horizontal.*

1. C. ALBUM, L. Erect, stoutish, more or less paniculately branching, 1—4 ft. high; *herbage pale green* or whitish with a mealy indument: leaves petiolate, ascending, rhombic-ovate, obtuse, acute or cuneate at base, sinuate-dentate or subentire, 1—2 in. long, whiter beneath than above: fl. densely clustered in *close spikes*, these forming a rather strict leafless panicle: sepals of fruiting calyx carinate, completely covering the fruit; seed smooth, shining, acutely margined.—A very common weed of fields, gardens and waste places. June—Oct.

2. C. MURALE, L. Stoutish and rather low, often with many decumbent or ascending branches from the base; *herbage dark green*, rather succulent, the growing parts very mealy: leaves petiolate, ascending, ovate-rhomboid, unequally and sharply toothed: fl. in rather *dense* axillary nearly *leafless cymes*: fruiting calyx nearly closed, the sepals slightly carinate: seed opaque, punctate-rugose, sharply margined.—A common weed flourishing at all seasons.

  * * *Herbage not mealy, glandular-pubescent and aromatic; seed horizontal (except in n. 6.).*

3. C. BOTRYS, L. Annual, erect, often widely branching, 1—2 ft. high, *glandular-pubescent and highly aromatic*: leaves ovate or oblong, 1—2 in. long, sinuate-pinnatifid, the lobes often toothed: fl. scattered in very numerous slender *axillary cymose panicles*: sepals acute, loosely investing the fruit: pericarp persistent: seed ⅓ line broad, thick-lenticular, black and shining.

4. C. ANTHELMINTICUM, L. Perennial, stems stoutish, decumbent, 1—2 ft. long; herbage light green, *glandular-puberulent*, pleasantly aro-

matic: leaves thin, oblong, narrowed at base, obtuse, sinuate-serrate or sometimes remotely dentate, 1 in. long or less: inflorescence a *terminal leafless panicle* of dense but slender spikes: sepals not carinate, completely enclosing the fruit: seed smooth and shining, obtusely margined.

5. C. AMBROSOIDES, L. Annual, erect or ascending, 2—3 ft. high, deep green, *glabrous or slightly scabrous*, the foliage occasionally puberulent: leaves oblong, attenuate at each end, acutish, remotely sinuate-toothed or entire, the uppermost and floral linear-lanceolate: *inflorescence loosely spicate and leafy:* fruiting perianth completely closed: seed smooth and shining, obtusely margined.

6. C. MULTIFIDUM, L. Prostrate, branching and leafy, aromatic: leaves pinnatifid into narrow lobes; flowers glomerate in the axils: *perianth deeply campanulate*, 3—5-toothed, *at length saccate* and contracted over the fruit and *reticulate-nerved:* pericarp whitish and with scattered glandular dots: seed subrostellate, obtusely margined, dark brown, minutely punctate-rugose.—Common in San Francisco.

* * * *Glabrous or slightly mealy; seed vertical more or less exserted.*

7. **C. Californicum**, Wats. Stems from a long fusiform *perennial root*, stout, decumbent, mostly simple, 1—3 ft. high; the young parts a little mealy: *leaves broadly triangular-hastate*, 2—3 in. long, truncate or with sinuses at base, acuminate, sharply, unequally, and often deeply sinuate-dentate: fl. in dense clusters in a long simple terminal spike: perianth campanulate, rather deeply 5-toothed, enfolding the utricle only loosely: pericarp persistent: seed somewhat compressed, ¾—1 line broad.— Common.

8. **C. rubrum**, L. *Annual*, stout and rather fleshy, erect with ascending branches, ½—1 ft. high: leaves *ovate to oblong-lanceolate*, 1 in. long, obtuse, petiolate, remotely and rather coarsely dentate, glabrous and green above, paler and mealy beneath: fl. in axillary spiked clusters: perianth small, with rounded lobes, not quite concealing the vertical, or as often horizontal utricle.—In the Suisun marshes on elevated and dry ground; apparently indigenous, though possibly introduced from Europe, where it is a common weed.

2. **BETA**, *Columna* (BEET). Rather coarse glabrous biennials, with alternate leaves, the radical large and long-petioled, the floral reduced and sessile. Flowers fascicled in the axils and spicate-congested along the paniculate branches, connate at base, perfect. Sepals 5, inserted on the margins of a concave receptacle, imbricate. Stamens 5, opposite the sepals, filaments subulate. Ovary partly inferior and encircled by a disk-like margin of the receptacle: style short, the 2 or 3 branches stigmatose on the inside. Fruit partly adnate to the receptacle, and enclosed by the thicked and somewhat fleshy sepals.

1. **B. VULGARIS, L.** Stout, 2—4 ft. high: radical leaves often 1 ft. long including the stout petiole, commonly with prominent nerves and a more or less undulate margin, the outline oblong or oval: inflorescence 1—3 ft. long.—Escaped from gardens; in some places a common weed.

3. ATRIPLEX, *Pliny* (ORACHE.) Herbs or shrubs, mealy or scurfy, monœcious or diœcious; inflorescence axillary and glomerate, or terminal and spicate or panicled. Staminate perianth bractless, 3—5-parted, enclosing as many stamens. Pistillate fl. bibracteate, without perianth or rarely with 2—4 distinct hyaline sepals; the bracts erect, appressed, distinct or more or less united, their margins often becoming dilated, the surface sometimes in age thickened, indurated and muricate. Fruit compressed, utricular. Seed vertical. Embryo annular.

* *Monœcious annuals, somewhat succulent and mealy; bracts distinct or nearly so, ovate-oblong to broadly triangular or hastate.*

1. **A. hastata, L. var. oppositifolia,** Moq. Rather slender, with divaricate and somewhat decumbent branches; 2—3 ft. long, or stouter and erect with ascending branches; herbage mealy, not very succulent: *leaves triangular-hastate or deltoid,* mostly entire, all the lower opposite: flower-clusters small, spicate: bracts small, triangular, entire or denticulate ¼ in. long: seed 1 in. long, dark-colored.—Common along the borders of brackish marshes at Petaluma, and elsewhere.

2. **A. patula, L.** Stout and succulent, mostly erect, 1 ft. high, with few ascending branches; herbage deep green, only the growing parts somewhat mealy: lowest *leaves* often opposite, *broadly lanceolate,* sometimes with hastate base: inflorescence more or less leafy at base: bracts rhombic-ovate, thick and subcoriaceous, often ½ in. long.—Very common in salt marshes and near beaches.

3. **A. spicata,** Wats. Stout, erect, 1—2 feet high, sparingly branching, mealy: *leaves alternate, rhombic-ovate, acute,* coarsely and irregularly sinuate-toothed, 2 in. long, attenuate to a short petiole: *fl. densely spicate,* the 4-sepalous calyx usually staminate, but not rarely pistillate and with a horizontal seed: bracts of pistillate fl. ovate, acute, little enlarged in fruit, partly coherent at base, 1½ lines long: seed black, ½ line broad; radicle inferior.—Alkaline soil among the foothills of the Mt Diablo Range.

* * *Herbs or shrubs seldom mealy, but silvery-scurfy; bracts mostly rounded and more or less completely united, naked or variously appendaged or winged, frequently hard and nut-like in fruit.*

← *Monœcious annuals.*

4. **A. verna,** Jepson. Only 3—4 in. high; the branches simple or

nearly so, two or three pairs opposite at base, the upper alternate; the plant loosely scurfy throughout; leaves oblong-lanceolate or ovate, sessile, 3—5 lines long; flowers from the axils of the leafy stems in clusters of two or three; calyx deeply 4-cleft, stamens 4; fruiting bracts orbicular, compressed, 2 lines long, the margins crenate-dentate.—Near Collinsville, on the lower Sacramento. May.

5. **A. depressa, Jepson.** Diffuse, grayish-scurfy, decussately branched throughout, the branches 1—4 in. long: leaves opposite, sessile, broadly ovate, acute, a line or two long; flowers in the axils of the opposite leaves in clusters of four, these and the leaves crowded on the branchlets, the internodes at time of flowering a line long or less; fruiting bracts ovate hastate, acute, wingless, or the pair of hastate lobes representing the wing.—In low saline spots, at the base of the Pelevo Hills, west of Vanden, Solano Co. Sept.

6. **A. cordulata, Jepson.** Rigid, virgate, 8—15 in. high, widely and oppositely branched at the base, alternately and sparingly so above; herbage scurfy throughout; leaves sessile, cordate-ovate, three or sometimes four lines long; flower-clusters in all the axils; calyx tomentosely scurfy and deeply 4-cleft; fruiting bracts semi-orbicular, 1½—2 lines broad; much compressed, sessile or shortly stipitate, the margin with acute teeth, the terminal tooth commonly the largest, the sides smooth or with one or two tooth-like projections.—Near Little Oak, Solano Co. August.

7. **A. nodosa, Greene.** Stout, branched from the base, 1 ft. high, mealy and apparently scabrous: leaves broadly rhomboid: fruit-clusters borne at the enlarged nodes of the widely and irregularly branching stem: pedicels stout, thickened under the bracts; these united and forming an almost globose fruit 2 lines in diameter, 3-lobed at summit, the sides covered with lichenoid spongy projections.—Near Antioch.

8. **A. trinervata, Jepson.** Stout, erect, 2—3 ft. high, closely and finely mealy-scurfy: leaves 1—3 in. long, broadly or deltoid-ovate, irregularly and sharply sinuate-toothed, the lower on stout petioles 9—10 lines long and strongly 3-nerved from the base: the upper reduced to sessile floral bracts as broad or broader than long; fruiting bracts sessile in the axils of the leaves, orbicular, 2 lines long, 2½—3 broad, usually emarginate, sharply toothed, partly distinct in the wing, and commonly bearing on one face a few irregular projections or crests.—Near the Araquipa Hills, Solano Co. Sept.

+ + *Monœcious perennials.*

9. **A. fruticulosa, Jepson.** Erect, suffrutescent at base, ½—1 ft. high: herbage gray-scurfy; leaves sessile, lanceolate or narrowly oblong,

¼ to ¾ of an inch long; staminate flowers in dense globose clusters on the terminal branchlets, naked or nearly so; pistillate flowers chiefly below, from the leaf-axils; calyx deeply 5-cleft, occasionally unequally parted and one lobe reduced; fruiting bracts orbicular, 1½—2 lines broad, the margins partly free, the sides tooth-crested; seed nearly a line broad.—In alkaline soil near Little Oak, Solano Co. ˙ Aug.

10. **A. Californica,** Moq. Branches many, slender and wiry, prostrate, from a *short and thick oblong or fusiform perennial root; herbage* densely *mealy;* leaves ovate- to linear-lanceolate, 3—8 lines long, entire, acute, the lowest opposite: flower-clusters all axillary, the upper ones more staminate, the calyx of these deeply 4-cleft: fruiting *bracts rhombic-ovate, membranous, distinct,* 1½ lines long, somewhat convex: seed ½ line broad.—On the seacoast, and along the edges of salt marshes, from near San Francisco and Alameda, southward. Sept.

+ + + *Diœcious shrubs.*

11. **A. leucophylla** (Moq.), Dietr. Stout, shrubby, but the stem and branches flexible and mostly reclining, 1—2 ft. long; plant hoary-scurfy throughout: leaves thick, broadly obovate, cuneate at base, sessile 3-nerved, ½—1½ in. long: staminate fl. in dense clusters in short terminal spikes; calyx large, 5-cleft: fruiting bracts in axillary clusters 2—4 lines long, rhombic-ovate, united, spongy, the sides 2-crested, the narrow margin entire or obscurely toothed. On sand beaches of San Francisco Bay, and along the seacoast southward. Oct.

4. **SALICORNIA,** *Tourn.* (SAMPHIRE). Herbs or shrubs with cylindrical fleshy jointed and apparently leafless branches. Flowers very simple, in threes at the joints of the spike-like ends of the branches; the lateral ones of each trio often only staminate. Perianth of 4 or 5 distinct or variously united sepals, at length spongy-thicked about the fruit. Stamens 1 or 2. Styles 2 or 3, short. Pericarp membranaceous, adherent to, or free from the vertical seed.

* *Branches and flowers opposite.*

1. **S. ambigua,** Michx. Perennial, decumbent, often rooting at the base, usually freely branching, ½—1½ ft. high: spikes not thicker than the sterile parts of the branches, ½—2 in. long: perianth sac-like, with an anterior opening (formed of 2 sepals united above and below), enclosing the fruit: pericarp membranous, adherent to the obovate-oblong seed, this ⅓ line long, pubescent.—Plentiful in salt marshes.

* * *Branches alternate and flowers spirally arranged in the spikes.*

2. **S. occidentalis** (Wats), Greene. Shrubby, diffusely branched, the main stem erect, often 5 ft. high, with a close and smooth gray bark:

scale like crowded and fleshy leaves broadly triangular and acute, amplexicaul, often nearly obsolete: fl. densely spiked: perianth of 4 or 5 concave carinate sepals more or less united: pericarp free from the oblong seed, this ¼ line long or less.—In alkaline soil at Byron Springs.

5. SUÆDA, *Forskaal* (SEA BLITE). Saline herbs or shrubs, with alternate fleshy linear entire leaves, and axillary sessile usually perfect flowers. Perianth minutely bracteolate, 5-cleft or -parted, fleshy; lobes unappendaged, more or less carinate, crested or winged, enclosing the fruit. Stamens 5. Styles 2, 3 or 4, short and thick. Pericarp membranous, free or slightly adherent to the vertical or horizontal lenticular seed. Testa shining, black and crustaceous. Embryo spiral.

1. **S. Californica,** Wats. Stout, 2—3 ft. high, very leafy, glabrous or somewhat pubescent: leaves broadly linear, subterete, not wider at base, ½—1 in. long, acute, crowded on the branchlets: fl. large, 1—4 in each axil: perianth cleft nearly to the base; lobes not appendaged: seed vertical or horizontal, nearly 1 line broad, faintly reticulate.— Vicinity of sand beaches about San Francisco Bay, but seldom seen.

ORDER XIX. PORTULACEÆ.

More or less succulent herbs, with entire leaves and regular complete flowers which open in sunshine only. Sepals 2 (in *Lewisia* 4—8), sometimes cohering at base. Petals 5 (in *Lewisia* 8—16), often united at base. Stamens commonly 5 (3—∞), opposite the petals, hypogynous, perigynous or epipetalous; filaments distinct; anthers versatile. Ovary 1-celled, with few or many ovules on a central placenta. Seeds commonly strophiolate; embryo curved or coiled around a mealy albumen.

1. **PORTULACA,** *Lobel* (PURSLANE). Fleshy annuals, with axillary and terminal yellow or rose-colored flowers. Sepals 2, united below and coherent with the base of the ovary; the limb free and deciduous. Petals 4—6. Stamens 7—20, perigynous with the petals. Style 3—8-cleft. Capsule circumscissile, opening by a lid. Seeds small.

1. P. OLERACEA, L. Prostrate, glabrous, the herbage usually reddish or purplish: leaves flat, obovate, obtuse: sepals acute, carinate: petals 1½—2 lines long, yellow: stigmas 5: capsule 3—5 lines long: seeds dull-black, minutely tuberculate.—Native of S. Europe.

2. **LEWISIA,** *Pursh.* Low acaulescent fleshy perennials, with thick fusiform roots, and short 1-flowered scapes. Sepals 4—8, broadly ovate, unequal, persistent, imbricate. Petals 8—16, large and showy. Stamens ∞. Style 3—8-parted nearly to the base. Capsule circumscissile at base, the upper and deciduous part more or less valvate-cleft. Seeds ∞, black and shining.

1. **L. rediviva, Pursh.** Leaves densely clustered on the short thick caudex, linear-oblong, glabrous, glaucous: scapes little exceeding the leaves, jointed at the middle, and with 5—7 subulate scarious bracts whorled at the joint: sepals 6—8, broadly ovate, scarious-margined, ½— ¾ in. long: petals 12—15, oblong, ½—1 in. long, pinkish or white: stamens 40 or more: capsule broadly ovate, ¼ in. long.—On Mt. Diablo, Mt. Hamilton, etc.

3. **CALANDRINIA,** *H. B. K.* Sepals 2 only, subequal, persistent. Petals 3—10. Stamens 3—25, apparently always hypogynous. Capsule 3-valved from the summit. Seeds several.

1. **C. Menziesii** (Hook), Torr. & Gray. Rather slender, diffuse, the branches 3—6 in. long: leaves linear-spatulate, mostly radical and long-peduncled; the upper and floral reduced and glandular-ciliate: sepals ovate, acuminate, the margins and sharp keel glandular-ciliate: corolla little exceeding the sepals, white or bright purple: stamens 3—10; seeds broadly ovoid, shining.—Coast Range. April, May.

2. **C. elegans,** Spach. Larger and stouter than the last, glabrous, the decumbent and ascending branches often 1 ft. long, flowering throughout: sepals ovate, acute or acuminate, less sharply carinate, the keel and margins entire or with a sparse short and flattened but in no wise glandular ciliation: stamens 10—15: corolla twice the length of the calyx, ¾ in. broad when expanded, bright rose-red: seeds larger, nearly orbicular.—Very common throughout the Bay region and elsewhere in the State. April—June.

3. **C. Breweri,** Wats. Habit of the preceding but still larger, the ascending branches often more than 1 ft. high, glabrous: pedicels rather remote, in fruit deflexed: sepals broadly ovate, truncate at base, surpassed by the long-conical (½ in. long) capsule: seeds dull, tuberculate. —Mt. Tamalpais.

4. **CLAYTONIA,** *Gronovius.* Glabrous herbs, often glaucous. Leaves radical except an involucral pair (sometimes united) under the racemose or subumbellate inflorescence of the usually scapiform peduncles. Sepals 2, persistent. Petals 5, equal, commonly united by their short claws. Stamens 5, each joined to the claw of its petal. Capsule membranaceous, ovate or globose, 3 valved, elastically dehiscent, each valve elastically involute, ejecting the rather few black and shining seeds.

Perennial; pedicels axillary to a bract.

1. **C. Sibirica,** L. Stems 1—1½ ft. high: herbage almost dark green, disposed to blacken in drying: radical leaves lanceolate to rhombic-ovate, acute or acuminate, 1—2 in. long, long-petioled; cauline sessile,

distinct, ovate or ovate-lanceolate, acute, not indistinctly parallel-veined: raceme very lax, the fl. on long pedicels: petals 4 lines long, rose-purple, retuse or emarginate at summit, at base narrowed to a distinct claw.—In open swamps of Marin Co.

* * *Annuals; bractless except at base.*

+— *Herbage light green, scarcely glaucous.*

2. **C. perfoliata,** Donn. Stems 2—16 in. high: radical leaves from deltoid-cordate, deltoid or rhomboidal to rhombic-lanceolate, 1—2 in. long, on long petioles; cauline pair joined into a more or less orbicular perfoliate nearly plane or strongly concave disk ½—4 in. broad: raceme short-peduncled or sessile, with an ovate acute or acuminate small foliaceous bract at base: petals 1—2 lines long, white; blade linear-oblong, retuse or emarginate; claws united at base and stamens epipetalous: fruiting calyx 2 lines long, twice the length of the subglobose 3-seeded capsule: seed ¾ line long, round-oval, black and shining but depressed granular (under a strong lens), with a small white strophiole. Var. (1) **carnosa,** Greene. Stout and low; the whole herbage very succulent: fruiting calyx ¼ in. long: seed nearly orbicular, 1½ lines broad. Var. (2) **angustifolia,** Greene. Quite like the type save that the lowest radical leaves are linear, almost without distinction of blade and petiole, the later ones somewhat broader and lanceolate; involucre truncate and with acute angles on the upper side (opposite the deflection of the pedicels) rounded on the other.—The most prevalent of Californian winter annuals, attaining its best development in the shade of oaks and laurels among the hills; in open grounds much smaller; in sandy soil near the sea usually reduced and depressed. The first variety is peculiar to the Mt. Diablo region, growing in open grounds, in fields and waste places. The second grows along with the type everywhere, and is remarkably different from it in that its early leaves are linear, only the later ones widening to the lanceolate, thus reversing the common order; for in the type the earliest leaves are broader than long, only the later ones being somewhat narrower.

3. **C. nubigena,** Greene. Habit of the preceding, but much smaller, the herbage pale and glaucescent, the white or pinkish flowers twice as large: leaves all linear: involucre orbicular.—On the summits of Mt. Hamilton, and Mt. Tamalpais.

+— +— *Herbage glaucous, in age flesh-colored.*

4. **C. gypsophiloides,** Fisch. & Mey. Pale and glaucous, 4—10 in. high: radical leaves linear, one-half or one-third as long as the slender scapes; cauline pair short and united on one side to form a quadrate disk-like involucre, or longer, lanceolate-acuminate and less perfectly

united: raceme peduncled, many-flowered; pedicels scattered, often 1 in. long: *petals rose-purple*, thrice the length of the calyx, cuneate-oblong, *deeply emarginate*, unguiculate at base and united around the ovary: seed dull to the unaided eye, under a lens roughened with a low and rounded but smooth and shining tuberculation.—On northward slopes in the Coast Range, from Tamalpais and Mt. Diablo northward.

5. **C. spathulata,** Dougl. Low, densely tufted and fleshy, 1—3 in. high: scapes little exceeding the linear leaves; involucral leaves lanceolate or linear, more or less dilated at base and there connate on one side, equalling or exceeding the short raceme: *petals white or purplish*, little longer than the sepals, *truncate or rounded at apex*: seed oval, ½ line long, black and shining, the polished low tuberculation appearing under a lens as a kind of reticulation.—Common on ledges of rock and gravelly summits of low hills along the seaboard.

5. **MONTIA,** *Micheli.* Annuals, or by stolons or bulblets perennial. Leave opposite or alternate. Flowers few or many in axillary racemose clusters, or in a single terminal raceme. Calyx, corolla, capsule and seeds as in *Claytonia*, but segments of corolla often unequal and stamens reduced to 3. Seeds sometimes 1 or 2 only, usually 3.

1. **M. fontana,** L. Annual, slender, erect, ascending or procumbent, 1—4 in. long: leaves opposite, narrowly oblanceolate or spatulate, dilated and somewhat connate at base, ¼—¾ in. long: corolla white, minute, little exceeding the calyx and seldom expanding, the petals unequal, united at base: seed minute, roundish, dull black, but under a lens shining and covered with an almost echinate murication.—Common and variable; the coarser form inhabiting the margins of streamlets and shores of muddy pools; the smaller and nearly prostrate state found on dry ground under growing grain in rather low fields.--March—May.

2. **M. parvifolia** (Moç.), Greene. Slender, succulent, 4—10 in. high: leaves alternate, on a short caudex ½—1 in. high, ovate or lanceolate, 1 in. long or less, including the slender petiole: racemose peduncles elongated, leafy below, the nodes in age bearing bud-like plantlets: calyx minute: petals rose-color, 2—4 lines long: seeds most solitary in the capsules, oval, shining.—Sonoma Co. *Bioletti.*

6. **CALYPTRIDIUM,** *Nuttall.* Glabrous and rather succulent herbs, with alternate leaves, and small ephemeral flowers in solitary or clustered scorpioid spikes. Sepals 2, broadly ovate or cordate-orbicular, scarious, usually persistent. Petals 2—4. Stamens 1—3. Style bifid. Capsule membranaceous, 2-valved, 6—12-seeded.

1. **C. tetrapetalum,** Wats. Branches erect or ascending from a more or less decumbent base, leafy up to the short dense spikes: leaves

broadly spatulate, 1—3 in. long: sepals round-reniform, conspicuously nerved and scariously margined, 2—4 lines broad, exceeding the 4 oblong or round-ovate petals: stigmas broad; nearly sessile: capsule oblong, 3 lines long, 12—20-seeded.—Lake Co. and Sonoma.

ORDER XX. ELATINEÆ.

Low annuals with opposite leaves, membranous stipules, and axillary regular symmetrical 2—5-merous flowers. Sepals, petals and stamens all distinct, hypogynous. Styles distinct; stigmas capitate; ovary 2—5-celled, becoming a 2—5-celled capsule with central placenta and a septicidal or septifragal dehiscence. Seeds straight or curved.

1. **ALSINASTRUM,** *Tourn.* (WATER-WORT). Glabrous dwarf and rather succulent plants of wet places, sometimes aquatic and floating. Flowers axillary. Sepals 2—4, nerveless, obtuse, persistent. Petals 2—4. Stamens as many or twice as many as the petals. Styles, or sessile stigmas, 2—4. Pod thin, globose, 2—4-celled, several- or many-seeded. Seeds cylindrical, straight or curved, striate-pitted.

1. **A. brachyspermum** (A. Gray). Commonly terrestrial: leaves oblong or oval, attenuate at base, sometimes lanceolate, ¼ in. long or less: fl. sessile, mostly dimerous; stamens 2 or 3: seed oval, nearly straight, ¼ line long, coarsely pitted in 6 or 7 lines of 10—12 pits.

2. **BERGIA,** *Linn.* Coarser plants, not succulent. Flowers pedicellate, often fascicled, 5-merous: Sepals with strong midrib, acute. Capsule crustaceous, more or less of the partitions remaining with the axis.

1. **B. Texana** (T. & G.), Seubert. Diffusely branched, the branches a foot long more or less; herbage glandular-pubescent: leaves oblanceolate, acute, serrulate, ½—1½ in. long, narrowed to a short petiole; fl. fascicled, pedicellate; sepals carinate, 1½ lines long, exceeding the petals and stamens: capsule globose: seeds smooth and shining.—Moist places along rivers and ditches.

ORDER XXI. HYPERICEÆ.

A small family, here represented by species of the genus

HYPERICUM, *Diosc.* (ST. JOHN's-WORT). Glabrous, the bright green herbage, punctate with pellucid or dark-colored dots. Leaves opposite, simple, entire exstipulate. Inflorescence cymose; flowers yellow. Sepals 5, imbricate in bud. Petals 5, convolute in bud, rotate in expansion. Stamens ∞, usually connate at base or into 3—8 clusters. Styles 2—5, nearly or quite distinct; ovary 1-celled with 3 parietal placentæ; or 3-celled by union of the placentæ with the axis. Seeds many, minute.

1. **H. conclnnum,** Benth. Erect, *wiry, very leafy, suffrutescent at base,* ½—1 ft. high: leaves thickish and somewhat conduplicate, linear or linear-oblong, acute: cyme few-flowered: fl. 1 in. broad: sepals ovate, acuminate: stamens ∞, in 3 fascicles.—Common on bushy hillsides in clayey soil, at middle elevations of the Coast Range. May, June.

2. **H. anagalloides,** Ch. & Schl. Diffusely branching, *slender, prostrate or assurgent, stoloniferous,* forming a mat a foot or more in breadth: leaves oval or elliptical, ¼—½ in. long, obtuse, clasping, only half as long as the internodes: inflorescence leafy-paniculate-cymose; fl. scarcely 2 lines long, the obovate- or linear-oblong sepals exceeding the petals: *stamens 15—20,* nearly or quite distinct.—In wet places.

3. **H. mutilum,** L. *Stem flaccid, erect, widely branching,* 6—20 in. high: leaves ovate to narrowly oblong, obtuse, partly clasping, 5-nerved; cymes numerous, leafy; fl. 2 lines broad: sepals narrow, erect: *stamens 5—12,* distinct.—Shores of the Sacramento in Solano Co., *Jepson.* Annual; common on the Atlantic slope; possibly introduced with us.

4. **H. Scouleri,** Hook. *Erect,* 1—2 ft. high, *mostly simple* up to the cyme of few and large flowers: leaves ovate to oblong, obtuse, 1 in. long or less: sepals ovate, 2 lines long: petals 4 or 5 lines: *stamens 60 or more, in 3 fascicles.*—On Howell Mountain, Napa Co., along streamlets.

ORDER XXII. MALVACEÆ.

Herbs or shrubs, with alternate stipulate leaves and a more or less stellate pubescence. Flowers usually perfect, complete and regular; the 5-cleft valvate and persistent calyx often subtended by a supplementary whorl of bracts and thus appearing double. Petals 5, hypogynous, at base commonly joined to each other and to the base of the tube of the monadelphous stamens, convolute in bud. Stamens 5—∞, more or less completely monadelphous and sheathing the styles; anthers usually reniform, 1-celled. Ovaries either distinct and forming a ring around a central columnar elevation of the receptacle, thus becoming achenes, or joined into one 5—10-celled organ and becoming more or less capsular. Seeds usually roundish, with little or no albumen.

Calyx with 3-lobed involucre at base..LAVATERA   1
Calyx 1—3-bracted at base:
    Fruit a whorl of 1-seeded carpels;
        Styles stigmatic lengthwise;
            Stamineal column single..........................MALVA   2
               "   "   double......................SIDALCEA   3
        Styles with terminal stigma;
            Seeds pendulous..................................SIDA   4
            " ascending..............................MALVEOPSIS 5
    Fruit a 5-celled capsule;
        Involucre wanting..ABUTILON   6
        " of many bractlets........................HIBISCUS

1. LAVATERA, *Tourn.* Stout shrubs with coarse flexible branches, ample palmately lobed leaves, and axillary showy flowers. Involucel 3-lobed. Stamineal tube divided at summit into numerous filaments. Style-branches stigmatose lengthwise, on the inside. Fruit a depressed whorl of 5—8 crowded achenes surrounding the angular column of the receptacle which scarcely exceeds them, and covered by the calyx.

1. **L. assurgentiflora,** Kell. Coarse, stout, soft-woody, flexuous-branched, 6—15 ft. high, the young branches, pedicels and calyx, rarely the leaves also, stellate-hairy or -tomentose: leaves long-petioled, 3—6 in. broad, angularly 5—7-lobed, the lobes coarsely toothed: fl. solitary, on a long deflexed and curved pedicel: petals 1—1½ in. long, cuneate-obovate, truncate or retuse, abruptly reflexed from near the base, rose-red, with crimson veins: stamineal column glabrous: styles exserted: fr. ½ in. broad; carpels not beaked, equalling the summit of the axis.— Native of the islands off Santa Barbara and San Pedro; long cultivated about San Francisco, where it is become spontaneous.

2. MALVA, *Pliny* (MALLOW). Herbs with broad angular or rounded leaves, and axillary solitary or glomerate flowers. Involucel 3-leaved. Stamens and pistils as in *Lavatera.* Column of receptacle short, seeming depressed below the whorl of achenes.

1. **M. PARVIFLORA,** L. Simple or branching, the branches depressed and a few inches long, or the main stem erect and 2—6 ft. high: herbage more or less pilose-hairy: leaves long-petioled, obsoletely 5—7-lobed, round-cordate, crenate, 1—3 in. broad: fl. glomerate, small, the pale blue corolla little exceeding the calyx: bractlets linear; *calyx accrescent, the broad-lobed limb rotately spreading away from the mature fruit: achenes* glabrous or pubescent, transversely and *sharply rugose on the back,* the acute winged *margins distinctly toothed.*—A homely weed, extremely common, often small and depressed when growing in the streets or along country waysides, but in good soil erect and tall.

2. **M. BOREALIS,** Wallm. Habit, aspect and foliage of the last, but herbage more conspicuously pilose and often a little stellate-hairy: bractlets lanceolate: *calyx-lobes deep, closed over the mature fruit:* corolla pale blue, ½ in. long, surpassing the calyx: achenes *reticulate-rugose, the acute margins entire.*—Rather common about Berkeley; easily distinguished from the foregoing by the larger flowers, connivent calyx-lobes, entire-margined and irregularly rugose achenes.

3. SIDALCEA, *A. Gray.* Herbs with rounded and commonly lobed or parted leaves; occasionally diœcious. Flowers in terminal racemes or spikes, rose-purple or white. Involucel 0. Stamineal column double; filaments of the outer series united into about 5 sets; of the inner distinct. Style-branches stigmatic lengthwise as in *Malva;* fruit the same, except that the achene is beaked.

* Annuals.

1. **S. diploscypha,** Gray. Erect, 1—2 ft. high, paniculately branching, pilose-hirsute with long spreading hairs: leaves long-petioled, rounded, the radical deeply crenate; cauline 7-parted with 2—3-cleft oblong segments: *inflorescence umbellate,* the umbels many, at the ends of the branchlets, 3—5 flowered: fruiting calyx ¾ in. long, deeply cleft, the segments lanceolate, acuminate: corolla 1 in. long, pale rose-color: *achenes cochleate and nearly orbicular,* scarcely a line in diameter, reticulate-rugulose on the back.—Mt. Diablo Range, both in the hills and upon the plains adjacent. March—May.

2. **S. secundiflora,** Greene. Pubescence and foliage as in the last, but plant less branching: fl. in terminal rather *lax spicate racemes:* petals oblique, purple, often with a very dark spot at base: *achenes* nearly 2 lines long, *semiobcordate,* strongly favose-reticulate.—Less common than the last; often associated with it.

3. **S. calycosa,** Jones. Stout, rather widely branching, 2 ft. high, glabrous below, sparingly hirsute above: inflorescence loosely spicate: calyx-lobes ovate-lanceolate, abruptly acuminate; corolla 1 in. long, deep or pale purple: *achenes* more or less perfectly *sulcate on the back,* by obliteration of the usual transverse ridges.—In Marin Co.

4. **S. hirsuta,** Gray. Stout, erect, simple or almost fastigiately much branched, 2—4 ft. high; the rather *densely spicate inflorescence and the growing parts densely hirsute:* lower leaves round-cordate, slightly crenate-lobed; cauline completely divided into 7—9 narrowly linear entire segments or leaflets: calyx ½ in. long, the lobes deep, acuminate: corolla rose-purple, 1 in. long: *achenes* rugose-reticulate, tipped *with a long rather soft but hispid erect beak.*—Coast range hills northward.

* * Perennials.

5. **S. humilis,** Gray. Stems clustered from a tuberous-enlarged root, simple, decumbent, virgate-racemose, 1—2 ft. long: herbage from hirsute to nearly glabrous: radical leaves rounded, crenate-incised: corolla rose-purple, 1 in. long or less, broad-funnelform: *achenes almost orbicular, rugose-reticulate.*—Very common in open hilly places.

6. **S. Oregana** (Nutt.), Gray. Stems solitary or few from the root, 2—6 ft. high, naked and paniculately branched above, leafy below; *inflorescence stellate-tomentose,* peduncles and lower part of stem sparingly hirsute, the plant otherwise glabrous: lower leaves orbicular, 7—9-lobed, the cuneate-obovate lobes 3-cleft at summit; upper 7—9-parted, narrowly and deeply cleft: spicate racemes usually dense but elongated: calyx-lobes broadly ovate, acute, not longer than the tube: corolla ½—1 in. long: *achenes small, straight* (semiorbicular), slightly beaked, *smooth and glabrous,* 1 line long.—Moist mountain meadows, in Sonoma Co.

4. SIDA, *Linn.* Herbs with undivided leaves. Involucel 0 (except in ours where it is 3-bracteate as in the preceding). Calyx 5-cleft. Stamineal tube simple. Stigmas capitate. Carpels 1-celled, 1-seeded, dehiscent or indehiscent, forming a short conical fruit. Seed pendulous.

1. **S. hederacea** (Dougl.), Torr. Perennial, stoutish, erect-spreading or prostrate, very leafy, ½—1 ft. high, hoary- or yellowish-tomentose throughout: leaves short-petioled, about 1 in. long, reniform, very oblique at base, plicate, serrate or crenate: fl. axillary, solitary or several, the pedicels slender: calyx subtended by 1 or 2 slender bractlets; lobes acuminate: corolla ¾ in. long, cream color: fr. short conical, smooth, glabrous; carpels 6—10, triangular, 1½ lines long.—A depressed hoary weed, very common in low and subsaline clayey soils.

5. MALVEOPSIS, Presl. Herbaceous or shrubby (ours mostly hoary-tomentose shrubs), with usually angular foliage, and solitary or racemose-panicled flowers. Calyx with an involucel of 1—3 bractlets, or none. Stamineal tube simple; free filaments terminal and distinct. Styles 5 or more; stigmas capitate. Carpels 1-seeded, bivalvate-dehiscent or indehiscent. Seed ascending.

1. **M. Fremonti** (Torr.). Suffrutescent, very stout, 2—3 ft. high, *densely white-tomentose: leaves very thick,* short-petioled, 1—3 in. long, broadly ovate, cordate at base, *slightly 3-lobed and crenate*: fl. in short axillary pedunculate racemose clusters: calyx ovate, ½ in. long, only the setaceous tips of its lobes visible amid the deep and dense white tomentum, almost equalled by the 3 linear setaceous involucral bractlets: corolla ¾ in. long, rose-color: carpels thin, smooth, promptly dehiscent. —Mt. Diablo and southward.

2. **M. fasciculata** (Nutt.), O. Ktze. Usually 6—8 ft. high, often larger and arborescent, the main stem a few inches thick; bark smooth, gray; branches long, wand-like, slender, racemose or amply racemose-paniculate above, these and the lower face of the leaves *canescently short-tomentose: leaves angularly 5-lobed* and coarsely toothed, 1½—3 in. long, and almost as broad: calyx-lobes triangular, as broad as long, acute: corolla rose-purple, ½ in. long: carpels smooth, tomentose above, promptly dehiscent: seed with a stellular-hairy minute reticulation.—A very handsome shrub or small tree, common in S. Calif., reaching Mt. Diablo, according to *Rattan.*

3. **M. arcuata.** Size and general habit of the last but stouter; leaves half as large, *obtuse and with inconspicuous rounded lobes, very strongly rugose-veiny and white-tomentose beneath:* interruptedly spicate flowering branches stout, strongly recurved, the flower-fascicles all on one side: corolla merely pinkish: carpels densely stellate-tomentose.— Eastern slopes of the Coast Range back of Belmont; wrongly referred to *M. marrubioides* in the Flora Franciscana.

6. ABUTILON, *Camerarius.* Herbs or shrubs with axillary solitary mostly yellow flowers. Involucel 0. Stamineal tube simple, antheriferous at summit. Styles 5 or more, with capitate stigmas. Fruit truncate-globose or -conical; carpels dehiscent, several-seeded.

1. A. THEOPHRASTI, Medic. A stout erect branching annual, 2—6 ft. high, the herbage green but velvety-pubescent and almost oily to the touch: leaves round-cordate, acuminate, crenate-dentate, 3—6 in. long, on petioles of 2—5 in.; peduncles axillary, erect, shorter than the petiole: fl. small, orange-yellow: carpels about 15, inflated, obliquely birostrate, pubescent, 3-seeded.—A common weed in cultivated grounds at the East; reported as established about Santa Rosa.

7. HIBISCUS, *Diosc.* Stout herbs, with large and showy axillary and solitary flowers. Involucel of many bractlets.. Stamineal column antheriferous below the summit; above naked and truncate or 5-toothed. Styles united; stigmas 5, capitate. Carpels united into a 5-celled loculicidal capsule; cells several-seeded.

1. **H. Californicus,** Kell. Perennial, stout, erect, branching, 5—7 ft. high, velvety-pubescent: leaves cordate-ovate, acuminate, coarsely but not deeply toothed, 3—5 in. long, exceeding the petioles: peduncle jointed above the middle, 2—3 in. long, 1-flowered: calyx 1 in. long, cleft to the middle, the lobes acute: corolla 3—4 in. long, yellowish or cream-color, with dark purple center: capsule 1 in. long, acute velvety-pubescent: seeds a line in diameter, globose, striate and tuberculate-roughened.—In swampy places along the lower San Joaquin.

ORDER XXIII. LINEÆ.

small order, comprising little besides the one genus

LINUM, *Vergil* (FLAX). Herbs with tough fibrous bark, alternate entire leaves without stipules or with glandular organs in the place of them, and cymose-panicled very regular and symmetrical 5-merous flowers. Sepals imbricate, persistent. Petals convolute, fugacious. Stamens monadelphous at the very base. Styles 2, 3 or 5, often united below. Ovary of as many carpels as styles, each more or less divided into 2 cells by a partition proceeding from the dorsal suture. Fruit capsular, septicidally dehiscent. Seeds 1 in each half-cell, ovate, compressed, mucilaginous when moistened; embryo large; albumen thin.

* Flowers ½—1 in. broad, blue; sepals not glandular-margined.

1. L. USITATISSIMUM, L. Annual, glabrous, glaucous, 1—2 ft. high, simple up to the ample inflorescence: sepals oval, short-acuminate, 3-carinate-nerved at base, the inner scarious-margined and ciliate: petals broad-cuneiform, blue, with deeper veins, ½ in. long: *capsule round-*

ovoid, equalling the calyx, tardily dehiscent, incompletely 10-celled, the *septa not ciliate.*—One of the cultivated flaxes; occasionally spontaneous.

2. **L. HUMILE**, Mill. Much like the last, but lower and more branching: *capsule more elongated*, promptly dehiscent, the *septa ciliate.*— Another flax of the Old World, sometimes found wild by waysides.

3. **L. Lewisii**, Pursh. *Perennial*, glabrous, glaucous, 1—2½ ft. high, densely leafy below, lax-corymbose above: sepals broadly ovate, not ciliate, 3—7-carinate-nerved: *petals large, deep blue:* capsule broadly ovate, obtuse, 3—4 lines long, twice as long as the sepals, the 10 valves dehiscing widely, the septa ciliate.—San Mateo Co.

* * *Annuals; leaves often with stipular glands; fl. small, white, rose-purple or yellow; sepals usually glandular-ciliate; petals commonly with lateral teeth and ventral appendages, pistils only 2 or 3.*

+— *Petals yellow.*

4. **L. Breweri**, Gray. Slender, 3—12 in. high, glabrous, glaucous, few-flowered: leaves linear-setaceous, 6—8 lines long; stipular glands conspicuous: sepals 1½ lines long, ovate, acute, glandular on the margin: petals spatulate, emarginate, ¼ in. long, 3-appendaged at base: capsule ovoid, acute, about equalling the calyx.—In the Vacaville and Mt. Diablo foot-hills; also on Lone Mountain, San Francisco.

+— +— *Flowers white or pink.*

5. **L. spergulinum**, Gray. Slender, loosely dichotomous-paniculate, 6—15 in. high, glabrous or with scattered hairs: leaves linear, obtuse, little narrowed at base, with or without stipular glands: *pedicels slender nodding:* sepals ovate, glandular-ciliate; petals white or rose-colored, obovate, 2—3 lines long, 3-appendaged at base: *capsule* ovoid, acute, *exceeding the calyx.*—Dry woods of the Coast Range; common in Marin and Sonoma counties.

6. **L. Californicum**, Benth. Glaucous, glabrate or puberulent, 5—15 in. high, with angular branchlets: leaves remote, linear, the stipular glands prominent: *pedicels short, erect*, not exceeding the rose-colored flowers, these clustered at the ends of the branchlets; sepals ovate-lanceolate, acute, carinate below, sparingly glandular-ciliate; petals obovate, ⅓ in. long, twice the length of the calyx, dilated and 3-appendiculate below; filaments not toothed: *capsule* ovoid, acute, little *shorter than the calyx*, the false partitions broad, gradually narrowed upwards. Var. **confertum**, Gray. Low, densely leafy, the inflorescence condensed; median appendage of petals obovate.—Eastern slope of Mt. Diablo Range, also about San Francisco.

7. **L. congestum**, Gray. Size of the last, glabrous except the calyx, the branches short and crowded: stipular glands small: *fl. rose-purple,*

in close terminal clusters; sepals pubescent, lanceolate, acuminate, not glandular; petals ¼ in. long, 2-toothed, 3-appendiculate, the median appendage long and hairy: *capsule sub-globose,* shorter than the calyx.— A rare species, to be sought at the north of Mt. Tamalpais.

ORDER XXIV. GERANIACEÆ.

Ours soft-herbaceous plants with acidulous, pungent or aromatic properties, and perfect mostly 5-merous flowers. Sepals and petals distinct, the later deciduous, their insertion, like that of the 5—15 stamens, hypogynous. Filaments distinct or slightly connate at base: anthers versatile, 2-celled, dehiscing lengthwise. Carpels as many as the sepals and alternate with them (or fewer), united around a central column, becoming distinct and 1-seeded in maturity, or else forming an elastically dehiscent 5—10-valved many-seeded capsule.

1. **GERANIUM,** *Diosc.* (CRANESBILL). Stems with enlarged joints. Leaves mostly opposite, palmately lobed; stipules scarious. Peduncles umbellately few-flowered, or 1-flowered. Flowers regular; sepals and petals imbricate in bud. Fertile stamens 10. Carpels 5, 2-ovuled, 1-seeded; styles persistent, coherent with the central column until the carpel is ripe, then splitting away from it elastically from below upwards, each forming a coil, not bearded within.

* Annuals.

1. **G. Carolinianum, L.** Erect, much branched from the base, 1 ft. high, the pubescent herbage light-colored: leaves 5-parted, the divisions cleft into many oblong-linear lobes: sepals awn-pointed, as long as the *pale flesh-colored emarginate petals:* carpels pubescent: seeds ovoid-oblong, blackish, minutely reticulate.—Common.

2. **G. DISSECTUM, L.** Taller than the last, the herbage of a darker green; leaves cut into narrower and more acute segments: *fl. larger, bright red-purple, the petals more deeply emarginate;* seed roundish, more strongly reticulate.—Rather common; preferring moist and partially shaded situations; continuing in flower until the end of June.

3. **G. MOLLE, L.** Low, slender, diffuse, the branches a few inches to 1 ft. long, the herbage softly and somewhat clammily villous: leaves 1 in. broad or more, cleft into oblong obtusish lobes: sepals ovate-oblong, not awn-pointed: *petals very small, rose-color: carpels glabrous, transversely rugose:* seed minutely striate.—About the U. S. Marine Hospital, San Francisco.

* * Perennial.

4. **G. RETRORSUM, L'Her.** Stouter than any of the foregoing, light green, glabrous except *a short stiffish retrorsely appressed pubescence on*

the stems and growing parts: leaves 2 in. broad, 5-parted, the segments obtusely and not deeply 3-lobed: petals 2 lines long, obtuse, purple, equalling the aristate sepals: carpels slightly hairy: seeds oblong, minutely striate-reticulate.—San Francisco and Alameda. Native of New Zealand.

2. ERODIUM, *L'Her.* (STORKSBILL). Vegetative characters of *Geranium*, but leaves often pinnate. Flowers and fruit almost the same; but fertile stamens 5 only, as many scale-like sterile filaments alternating with them. Beak of ripe carpel silvery-bearded within, and spirally twisted.

* Naturalized species; leaves pinnate.

1. E. CICUTARIUM (L.), L'Her. Leaves chiefly radical, in a depressed rosulate tuft, usually 6—10 in. long, the many *leaflets laciniately pinnatifid with narrow acute lobes;* cauline leaves reduced: peduncles exceeding them and bearing an umbel of 4—8 small bright bright purple flowers: beak of carpels 1—2 in. long.—Frequent in the Bay region; perhaps more common in the interior and southward. This is one of the pasture plants commonly called *Pin-clover* and *Alfilerilla;* but the next is the important one.

2. E. MOSCHATUM (L.), L'Her. Coarser and larger, the radical leaves ascending, 1 ft. long or more; cauline more ample; *leaflets unequally and doubly serrate:* corolla pale and rather dull purple or rose-color: herbage with a delicate musky odor.—The prevalent *Pin-clover* of middle Calif.

3. E. BOTRYS, Bertol. Radical leaves rosulate, closely depressed, shining above, of oblong obtuse outline, the segments coarsely dentate: stems short: sepals 4 lines long; pale purple or lilac petals longer: beak of carpels 2—3 in. long.—Common in Marin Co.

* * Native species; leaves simple, rounded.

4. E. macrophyllum, Hook & Arn. Subacaulescent, 4—10 in. high, soft-pubescent and with some gland-tipped pilose hairs, leaves 1—3 in. broad, *reniform-cordate with a broad open sinus,* crenate-serrate: peduncles exceeding the leaves: sepals oblong, accrescent, at length ½ in. long; *petals* equalling them, *dull white:* carpel clavate, ⅓ in. long (excluding the 1 in. beak), densely velvety-pubescent: seed oblong linear, ¼ in. long, dull, smooth.—Plains of the interior; also toward the seaboard in Marin Co. March, April.

5. E. Californicum, Greene. Caulescent, the stem exceeding the rather few radical leaves, 1—2 ft. high; herbage without soft pubescence, but upper part of stem and growing parts with abundant spreading hairs tipped with purple glands: leaves broadly *cordate-ovate with closed sinus,* slightly 5-lobed, rather coarsely crenate, the teeth obtuse, mucronulate: fl. much as in the preceding but *petals deep rose-red:* fruit as in the last.—Berkeley hills and eastward. April—June.

3. CARDAMINDUM, *Tourn.* (NASTURTIUM. TROPÆOLUM). Tall leafy climbing plants, the succulent herbage with a pungent juice. Leaves alternate, simple, extipulate. Flowers large, axillary, irregular. Sepals not quite distinct; the 3 upper somewhat conjointly produced at base into a long spur. Petals 5, unequal; the 3 lower often shorter. Stamens 8, distinct. Carpels 3, becoming large corky sulcate achenes.

1. C. MAJUS (L.), Mœnch. Leaves orbicular, peltate, repandly lobed: petals usually orange-red, 1—2 in. long, broad and obtuse, unguiculate, the 3 lower fimbricate lacerate at the base of the blade: achenes ⅓—½ in. in diameter.—Native of Peru; escaped from cultivation in many places in California.

4. FLŒRKEA, *Willd.* Low annuals, slightly succulent, the juice pungent. Leaves alternate, pinnately cleft, exstipulate. Flowers axillary, solitary, regular, 3—5-merous, (all ours 5-merous, or by exception 4-merous). Sepals valvate in bud. Petals convolute, as many hypogynous glands alternating with them. Stamens 10, distinct. Style 5-cleft; carpels distinct, subglobose, fleshy when young, becoming soft variously roughened achenes separating from their short axis.

1. F. Douglasii (R. Br.), Baillon. Glabrous throughout, 6—18 in. high: leaflets narrowly cuneiform, incisely lobed or parted: peduncles 2—4 in. long: sepals lanceolate, ¼—⅓ in. long: petals yellow, ¾ in. long, obovate, emarginate; achenes obovate-pyriform, more or less tuberculate.—In very wet places. April, May.

5. OXYS, *Tourn.* (WOOD-SORREL). Herbs with sour juice (containing oxalic acid), alternate palmately 3-foliolate leaves, and cymose or umbellate regular 5-merous flowers. Sepals imbricate, distinct or slightly coherent at base, persistent. Petals convolute, deciduous. Stamens 10, more or less monadelphous, those opposite the petals longer than the others. Ovary of 5 united carpels; styles distinct. Fruit an ovoid or columnar loculicidally dehiscent capsule; the valves remaining attached to the central axis; cells 2—several-seeded. Seeds pendulous, the testa aril-like, at length splitting and becoming recurved.

1. O. Oregana (Nutt.). *Acaulescent*, perennial by simple or sparingly branched scaly rootstocks; herbage rusty-pubescent: leaves 1 ft. high: leaflets broadly obcordate, ciliate, 1 in. long, 1¼ in. broad: *scapes 1-flowered*, shorter than the leaves, bibracteolate above the middle: *petals* oblong obovate, emarginate, *white, with purple veins:* capsule ovoid.— Shaded slopes in the Coast Range.

2. O. lutea, J. Bauh. (1651); Tourn. (1700); Lam. (1778). *Oxalis corniculata*, L. Perennial, erect or decumbent, 3—10 in. high, *branching*, pubescent: leaflets broadly obcordate: peduncles mostly 2-flowered: *fl. small, yellow:* capsule columnar, ¾ in. long, densely pubescent, many-seeded.—Not common in California.

Order XXV. RUTACEÆ.

Represented by a single species of the genus

PTELEA, *Linn.* (HOP-TREE). Shrubs or small trees with alternate 3-foliolate aromatic pellucid-dotted leaves, and corymbose regular flowers. Sepals, petals and stamens each 4 or 5, the latter inserted outside of a disk encircling the ovary. Ovary 2-celled, surmounted by a short style, and becoming an orbicular broadly winged 2-seeded samara.

1. **P. crenulata,** Greene. Tree 10—25 ft. high, strongly aromatic when fresh; glabrous except the tomentulose flowers, and a sparse pubescence on the lower face of the leaves and on the fruit: leaflets cuneate-obovate, obtuse or acute, 1—3 in. long, crenulate or crenate-serrate: filaments villous near the base: samara ¾ in. long and as broad, truncate or emarginate at both ends, often triquetrous and 3-seeded.—In the Coast Range, from Lake Co. southward through Contra Costa, etc. May.

Order XXVI. SAPINDACEÆ.

Trees or shrubs with opposite compound, or at least deeply lobed leaves, without stipules. Inflorescence compound, usually racemose or thyrsoid. Sepals 5, nearly distinct, or joined into a tubular calyx. Petals 4 or 5, distinct, and, with the few and definite stamens, inserted hypogynously, or around a hypogynous disk. Fruit a 3-celled capsule, or a double samara. Seeds large; without albumen.

1. **ACER,** *Pliny* (MAPLE. BOX-ELDER). Trees or shrubs with opposite palmately lobed or pinnately compound leaves without stipules. Flowers small, greenish or reddish, in terminal racemes, umbel-like corymbs, or fascicles, perfect or unisexual. Calyx usually 5-lobed. Petals 5 or 0. Stamens usually 8 (3—12), in the perfect flowers inserted with the petals upon a lobed disk. Ovary 2-lobed, 2-celled; styles 2.

* Leaves simple; tree not diœcious.

1. **A. macrophyllum,** Pursh. Tree 50—90 ft. high, 2—3 ft. in diameter: leaves ½—1 ft. broad, deeply 5-lobed, the sinuses rounded, the segments often 3-lobed, coarsely toothed: fl. large, in large crowded pendulous racemes which appear with the unfolding leaves, greenish yellow or reddish: stamens 9 or 10; filaments hairy: fruit densely hirsute or almost hispid, the glabrous wings 1 in. long or more, divergent. —Along mountain streams or on hillsides.

* * Leaves unequally pinnate; tree diœcious.

2. **A. Californicum** (T. & G.), Dietr. Tree 30—70 ft. high, the young twigs and partly developed leaves villous-canescent: leaflets 3, ovate, or

the lateral ones oblong, acute, 3—4 in. long, the terminal largest and 3—5 lobed, or coarsely serrate: fl. of sterile tree umbellately clustered, the pedicels long and capillary, those of the fertile in drooping racemes: fruit pubescent 1—1½ in. long, including the nearly erect wings.—In the Coast Ranges from San Luis Obispo northward.

2. HIPPOCASTANUM, *Tourn.* (BUCKEYE. HORSE-CHESTNUT). Trees with opposite palmately compound exstipulate leaves, and a large thyrsoid inflorescence, the flowers on jointed pedicels. Flowers polygamous. Calyx tubular, unequally 5-toothed. Petals 4 or 5, unguiculate. Stamens 5—8, exserted, often unequal. Ovary 3-celled: ovules 2 in each cell, 1 abortive. Fruit a large coriaceous 3-valved capsule. Seed very large; testa chestnut-brown, showing a large white hilum. Cotyledons large, fleshy, somewhat coherent.

1. H. Californicum (Spach). A low spreading tree, glabrous, except the petiolules and inflorescence which are minutely pubescent: leaflets 5, on distinct stalklets, oblong or elliptic-oblong, mostly rounded at base, acute or acuminate at apex, serrulate, 3 –5 in. long: thyrsus cylindrical, often 1 ft. long: calyx 2-lobed, the lobes scarcely toothed: corolla white with a faint tinge of rose, ½ in. long: stamens 5—7, long-exserted: fruit smooth, usually 1-seeded: seed 1 in. thick.—Tree often 25 or 30 ft. high, the rounded or depressed head of still greater breadth: very common throughout middle California. Admirable specimens are seen at Shell Mound, and on Point Isabel. Fl. May: fr. November.

ORDER XXVII. ANACARDIACEÆ.

Shrubs or trees with resinous and often acrid juice, alternate exstipulate leaves, and small variously clustered regular flowers. Stamens definite in number, as many or twice as many as the petals. Pistil 1: ovary free from the calyx. Fruit drupaceous.

1. RHUS, *Theophr.* Ours deciduous shrubs with trifoliolate leaves and small perfect or unisexual flowers in axillary bracted panicles or spikes. Sepals and petals usually 5. Stamens inserted under the edge of a disk lining the base of the calyx. Pistil 1; styles 3, distinct or united. Fruit a small compressed drupe with thin flesh and ligneous putamen. Seed erect; albumen 0.

* *Flowers greenish, in small axillary panicles, appearing with the leaves; drupe white; putamen striate.*

1. R. diversiloba, Torr. & Gray. Erect and 3—6 ft. high, or ascending trees by aerial roots to the height of 15 ft. or more: leaflets ovate, obovate or elliptical, 1—4 in. long, variously lobed or toothed, the indentations obtuse, or the leaflet rarely entire: panicles short-peduncled, more or less pendulous: fl. 1½ lines long: fr. 2—3 lines broad.—Copious in the Coast Range hills, preferring cool northward slopes and the banks of streams; the terror of many excursionists and of some botanists, and commonly called *Poison Oak.*

* * *Flowers yellow, in small dense spikes, appearing before the leaves; drupe red, hairy; putamen smooth.*

2. **R. trilobata,** Nutt. Diffusely branching, 2—5 ft. high, aromatic-scented, more or less pubescent when young: terminal leaflet thrice as large as the lateral, cuneate-obovate, 1—2 in. long, 3-lobed and coarsely toothed above the middle; lateral pair round-obovate, scarcely lobed, but coarsely crenate: spikes ½—¾ in. long, short-pedicelled: fr. viscidly hirsute, the thin pulp keenly and pleasantly acid. Var. **quinata,** Jepson. Leaves apparently quinate, the terminal leaflet being 3-parted, the lateral ones little larger than the lateral divisions of the terminal one.—Perhaps only the variety is found within our limits; and this plentiful in the Vaca Mts. and northward.

Order XXVIII. CELASTRINEÆ.

Shrubs with simple exstipulate leaves, and small perfect regular flowers. Sepals and petals 4 or 5, imbricate in bud. Stamens as many as the petals, inserted alternately with them on or under the edge of a perigynous disk. Ovary free from the calyx, but immersed in the disk or encircled by it, 3- or 4-celled; cells 1- or several-ovuled. Fruit capsular, loculicidal: seed without albumen.

1. **EUONYMUS,** *Theophr.* (BURNING-BUSH). Deciduous shrub with 4-angular green branches, opposite leaves, and flowers in loose axillary cymes. Sepals and petals 4 or 5, widely spreading. Stamens very short, on a broad angled disk. Ovary immersed in the disk, 3—5-celled; style short or 0. Capsule coriaceous, 3—5-lobed and -valved. Seeds 1—4 in each cell, covered with a fleshy red aril.

1. **E. occidentalis,** Nutt. Erect, slender, 7—15 ft. high: leaves ovate or oblong-lanceolate, acuminate, serrulate, short-petioled, 2 -4 in. long: peduncles slender, 2—4-flowered: fl. 5-merous, dark-brown purple, 4—6 lines wide: fr. smooth, deeply lobed.—Marin and San Mateo counties, along mountain streams.

Order XXIX. RHAMNEÆ.

Shrubs with simple leaves; stipules minute, mostly caducous. Flowers 4—5-merous, small, perfect or unisexual, regular. Calyx 4—5-cleft, valvate in æstivation. Petals cucullate or convolute, sometimes 0. Stamens as many as the calyx-lobes and alternate with them, *i. e.,* opposite the petals. Ovary more or less free, surrounded by a fleshy disk, 2—3- or 4-celled; ovules solitary, erect. Fruit baccate or capsular. Seeds erect; albumen fleshy or 0.

1. **RHAMNUS**, *Nicander* (BUCKTHORN). Shrubs (all ours evergreen), with alternate leaves and axillary clusters of small greenish 4—5-merous flowers. Disk thin, lining the tube of the calyx. Fruit sub-globose, the juicy pulp enclosing 2 or 3 large nut-like seeds.

1. **R. crocea**, Nutt. *Low, spreading, slender and spinescent,* 2 ft. high; leaves coriaceous, ½ in. long, bright-green above, yellow beneath, roundish ovate in out-line, glandular-denticulate: *fl. 4-merous,* apetalous, often unisexual, short-pedicelled, solitary or few in a fascicle: fr. small, obovoid, scarlet, 1- or 2-seeded.—Mission Hills and islands in the Bay.

2. **R. ilicifolia**, Kell. *Erect,* bushy or arborescent, *6—15 ft. high, not spinescent:* leaves firm-coriaceous, 1 in. long more or less, oval, spinu-lose-dentate; very short-petioled: *fl. 5-merous:* fr. as in the last, though larger.—Mt. Diablo Range.

3. **R. Californica**, Esch. Bushy or arborescent, 4—20 ft. high, *nascent parts pubescent, otherwise glabrous:* leaves thin-coriaceous, elliptic-oblong, acute or obtuse, denticulate or entire, 1—4 in. long: fl. subumbellate, 5-merous: petals small, ovate, emarginate: filament long; anther exserted from the cucullate petal: fr. globose, ⅓—½ in. in diameter, copiously pulpy, black: seeds usually 2, hemispherical, as broad at base as at summit.—Along the seaboard, on sandy plains near the shore, where it is a low shrub; also among the hills in the form of a small tree with ample spreading branches.

4. **R. tomentella**, Benth. Near preceding, of similar habit, but never either low-bushy or aborescent: *leaves* 2 in. long, narrowly oblong or elliptical, abruptly acute or acuminate, entire, the margin narrowly revolute, *glabrate above, minutely and very densely silvery- or yellowish-tomentose beneath:* fl. and fr. as in the last.—Eastern slope of the Mt. Diablo Range.

2. **CEANOTHUS**, *Linn.* (CALIFORNIA LILAC). Arborescent, shrubby or suffrutescent, unarmed or spinescent, with petioled leaves and mostly thyrsoidly arranged, caducous-bracted fascicles or cymes of small perfect blue or white flowers. Calyx campanulate, 5-cleft, the lobes acute, con-nivent; disk thick, adnate to the calyx and base of the ovary. Petals 5, cucullate and arched, on long claws. Stamens 5; filaments filiform, long-exserted. Ovary 3-lobed; style short, 3-cleft. Fruit 3-lobed and capsular, though coated with a thin pulp; ultimately separating into 3 unilocular 1-seeded carpels which are elastically dehiscent by the ventral suture. Seeds obovate, without a furrow.

* *Leaves alternate, membranous or thin-coriaceous, glandular-toothed or entire; fruit unappendaged or slightly crested.*

+— *Branches flexible, not spinescent.*

++ *Leaves plane, glandular-toothed, except in n. 1.*

1. **C. integerrimus,** Hook. & Arn. Tall, loosely branching and sometimes arborescent, 5—12 ft. high, the branchlets green, more or less angular when young, and warty in age: *leaves ovate, 1—3 in. long,* prominently triple-veined, pubescent or glabrate, *entire* or very slightly glandular-serrate: thyrse long and dense, terminating leafy branchlets: fl. from deep blue to white.—One of the most common species of the Coast Range northward, but mostly beyond our limits.

2. **C. velutinus,** Dougl. Stout, diffusely branching, 2—4 ft. high: *leaves subcoriaceous, broadly oval,* 1½—3 in. *long, shining and thick-glutinous above,* more or less velvety-pubescent and strongly 3-ribbed beneath; petioles stout, ½ in. long: thyrse compound, loose and broad, rather short-peduncled: fl. white.—Higher parts of the Coast Range, as far south as Mt. St. Helena. June.

3. **C. thyrsiflorus,** Esch. Arborescent, 6—15 ft. high, glabrous or nearly so, branches angular, foliage firm-membranous, bright and shining; *leaves 1—2 in. long,* short petioled, *ovate oblong, strongly 3-ribbed:* thyrse dense, sometimes broader than long, on short leafy peduncles: fl. deep blue: fr. small, smooth.—Frequent, preferring northward slopes and cool ravines.

++ ++ ++ *Leaves pinnate-veined; margins glandular-toothed, undulate or revolute; surface mostly papillose or rugose.*

4. **C. Parryi,** Trel. Arborescent, 6—10 ft. high; branches sparingly villous or glabrate, angular, more or less papillose: *leaves oblong, obtuse.* 1—1½ in. *long,* the pinnate veins supplemented by a pair of laterals which run near the more or less strongly *revolute margin;* surface of leaf glabrate, lower face more or less tomentose-canescent: thyrse narrowly oblong, umbels subsessile: fl. blue: fr. small, smooth.—In the hill-country between Napa and Sonoma counties. May, June.

5. **C. papillosus,** Torr. & Gray. Stouter than the last, less arboreous, 4—6 ft. high; branchlets and stalklets hirsute-pubescent: *leaves* narrowly oblong, 1—2 in. long, *glandular-serrate, the surface rugose and glandular-papillose:* fl. blue, in short, mostly simple and short-stalked racemes: fr. small, smooth.—Hills along the seaboard.

6. **C. foliosus,** Parry. Low, slender, the erect stems 2—3 ft. high, with many ascending very leafy branches; nascent parts pubescent: *leaves subcoriaceous,* often fascicled, *glaucous beneath,* deep but dull

green above, *2—5 lines long, oborate, or oval*, obtuse, short-petioled, closely denticulate, the mucronate teeth having very large rather deciduous resin-glands: fl. few, light blue, in a simple usually capitate raceme on a slender more or less leafy-bracted peduncle: capsule sharply crested at summit.—Wooded hills of Napa, Sonoma and Marin counties.

✚ ✚ Branches spinescent; flowers in smaller clusters.

7. **C. divaricatus**, Nutt. Rigidly and diffusely branched, the *branches spinescent* and divaricate, *nearly glabrous:* leaves ovate to oblong, $\frac{1}{3}$—$1\frac{1}{4}$ in. long, rounded at base, acute or obtuse at summit, not tomentose beneath, entire or minutely glandular-serrulate: racemes rather lax, often leafy; fl. blue or white: fr. of middle size, very resinous. —In the Coast Range; very common.

8. **C. incanus**, Torr. & Gray. Spinescent *branches thick and stout, minutely canescent*, the foliage also cinereous-velvety and pale: leaves coriaceous, tomentose beneath, broadly ovate or elliptical obtuse, subcordate at base or somewhat cuneate, $\frac{3}{4}$—2 in. long: fl. white, in short racemes from thick spurs or axillary branchlets: fr. 2 lines in diameter, resinous and warty.—In the Coast Range.

9. **C. sorediatus**, Hook. & Arn. Shrubby or arborescent, 5—10 ft. high, nearly glabrous: *branches spreading or recurved*, and with short stiff branchlets: *leaves subcoriaceous, glossy above*, glabrous or somewhat tomentose beneath, but silky along the veins, *oblong-ovate*, $\frac{1}{2}$—$1\frac{1}{2}$ in. long, rounded or subcordate at base: racemes of deep blue, $\frac{1}{2}$—2 in. long, usually not longer than broad.—Plentiful on Mt. Tamalpais, on the northern slope; common in the Berkeley Hills. March—May.

* * Evergreen shrubs; branches small-leaved, with warty stipules; leaves opposite, coriaceous, closely pinnate-nerved; fruit with 3 horns.

✚ ✚ Shrubs erect, with short rigid branchlets.

10. **C. cuneatus** (Hook.), Nutt. Stems clustered, 6—12 ft. high, the branchlets short and remote, glabrous or nearly so: *leaves cuneate-obovate, or oblong, obtuse or retuse, entire*, $\frac{1}{2}$ in. long or less, exceeded by the profuse simple subsessile umbellate clusters of rather large dull-white heavy-scented flowers: fr. rather large; horns short, erect. Var. **ramulosus**, Greene. Smaller, the branchlets more numerous and more leafy: *leaves narrower and longer*, more tomentose beneath: fl. half as large, scentless, deep blue: fr. smaller and more elongated.—The type on Mt. Diablo and near Los Gatos: the variety in the Coast Range only. February—April.

11. **C. divergens**, Parry. Low, almost diffuse, the *long rigid divergent branches sometimes almost trailing*, pubescent when young: *leaves 1*

in. long or less, rigidly coriaceous, *cuneate ånd entire below the middle*, above bearing 2 or 3 opposite pairs of coarse spinescent serrate teeth, the broad truncate apex with or without a similar tooth: umbels peduncled or subsessile: fl. purplish or even blue: fr. large, elongated, with 3 prominent horns and as many alternating crests.—Higher mountains of Marin and Sonoma counties.

12. **C. Jepsonii.** Low bush *rigidly erect* and intricately branching, 2—4 ft. high, the branches and branchlets short and very stout, divaricate, puberulent when young: leaves ¾ in. long, hard-coriaceous, *oblong, obtuse*, or even truncate at both ends, the whole *margin coarsely and saliently spinose-toothed:* fl. in short-peduncled simple clusters at the ends of all the branchlets, large, dark blue, varying to white: fr. large, prominently 3-horned. — Open hills in Marin County, near San Geronimo, and northward. Confused with the preceding by Parry.

Order XXX. SARMENTOSÆ.

A small family, important as containing the *Grape*.

1. **VITIS,** *Varro* (GRAPE). Shrubs with watery juice, climbing by branching tendrils opposite the leaves. Flowers small, greenish, numerous, in thyrsiform clusters opposite the leaves. Calyx minute, cup-like, with or without traces of 4 or 5 teeth. Petals 4 or 5, united at apex, and falling off like a calyptra. Stamens as many as the petals and opposite them, on a perigynous disk or elevation of the torus; filaments slender; anthers introrse. Pistil with a short style or none; stigma slightly 2-lobed. Fruit baccate, 1—4 seeded. Seeds bony, rather large, grooved on one side; embryo small in a hard albumen.

1. **V. Californica,** Benth. Stem often 1—2 in. thick below, climbing trees to the height of 20—50 ft.: leaves 3 in. long, nearly as broad, round-cordate with deep and narrow sinus, obtuse, rather coarsely serrate, sometimes 3-lobed, canescently tomentose beneath, and when young more or less so on both faces: fr. 4 lines thick, in large clusters, purple, glaucous: seeds broad.—Along streams back from the seaboard.

2. V. VINIFERA, L., the wine grape, native of the Old World, has escaped from cultivation, and will be occasionally seen in a wild state.

Order XXXI. TITHYMALOIDEÆ.

Herbs, shrubs or trees, often with milky acrid juice, the leaves simple, stipulate. Flowers axillary or terminal, bracted, imperfect, monœcious or diœcious, in all ours apetalous. Stamens 1—∞. Pistil 1; ovary superior, 1—3-celled. Fruit a 1—3-celled capsule with as many lobes as cells; the lobes in maturity separating from a central axis as a 1-celled 1-seeded carpel; this elastically dehiscent by two sutures and exposing or ejecting the usually arilled or strophiolate seed. Ovules and seeds pendulous. Embryo embedded in fleshy albumen; cotyledons flat.

1. CROTON, *Linn.* Pale scurfy or stellate hairy plants with alternate exstipulate entire leaves and racemose, cymose or solitary unisexual apetalous flowers. Staminate calyx 4—6 parted, slightly imbricate in bud. Stamens 5—7, on a hairy receptacle; anthers inflexed in bud. Pistillate calyx when present 5-parted. Ovary simple and 1-celled, or 2 —3-lobed with as many cells; styles as many as the ovary-cells, simple. or once or twice forked. Seed grayish, smooth and shining.

* Fruit 3-lobed; styles forked.

1. C. Californicus, Müll. Arg. Suffrutescent, weak, decumbent or prostrate: leafy branches erect, 1 ft. high; these and the foliage silvery-canescent with a fine scurf and a minute stellate pubescence: leaves narrowly oblong or elliptical, obtuse at each end, 1 –2 in. long, on petioles half as long: staminate flowers greenish, in short subsessile racemes; calyx-lobes about 1 line long; filaments hairy: pistillate fl. mostly solitary, on short pedicels; styles twice forked: capsule deeply 3-lobed, ¼ in. thick: seed 2½ lines long, with a small appressed caruncle.—Plentiful among the sandhills at San Francisco and southward.

* * Fruit of a single 1-seeded carpel, style simple.

2. C. setigerus, Hook. A stout low annual with short but widespread leafy branches, the heavy-scented ·herbage with a spreading hispid and an appressed stellate pubescence: leaves ovoid or rhomboid, ½—2 in. long, on slender petioles, the upper crowded and appearing opposite or whorled: staminate fl. few in corymb, long-pedicelled; calyx with oblong obtuse segments a line long: pistillate fl. 1, 2 or 3 in an axil; ovary and style densely pubescent: capsule and seed 2 lines long. —Abundant in autumnal fields, mostly in the interior.

2. EUPHORBIA, *Linn.* (SPURGE). Herbs with milky juice, alternate or opposite toothed or entire leaves, and inflorescence either terminally clustered, or solitary in the forks of the many branches. Both staminate and pistillate flowers within the same involucre; this cup-shaped and like a calyx, the 4 or 5 lobes minute, usually alternating with as many glands which have often a colored margin resembling a petal. Staminate flowers many, of a single naked stamen jointed upon a short pedicel which has often a minute bract at base. Pistillate flower 1, in the center of the involucre, pedicellate and soon exserted from it, consisting of a single 3-celled ovary, 3 forked styles and 6 stigmas each 2-lobed. Capsule 3-seeded. Seeds smooth, reticulate, rugose or pitted.

* *Stems erect; stipules 0; involucres in forked or umbellate terminal cymes; glands flattened or convex; seed carunculate.*

1. E. LATHYRIS, L. Biennial or perennial, erect, stout, 1—3 ft. high, glabrous throughout: leaves opposite, 4-ranked, linear-lanceolate, sessile, entire, obtuse, cuspidate, 3—4 in. long: inflorescence bracted, the branches twice or thrice dichotomous, the leaf-like bracts oblong-ovate: glands crescent-shaped, with broad obtuse horns: capsule ⅓ in. thick, the lobes rounded, in age wrinkled: seeds reticulate rugose.—Native of the Mediterranean region; with us, an escape from gardens.

2. E. EXIGUA, L. Annual, slender, glabrous, 3—10 in. high: leaves alternate, linear, entire, acute or obtuse, the floral diliated at base and subcordate: inflorescence lax, repeatedly dichotomous: glands semilunate, the horns divergent: capsule smooth, scarcely a line wide: *seed ovate-quadrangular, whitish, minutely tuberculate.*—A weed of the grain fields in Europe; reported as occurring in Santa Clara.

3. E. PEPLUS, L. Annual, slender, with light-green delicate herbage; simple or with a few erect branches from near the base: leaves alternate, the lowest round-ovate, the others cuneate-obovate, slenderly petiolate, the floral broadest at the unequal and sessile base: glands sublunate, the long slender horns little divergent: *capsule with a pair of wing-like crests on each lobe: seeds ash-color, obscurely 6-sided,* with a few large dark-colored depressions.—Common garden weed at Berkeley.

4. **E. leptocera,** Engelm. Annual or biennial, 1 ft. high: leaves alternate, obovate-spatulate, obtuse, ½—1½ in. long, entire or erose-denticulate; the floral opposite or ternate, broadly rhombic-ovate, sometimes connate, acute, ¼—¾ in. broad: involucre turbinate, the oblong lobes nearly entire; glands large, crescent-shaped, the slender horns entire or cleft: styles long, bifid: capsule 2 lines broad: *seeds ash-colored, oblong-ovate, dark-pitted,* about 1½ lines long, prominently carunculate.—Common in bushy places either in sandy or clayey soil.

5. **E. dictyosperma,** Fisch. & Mey. Annual, erect, ½—1½ ft. high, glabrous; stem simple below, or branched from the base: cauline leaves alternate, oblong- to obovate-spatulate, obtuse or retuse, obtusely serrulate, ½—1½ in. long; floral opposite, round-ovate, subcordate, mucronate, 2—6 lines long: involucres and glands small: styles deeply bifid: *capsule rough with small warty protuberances: seeds subglobose,* dark-colored, delicately net-veined, the caruncle thin and flat.—Of wide dissemination in the State, but less common than the last. March—June.

* * *Stems diffusely branched, often prostrate; leaves all opposite, unequal at base, stipulate; glands with petaloid appendages; seeds angular, not carunculate.*

6. **E. serpyllifolia,** Pers. Var. **consanguinea,** Boiss. Diffuse annual, with ascending or horizontal but seldom prostrate slender

branches: herbage glabrous, deep green, reddening in age: leaves obovate- to spatulate-oblong, 1—3 lines long, obscurely pinnate-veined, sharply serrate above the middle: stipules setaceous, lacerate or sub-entire: *glands of involucre minute, transversely oblong,* reddish and with *narrow 2—3-lobed or entire* white or rose-colored *appendages:* seed quadrangular, the length scarcely twice the breadth, the *sides more or less rugose-pitted,* the angles somewhat prominent.—Not common.

7. **E. occidentalis,** Drew. Habit of the last, but the glabrous herbage of a dull rather yellowish green: leaves oval or broadly oblong, only slightly unequal, very obtuse at each end, serrate above the middle or quite entire, mucronulate, 2—4 lines long; stipules setaceous lacerate: *appendages of involucre crenate-lobed:* seed ½ line long, whitish, the *faces more or less distinctly sinuate-rugose* between the rather prominent angles.—On Mt. St. Helena.

8. **E. rugulosa,** Greene. Wholly prostrate and very closely depressed, rather succulent, much branched and in age forming a very close mat a foot broad or more: herbage glabrous, pallid and glaucescent: leaves veinless, sharply serrate or almost entire: stipules, involucre, etc., as in the preceding: seeds whitish, finely *transverse-rugose between the scarcely prominent angles.*—Native of the southern extremity of the State, but well established along our railroads.

9. **E. maculata,** L. Prostrate, puberulent or hairy: leaves *oblong-linear,* very oblique at base, senulate upward, 4—6 lines long, usually with a brown-red spot in the centre; stipules lanceolate, fimbricate: glands of the small involucre minute, with narrow slightly crenate reddish appendages: pods acutely angled, puberulent: seeds ⅔ line long, *sharply 4-angled and with about 4 shallow grooves across the concave sides.*—An immigrant from the Mississippi Valley; not rare.

Order XXXII. POLYGALEÆ.

Herbs or shrubs often with milky juice. Leaves simple, entire, exstipulate. Flowers, except as to the pistil, simulating the papilionaceous; but the affinities apparently with certain allies of *Euphorbia.* We have but the genus

POLYGALA, *Diosc.* Ours low undershrubs with alternate leaves and few irregular flowers in terminal cymes. Sepals 5, two larger than the others, lateral and petal like. Petals 3, joined to each other and to the stamen-tube, the middle one hooded above and beaked or crested. Stamens 6--8, unequal, monadelphous, forming a sheath, this open on one side; anthers 1-celled, opening at top. Ovary short, 2-celled; ovules solitary, pendulous, style long, curved dilated. Capsule membranaceous, flattened contrary to the narrow partition, rounded and notched at summit, dehiscent at the margin.

1. **P. Californica,** Nutt. Stems many, slender, 2—8 in. high, from a woody base, mostly simple: leaves oblong-lanceolate or ovate-elliptical, ½—1 in. long: fl. rose-purple on bractless pedicels 1—3 lines long: outer sepals 2½ lines long, rounded, saccate at base; inner ones broadly spatulate, ⅙ in. long or less: lateral petals linear-lanceolate, somewhat ciliate, equalling the broad obtuse somewhat curved beak of the rounded hood: fr. mostly from apetalous fl. near the root; capsule glabrous, broadly ovate, ¼ in. long, retuse narrowly margined: seed 2 lines long, pubescent; caruncle wrinkled and bladdery, calyptriform, half the length of the seed.—In the Coast Range.

Division II, CHORIPETALÆ PERIGYNÆ.

Petals mostly distinct. Calyx more or less distinctly synsepalous, and stamens perigynous.

Order XXXIII. LEGUMINOSÆ.

Herbs, shrubs or trees with alternate, stipulate, compound (in *Siliquas-trum* simple) leaves; leaflets mostly entire, the upper, in some genera, converted into tendrils. Sepals more or less united and forming a 2—5-toothed or -cleft cup, the odd tooth or segment inferior. Petals 5 (sometimes by abortion fewer), more or less united above the base; the two lowest joining to form the *keel;* the two lateral enfolding this and called the *wings;* the uppermost one broader than the others, usually erect, but in the bud folded over the other (except in *Siliquastrum*) and called the *banner;* the corolla as a whole papilionaceous, or butterfly-shaped. Stamens usually 10, distinct or diadelphous (9 and 1), or monadelphous. Pistil 1, usually becoming a *legume, i. e.*, a 2-valved 1-celled pod with 1 row of seeds; these attached to the upper suture, and containing no albumen, the large embryo filling the integuments.

* *Unarmed shrubs.*

Leaves broad, simple.... ....SILIQUASTRUM 1
" unequally pinnate, leaflets many.............................AMORPHA 6
" few, 1-foliolate; branches reedy.............................SPARTIUM 15
" 3-foliolate, not aromatic....................................CYTISUS 13
" " glandular and aromatic........................PSORALEA 8
" palmately 5—9-foliolate.... ....LUPINUS 16

* * *Spinescent or prickly shrubs or trees.*

Leaves unequally pinnate; leaflets many................. ....PSEUDACACIA 7
" all in the form of spine-like green organs................ ...ULEX 14
" 1—3-foliolate; branchlets spinescent........................XYLOTHERMIA 17

* * * *Herbaceous plants.*

Leaves equally pinnate, tendril-bearing;
   Style villous all around at apex............................... ...VICIA 2
   " " lengthwise on one side........................LATHYRUS 3
Leaves unequally pinnate; fl. capitate or racemose;
   . Pods commonly inflated, not prickly......................ASTRAGALUS 4
   " oblong, prickly.....................................GLYCYRRHIZA 5
Leaves of 1—∞ often unequally distributed leaflets;
   fl. in an umbel often with a bract at its base..............LOTUS 9

Leaves pinnately 3-foliolate ⎰ fl. in slender racemes.................MELILOTUS 11
                 pods coiled or curved.. ....MEDICA 12
                 fl. capitate, yellow..................TRIFOLIUM 10
                ⎱ herbage dotted, aromatic...........PSORALEA 8

Leaves palmately 3-foliolate ⎰ fl. capitate or umbellate.......TRIFOLIUM 10
                ⎱ fl. racemose, yellow...................THERMOPSIS 18

Leaves palmately 5—9-foliolate ⎰ calyx bilabiate....................LUPINUS 16
                ⎱ herbage dotted, aromatic.........PSORALEA 8

1. SILIQUASTRUM, Tourn. (RED BUD). Shrubs with simple leaves; the flowers in axillary fascicles, appearing in spring before the leaves. Calyx campanulate, with 5 broad obtuse teeth. Petals 5, the banner small, enfolded by the wings: keel-petals distinct, larger than the wings. Stamens 10, distinct. Pod thin, flat, oblong, wing-margined along the upper suture.

1. **S. occidentale** (Torr.). Widely branching, 6—20 ft. high: leaves round-cordate, entire, obtuse or emarginate, 2 in. broad, on petioles of 1 in. or less: fl. ½ in. long, rose-purple: pod 2 in. long, ⅓ in. broad, acute at each end.—From near Suñol, *Behr*, northward. April.

2. VICIA, *Varro* (VETCH). Weak herbs with angular stems, climbing by tendrils which terminate the pinnate leaves. Peduncles axillary, 1—∞-flowered. Calyx 5-cleft or -toothed, the upper teeth shorter. Stamens diadelphous (9 and 1). Style filiform, bent upward at apex and villous all around, under the stigma, or else on the outside only. Pod oblong, several-seeded.

* Racemose-flowered perennials.

1. **V. gigantea,** Hook. Stout, 5—10 ft. high: *leaflets 10—13 pairs, linear-oblong*, obtuse, mucronulate, 1—2 in. long; stipules 1 in., semi-sagittate, toothed at base; peduncles much shorter than the leaves; the dense raceme 1-sided, 5—18-flowered; fl. dull red: pod glaucous, black when ripe.—Common along streams. May, June.

2. **V. Americana,** Muhl. Weak, 2—5 ft. high, climbing by branched tendrils, nearly glabrous: *leaflets 4—6 pairs*, thin-membranaceous, vivid green above, paler beneath, closely but delicately feather-veined, *elliptic-lanceolate, entire, obtuse*, mucronulate, 1 in. long: *peduncles* shorter than the leaves, *3—8 flowered:* fl. ¾ in. long, bright purple: upper calyx-teeth very short, lower well elongated: pods 1 in. long, glabrous. Var. **truncata,** Brewer. Lower and stouter than the type; leaflets linear to oblong-linear, usually dentate or even serrate toward the truncate apex: fl. larger and paler. Var. **linearis,** Wats. Leaflets firmer in texture, narrowly linear, the veinlets confluent along the margin.—Common and variable.

* * Few-flowered annuals.

3. **V. exigua,** Nutt. Slender, 1—2 ft. high: *leaflets 4 or 6, oblong-linear, obtuse:* peduncles filiform, shorter than the leaves, 1—2-flowered: calyx-teeth lanceolate from a broad base: corolla white or purplish, 2 lines long: pod glabrous, 4—5-seeded.—Hillsides or plains, preferring stony or sandy soil. March—May.

4. **V. Hassei,** Wats. Taller and less delicate than the last, the *leaflets* ampler, more numerous, *deeply notched at apex:* fl. 3 lines long:

pod shortly stipitate, 5—8-seeded.—Of more southerly distribution than the preceding, but found at Benicia, *Bigelow.*

5. V. sativa, L. Stoutish, suberect, 2—3 ft. high: leaflets 8 or 10, obovate-oblong, truncate or retuse, mucronate: fl. 1 or 2, subsessile, ½ in. long, red-purple.—The Vetch or Tare, cultivated from time immemorial as a food and fodder plant; of frequent occurrence by way-sides and in old fields.

3. **LATHYRUS**, *Theophr.* (WILD PEA). · Coarser plants than *Vicia*, with broader leaves and flowers, the style villous in a line up and down the inside (next the free stamen).

  * *Tendril-bearing; the racemes many-flowered.*

1. **L. Bolanderi**, Wats. Often shrubby below, 3—5 ft. high: leaflets 3—5 pairs, ovate, obtuse or retuse, mucronate, 1—1½ in. long, thin, on very short petiolules: *stipules broadly semisagittate,* acute, more or less toothed: peduncles equalling the leaves: *lower calyx-teeth lanceolate-acuminate, longer than the tube; upper very short,* broadly triangular, all glabrous along the margin, or nearly so: corolla ¾ in. long, rose-purple, fading yellowish.—Frequent on wooded slopes.

2. **L. Jepsonii**, Greene. Nearly or quite glabrous; *stem* 5—8 ft. high, *strongly winged* along the angles and striate between them: leaflets 8—12, linear-lanceolate, acute, 2—3 in. long, subcoriaceous, venulose: *stipules small,* setaceously acuminate: peduncles stout, about as long as the leaves: fl. rose-purple; *calyx-teeth ovate-lanceolate, the lowest not much longer than the others;* corolla ¾ in. long, relatively broad: pod 2—3 in. long, sessile in the calyx, 12—16-seeded.—Muddy margins of sloughs, within reach of tide-water in the Suisun marshes.

3. **L. puberulus**, White. Low, herbaceous, or, 8—15 ft. high and shrubby at base, soft-pubescent or nearly glabrous, the *stems angled:* leaflets 5—7 pairs, ovate-oblong to linear, cuspidate, subcoriaceous: stipules broadly or narrowly semisagittate, toothed or entire: peduncles about equalling the leaves; fl. ¾ in. long, broad, purplish; lower calyx-teeth lanceolate, acuminate, as long as the tube: ovary and pod appressed-pubescent.—From Sonoma Co. southward. Feb.—May.

  * * *Without tendrils; peduncles 1--3 flowered.*

4. **L. Torreyi**, Gray. Erect, slender, 1—2 ft. high, the herbage thin, light green, fragrant: leaflets 4—6 pairs, with or without a reduced terminal odd one, round-ovate or oblong, ½ in. long, mucronate: stipules narrow, acuminate, the lower lobe short or almost obsolete: fl. 1 or 2, short-peduncled, white or pinkish: calyx-teeth narrowly subulate, the upper a little shorter: pod 1 in. long, pubescent, 3—6-seeded.—From Santa Clara Co., to Napa and northward, in dry woods. Herbage remarkable as exhaling the fragrance of *Asperula odorata.* May.

* * * *Rachis dilated, ending in a rudimentary odd leaflet.*

5. **L. littoralis** (Nutt.), Endl. Stout and low, decumbent, densely silky-villous: stipules large, ovate or semihastate; leaflets 1—3 pairs, cuneate-oblong, ½ in. long or more: peduncles exceeding the leaves; calyx-teeth nearly equal, about as long as the tube: corolla ½—¾ in. long, banner bright purple, wings and keel white: pod large, oblong, obtuse, villous, 3—5-seeded.—Strictly maritime, in sandy or clayey soil within reach of the sea-spray.

4. **ASTRAGALUS**, *Diosc.* (RATTLE-WEED, LOCO-WEED). Herbs either erect or decumbent, with unequally pinnate leaves, no tendrils, persistent stipules, and axillary spikes or racemes of flowers which are usually small for the size of the plant, and rather narrow. Calyx 5-toothed. Petals with slender claws, the keel obtuse. Stamens diadelphous, (9 and 1); anthers uniform. Stigma terminal, minute. Pod various, seldom or never promptly dehiscent, often coriaceous and turgid, or thin and bladdery-inflated, or thin and flat; 1-celled, or partly 2-celled by intrusion of one or both sutures. Seeds few or many, small for the size of the pod, commonly reniform, on slender funiculi.

* *Annuals.*

1. **A. didymocarpus**, Hook. & Arn. Slender, pubescent, 1 ft. high: leaflets 9—15, cuneate-oblong to linear, emarginate, 3—5 lines long: spikes long-peduncled, dense, ovate or oblong: fl. small, dull purplish: *pods erect, 2 lines long, and about as broad, scarcely exserted* from the calyx, *strongly wrinkled*, 2-celled, 2-seeded.—Abundant along the eastern base of Mt. Diablo Range and far southward.

2. **A. nigrescens**, Nutt. Smaller than the last, more slender, less pubescent, the less dense spikes cylindrical: *pods deflexed, well exserted from the calyx, slightly wrinkled*, strongly obcompressed.—Common on sterile gravelly hill-sides of the Bay region; the flowers commonly minute and dull, but on the flanks of Mt. Tamalpais and northward larger and violet. April—June.

3. **A. tener**, Gray. Slender, sparsely pubescent, 6—10 in. high: leaflets 9 —15, linear or cuneate, acute or retuse: fl. many, capitate on a slender peduncle, purple: *pod* ¾ in, long, *slender, incurved*, 2-celled, 5—10-seeded.—In moist lands, either sandy or alluvial. A handsome species; the heads of purple and white recalling those of some kinds of clover. April, May.

4. **A. Breweri**, Gray. Smaller than the last, relatively stouter, leaflets broader, heads few-flowered: *pods with a short body and* a very long *incurved beak.*—Common in fields of the Sonoma valley.

* * *Perennials; pods bladdery-inflated.*

5. **A oxyphysus,** Gray. Erect, 2—3 ft. high, stoutish, canescent with a minute pubescence: leaflets 9—21, oblong, 1 in. long; peduncles exceeding the leaves, raceme elongated: calyx-teeth subulate, half as long as the oblong tube: corolla greenish-white, ⅔ in. long: pod compressed, oblique (semiobovate), *acuminate at both ends,* 1½ in. long, on a *stipe little exceeding the calyx.*—Dry hills of the Mt. Diablo range.

6. **A. leucophyllus,** Torr. & Gray. Erect, tall, growing parts silvery-canescent, when older glabrate: leaflets 27—37, broadly linear, acutish, ¾ in. long: peduncles long, racemes short: calyx-teeth subulate, half as long as the oblong tube; corolla yellowish: *pod* obliquely oval, 1½ in. long, *on a long filiform pubescent stipe.*—Low hills skirting the interior valley; very common between Livermore and Niles.

7. **A. crotalariæ** (Benth.), Gray. Stout, decumbent, glabrous, except the canescent growing parts: leaflets very many, oblong-linear to obovate, sometimes retuse, ⅓—1 in. long: stipules broadly triangular, distinct: calyx-teeth subulate, half as long as the short-campanulate tube: fl. white: *pod thin, ovoid, 1—1½ in. long, sessile in the calyx.*— Plains and hills.

8. **A. Menziesii,** Gray. Stout, erect, 2—4 ft. high, glabrous or nearly so: stipules broad, not pointed, continued around the stem, sometimes nearly meeting or even cohering opposite the base of the leaf: *raceme long and dense; fl. greenish:* pod thin, large as in the last.—Plentiful in sandy soils along the seaboard, at Alameda, San Francisco, etc.

9. **A. macrodon** (H. & A.), Gray. Erect, tall, glabrous in age, the nascent parts canescent: leaflets 23—27, oblong-lanceolate, obtuse, mucronulate: stipules small, lanceolate-acuminate: peduncles rather shorter than the leaves; racemes long: *calyx-teeth* slender-subulate, equalling the campanulate tube, and *almost as long as the corolla:* ovary silky; pod unknown.—This plant was collected by Douglas only, some sixty years ago, somewhere between Monterey and Sonoma, probably near the former place. It should be carefully sought, though it may have become extinct. The long and slender calyx-teeth, according to the original description, so distinguished it from its allies, as to make its recognition easy in case it should be rediscovered.

10. **A. Douglasii** (T. & G.), Gray. Ascending, 1 ft. high, cinereous-puberulent when young, otherwise nearly glabrous: leaflets very many, linear or linear-oblong, ⅓—¾ in. long: spike short, dense, 10—20 flowered: *calyx-teeth subulate, shorter than the campanulate tube:* pod thin, obliquely ovoid, 1½—2 in. long.—In gravelly places along streams, from San Francisco southward.

* * * Perennials; pods not bladdery.

11. **A. pycnostachyus,** Gray. Stout, 2 ft. high, more or less villous-hoary: leaflets about 21, oblong, ½ in. long: fl. yellowish, in dense cylindrical short-stalked spikes: *pods crowded, retrorsely imbricate,* ovate, acute, *laterally flattened,* thin-coriaceous, glabrous, coarsely reticulate, 1-celled.—In moist subsaline grassy land near the entrance to Bolinas Bay, *Bolander,* 1863, *Greene,* 1888, also in a similar locality not so near the sea southwest of Mt. Tamalpais.

5. **GLYCYRRHIZA,** *Diosc.* (LICORICE). Glandular-viscid perennials with unequally pinnate leaves, and flowers in axillary peduncled spikes; calyx 5-cleft. Stamens monadelphous or diadelphous; the alternate anthers smaller. Pod short, compressed, prickly, indehiscent.

1. **G. glutinosa,** Nutt. Two or three ft. high, erect or decumbent, either nearly glabrous and viscid with minute sessile resinous dots, or more decidedly glutinous by a villous or hirsute glandular pubescence, never scurfy: leaflets 13 to 19, oblong-lanceolate, 1 or 2 in. long; stipules ovate-acuminate to lanceolate, persistent: spikes merely oblong, 1 to 1½ in. long, on peduncles of 1 in.: pod bur-like.—Common in the Mt. Diablo Range, and on the plains eastward; also at Alameda.

6. **AMORPHA,** *Linnæus.* Shrubs with unequally pinnate leaves which, with the young twigs and inflorescence, are pellucid glandular and heavy-scented, the glands in age dark brown and opaque. Leaflets many; stipules and stipels caducous. Flowers very small, dark purple, in long and narrow terminal spikes. Calyx obconic-campanulate, 5-toothed, persistent. Banner (the only petal present) erect, concave, unguiculate. Stamens monadelphous at the very base. Pod short, lunulate, glandular, scarcely dehiscent, 1- or 2-seeded.

1. **A. hispidula,** Greene. Two to four ft. high, pubescent or glabrous, the glandular dots supplemented on the twigs, stalklets and leaf-rachis by acute prickle-like glands with tips more or less recurved: leaflets 8—12 pairs, oval to linear-oblong, an inch long, retuse or emarginate: calyx-teeth triangular-lanceolate, more than half the length of the tube: petal red-purple: pod half obcordate, very glandular, twice the length of the calyx.—Marin and Napa counties.

7. **PSEUDACACIA,** *Tourn.* (LOCUST-TREE). Trees or shrubs with odd-pinnate leaves and stout prickles in place of stipules, the leaflets prickly-stipellate. Flowers showy, in pendulous racemes. Calyx slightly bilabiate, 5-toothed. Banner large, roundish, reflexed, little longer than the wings and keel. Stamens diadelphous. Pod linear, flat, several-seeded, margined along the upper suture, readily dehiscent.

1. **P. VULGARIS,** Tourn. *P. odorata,* Mœnch. *Robinia Pseudacacia.* L. Tree 30—50 ft. high: prickles on older branches small, straight, on

the younger larger and somewhat curved: leaflets 9—17, oblong-ovate or elliptical: racemes pendulous, oblong: fl. white, very fragrant.—Native of the Atlantic states; long cultivated in California for shade and ornament; now spontaneous in many places.

8. PSORALEA, *Royen.* Perennials (one adventive species shrubby), punctate with dark dots and heavy-scented: leaves pinnately 3-foliolate (in n. 6 palmately 5-foliolate): stipules free from the petiole. Calyx-lobes nearly equal, the two upper sometimes connate. Keel broad, obtuse, joined to the wings. Stamens monadelphous or diadelphous: anthers uniform. Pod ovate, indehiscent, 1-seeded.

1. **P. orbicularis,** Lindl. *Stem prostrate, creeping,* the leaves and racemes erect, long-stalked; *leaflets 2—3 in. long, the terminal one nearly orbicular,* the lateral pair obovate: raceme a few inches to a foot long, the flowers subtended by large deciduous bracts: calyx villous and pedicellate-glandular, cleft almost to the base, the lowest tooth as long as the purplish corolla: stamens diadelphous: pod ovate, acute, 3 lines long.—Moist grassy places. July.

2. **P. strobilina,** Hook. & Arn. *Erect,* 2—3 ft. high, villous throughout; the stem and stalklets glandular: *leaflets rhombic-ovate, 2 in. long; stipules large, broadly ovate,* acuminate: peduncles shorter than the leaves: spike oblong, the bracts very large, deciduous: calyx ½ in. long, the lower tooth much the longest, equalling the purple corolla: stamens monadelphous: ovary pubescent.—Mountains of Contra Costa and Santa Clara counties.

3. **P. macrostachya,** DC. Erect, stout, 3—12 ft. high: *leaflets ovate-lanceolate: stipules small, lanceolate:* peduncles greatly surpassing the leaves: spikes cylindrical, silky-villous: bracts acuminate, as long as the flowers: lower calyx-tooth longest, scarcely as long as the corolla: tenth stamen almost free: pod ovate-oblong, acute, 3 or 4 lines long, compressed, villous.—Very common, either on hillsides or in low ground, but always in moist places. June—Oct.

4. **P. physodes,** Dougl. Erect, 2—3 ft. high, nearly glabrous: *leaflets ovate, acute,* 1 in. long; *stipules linear-lanceolate:* peduncles about as long as the leaves; raceme short, dense, the bracts small: calyx covered with sessile glands and somewhat black-hairy, at length much enlarged and inflated, becoming 4 or 5 lines long, its teeth short, subequal: corolla scarcely ½ in. long, ochroleucous, often with a deep purple tinge: stamens monadelphous: pod rounded, compressed, 3 lines long.—Common in both the Coast and Contra Costa Ranges, in open places among thickets and trees. May—July.

5. P. GLANDULOSA, L. *Shrubby or arborescent,* with loose elongated branches; glabrous, but roughish with elevated glands: leaflets ovate-

lanceolate, acuminate, 2 or 3 in. long; *stipules subulate-setaceous, deciduous:* racemes longer than the leaves, the bluish flowers more or less
verticillate.—Native of Chile; frequent in cultivation, occasionally wild.

6. **P. Californica,** Wats. *Low, tufted; pubescence short, silky, appressed: leaves palmately 5-foliolate;* stipules scarious, lanceolate, deciduous; leaflets broadly oblanceolate, acutish, ¾ –1¼ in. long: racemes
shorter than the leaves, short-peduncled, rather loose; pedicels slender:
calyx silky-villous, ½ in. long, the linear acuminate lobes a little
exceeding the petals: pod thin, villous, oblong with a lanceolate beak:
seed compressed, 2 lines long or more.—Summit of Mt. Diablo.

9. LOTUS, *Tourn.* (Lotus. Hosackia). Herbaceous or suffrutescent,
with pinnately 3—∞ - foliolate (in the first species often 1-foliolate)
leaves; leaflets sometimes of even number but unequally distributed on
the two sides of the rachis; stipules foliaceous, scarious, or more commonly reduced to dark glands. Flowers solitary, or in umbels or heads
which are naked or subtended by a 1—5-foliolate bract. Calyx 5-toothed
or -cleft. Corolla whitish, yellowish or purplish, changing to orange or
red; petals free from the stamens; banner ovate or rounded: wings
commonly meeting imperfectly and (by a twist in the claw) obliquely in
front of the obtuse or acute, sometimes rostrate keel. Stamens diadelphous; the alternate filaments dilated or thickened under the anthers.
Pod linear, compressed or terete, straight or arcuate, promptly or tardily
dehiscent, or indehiscent, 1—∞ - seeded. Seeds variously rounded or
elongated, sometimes quadrate, smooth, tuberculate or rugose.

    * *Annuals with gland-like traces of stipules; leaflets 1—4, on a linear
       rachis; pods straight, readily dehiscent.*

    1. **L. Americanus** (Nutt.), Bisch. Erect or decumbent, 1—2 ft. high,
more or less villous: leaflets (rarely 5) ovate or oblong, acutish, ¾ in.
long: peduncles slender, exceeding the leaves, the solitary salmon-
colored or whitish flower subtended by a bract 3—6 lines long: calyx-tube
very short, the linear teeth equalling the corolla: pod 1—1½ in. long:
seeds oblong, smooth, dark-colored.—On sunny banks, or in the dry
gravelly beds of streams, or even in moist meadow lands. May—Dec.

    * * *Stipules gland-like; leaflets 4—10, unequally distributed on opposite
       margins of a dilated rachis; pods readily dehiscent.*

    ←*Annuals; flowers solitary, short-pedicelled, not bracted; claws of petals
       approximate; keel pointed.*

    2. **L. Wrangelianus,** Fisch. & Mey. Less than 1 ft. high, ascending,
much branched, densely leafy, *sparsely or canescently villous:* leaflets
about 4, cuneate-obovate to oval or oblong, 3—6 lines long; *calyx-teeth
broadly subulate, equalling the tube:* corolla 3 lines long, bright yellow,

the broadly obovate banner erect: wings meeting above the keel, not enfolding it: pod pubescent, straight, 7—10 lines long, 5—7 seeded.— Common throughout middle California, especially toward the seaboard.

3. **L. humistratus,** Greene. Low and diffuse, the branches 5—8 in. long, *herbage soft-villous:* fl. nearly sessile, yellow; *calyx-teeth linear, much longer than the tube:* pod oblong. ⅓ in. long, pilose, 2—3-seeded. —Clayey banks and hillsides; as widely dispersed as the preceding, but less common. May, June.

4. **L. denticulatus,** Greene. Erect, 1—2½ ft. high, fastigiately branching, pale green and glaucous, sparingly pilose; *calyx-teeth longer than the tube,* and, with the margins of the upper leaves, *somewhat denticulate:* corolla 2 lines long, pale yellow or salmon-color, changing to red: pod pubescent, short, 3-seeded.—A weed in grain fields of the Sacramento. April—June.

+—+—Flowers 1 or many, on an elongated, usually bracted peduncle; claw of the banner commonly remote from the others, keel mostly obtuse.

++Annuals; few-flowered.

5. **L. micranthus,** Benth. *Erect, slender,* 4—10 in. high, *glabrous, glaucous:* peduncle filiform, bracted, 1-flowered: fl. minute, pale salmon, turning red; pod 1 in. long or less, compressed, constricted between the seeds, these oval or roundish, little compressed, smooth.—April, May.

6. **L. salsuginosus,** Greene. Ascending or depressed, *slightly strigose, somewhat succulent,* the branches 8—18 in. long: *leaflets 4—6, obovate, obtuse:* peduncles 1 in. long, 1—4 flowered, naked or with a conspicuous 1—3-foliolate bract: corolla yellow, 3 lines long, the banner and wings equalling the straight keel: pod scarcely compressed, 10—12-seeded: seeds obliquely oval, smooth.—From San Jose southward, either toward the sea, or on subsaline flats of the interior. March—June.

7. **L. rubellus** (Nutt.), Greene. Prostrate, slender, *not succulent,* strigose-pubescent or nearly glabrous: *leaflets 6—10, linear-oblong,* mostly acutish: early peduncles shorter than the leaves, bractless, 1-flowered, the later longer, bracted, 2-flowered: corolla reddish, scarcely twice as long as the calyx: pod slender, straight, 7—10-seeded: *seeds* quadrate, *minutely granulate.*—Plentiful in sandy soils, San Francisco, Alameda and southward; apparently only along the seaboard. April—July.

8. **L. nudiflorus** (Nutt.), Greene. Near the last, but *leaflets smaller and broader:* fl. thrice as large: pod broader, more flattened, slightly curved upward at apex: *seeds larger,* quadrate, *faintly tuberculate.*— Eastern base of Mt. Diablo Range, near Byron, etc., on gravelly hill-tops; thence southward. March—May.

9. **L. strigosus** (Nutt.), Greene. Strigose-pubescent, decumbent or prostrate: peduncles long, commonly 1—2-flowered and 3-foliolate-bracted: fl. 4—5 lines long, yellow: pod pubescent, slightly curved upwards: *seeds* quadrate: but *somewhat cruciform*, being deeply notched at each end and at the hilum, the surface *closely sinuate-rugose.*—Same range as the last, and readily distinguished by its seeds which have something of the outline of a Maltese cross. March—June.

10. **L. hirtellus**, Greene. Stoutish, depressed, *canescently hirsutulous*, not at all strigose: leaflets 5—7, cuneate-oblong or -obovate, obtuse: peduncles stoutish, bracted, surpassing the leaves, 2-flowered: pod 1 in. long, subterete, straight, 7—10-seeded; *seeds* quadrate, *notched at the hilum only, faintly rugose and coarsely granulate.*—The Mt. Diablo Range near Livermore.

++ ++ *Perennials; flowers capitate-umbellate.*

11. **L. leucophæus,** Greene. *Low; ascending*, less than a foot high, internodes short, leaves ample, herbage *velvety-pubescent:* leaflets 5—7, obovate, 6—9 lines long, acute: peduncles equalling or exceeding the leaves; umbel 1-foliolate-bracted, 5—8-flowered: fl. more than ½ in. long, ochroleucous, becoming red-purple.—Dry ridges of the inner Coast and Mt. Diablo Ranges. May.

12. **L. grandiflorus**, (Benth.), Greene. *Tall, slender* with few leaves and long internodes, *nearly glabrous:* peduncles slender, elongated, small-bracted, 5—8-flowered: fl. nearly 1 in. long, deep yellow, the petals broader than in the last, turning orange.—Same range as the last; but less frequent.

* * * *Perennials with true stipules; leaflets never inequilaterally distributed; flowers in bracted umbels; pods long, straight.*

13. **L. formosissimus**, Greene. Slender, glabrous, the *decumbent stems several, 1 ft. long:* leaflets 5—7, from broadly obovate to obovate-oblong, obtuse, the lowest truncate or retuse; stipules thin, ovate: umbels equalling the leaves, or shorter, the bract 3-foliolate; calyx-teeth unequal, triangular, acute or acuminate, shorter than the campanulate tube: corolla 7 lines long, the *wide-spread wings and much shorter keel rose-red, the banner yellow.*—In moist ground along the seaboard. May.

14. **L. pinnatus**, Hook. *Stoutish, glabrous, the erect stems 2 ft. high:* leaflets 5—9, obovate or oblong, acutish: stipules scarious, triangular: peduncles longer than the leaves, 3—7-flowered, naked or with a small scarious 1—3-foliolate bract: calyx-teeth triangular, half as long as the tube: corolla as in the last, but *keel and wings white, banner yellow.*—Said to inhabit the seaboard districts from San Francisco northward.

15. **L. Torreyi** (Gray), Greene. Habit of the last, but *slender, more or less silky-pubescent;* leaflets narrower, acute or obtuse: bract of the

umbel sessile: fl. smaller, the keel and wings white, the latter not spreading.—Along streamlets in the middle or higher Coast Range.

16. **L. Crassifolius** (Benth.), Greene. Erect, stout, 2—3 ft. high, *of a dull green hue*, as if glaucous, but *minutely pubescent:* leaflets 9—15, thickish, obovate or oblong, obtuse, mucronulate, ½ in. long or more: peduncles nearly equalling the leaves; umbel many-flowered, the 1—3-foliolate bract a little below it; calyx-teeth triangular, short; corolla purplish marked with green spots: pods thick, 2 in. long.—Common in the mountain districts. May, June.

17. **L. stipularis** (Benth.), Greene. Not as tall as the last, more slender, *villous with spreading hairs* and often somewhat glandular; leaflets 15—21, obovate-oblong, acute, mucronate, ½—1 in. long; stipules large, ovate: peduncles short, 4—8-flowered, the leaf-like bract near the middle, 3—9-foliolate: calyx 2 lines long, the subulate teeth short: corolla purple; pod straight, 1—1½ in. long.—Contra Costa and Sonoma counties. Seldom seen.

18. **L. balsamiferus** (Kell.). Stoutish, erect, 2 ft. high, with the foliage and inflorescence of *L. crassifolius* nearly; but *herbage of a vivid green*, the stem and growing parts very *glutinous* from abundant *glandular-hispidulous short hairs.*—Hood's Peak, Sonoma Co., *Bioletti.* Doubtless a rediscovery of Dr. Kellogg's *Hosackia balsamifera.*

* * * * *Stipules gland-like; leaflets few, unequally distributed; pods small, indehiscent, usually arcuate and long pointed.*

+—*Perennials (sometimes woody at base).*

19. **L. glaber** (Vogel.), Greene. Suffrutescent, 2—8 ft. high, erect or decumbent, nearly glabrous; leaflets mostly 3, on young shoots 4—6, oblong to linear-oblong, ¼—½ in. long, obtuse or acute: umbels many, sessile; fl. 3—4 lines long, yellow, turning red: *calyx-teeth subulate, erect, rather less than half as long as the tube.* Usually tufted and reedy-looking, the foliage sparse, the flowers profuse.—Common about San Francisco, and southward throughout the State, in the Coast Range chiefly; flowering almost all the year round.

20. **L. Benthami**, Greene. Resembling the last, but smaller and mostly prostrate: umbels on peduncles which equal or exceed the leaves and are 1—3-foliolate-bracted at top: *calyx-teeth more slender, stellate-spreading in the bud and recurved in flower.*—Common on low hills near the sea in San Mateo Co. June, July.

21. **L. junceus** (Benth.), Greene. Nearly glabrous, erect, shrubby, with slender branches reedy and sparsely leafy: leaflets obovate to oblong, 2—4 lines long: fl. 3 lines; *calyx 2 lines long or less; teeth very short and blunt.*—A more southerly species than either of the two preceding; but said to have been found near San Francisco.

22. L. Biolettii, Greene. Slender, the somewhat wiry prostrate branches 1—2 ft. long: herbage *cinereously or canescently pubescent with short appressed hairs:* leaflets usually 4, cuneate-obovate, obtuse, 2—5 lines long: umbels on slender peduncles little surpassing the leaves, unifoliolate-bracted, 6—10-flowered: calyx a line long or less, narrowly funnelform, the *triangular pointless teeth* a third as long, *erect:* corolla 2 lines long, yellow, turning dark-red: pod strongly arcuate, slender-braked.—Dry ridges above Mill Valley, Marin Co.

+—+ *Annuals.*

23. L. eriophorus, Greene. The numerous branches a foot or two long, *flexuous, weak and prostrate: pubescence dense, somewhat tomentose:* leaflets 5—7, obovate or cuneate-oblong, acute, 3—6 lines long: umbels short-peduncled or subsessile, bracted: fl. 3—4 lines long; calyx half as long, very villous; the filiform teeth about equalling the tube.—In sandy grounds near the sea, from San Francisco southward.

24. L. Heermani (Dur & Hilg.), Greene. Near the last, but less pubescent, neither the leaflets nor the flowers more than half as large, the leaflets broader and rounded.—Same range as the last.

10. TRIFOLIUM, *Pliny* (CLOVER). Herbs with palmately (in one pinnately) 3-foliolate leaves and adnate stipules; leaflets commonly denticulate. Flowers in roundish or ovoid or somewhat depressed capitate or umbellate clusters, on axillary or terminal peduncles. Calyx 5-cleft or -toothed. Corolla persistent; banner and wings commonly coherent with the stamineal tube; keel mostly obtuse and shorter than the wings. Stamens diadelphous. Pod concealed within or little exserted from the calyx, 1—6-seeded, dehiscent or indehiscent.

* *Heads or spikes not involucrate.*

+ *Flowers pedicellate, at length reflexed; calyx-teeth subulate, not plumose.*

1. T. gracilentum, Torr. & Gray. Erect, slender, 1—2 ft. high, wholly glabrous: stipules ovate- or linear-lanceolate, acuminate: *leaflets cuneate-obcordate, spinulose-serrulate,* ½ in. long: heads 15—25-flowered: calyx-teeth lanceolate-subulate, setaceously acuminate, thrice as long as the tube, shorter than the usually deep reddish corolla: pod exserted, 2-seeded: seeds obliquely oval, straw-colored, very smooth.—Open plains and hillsides. April—June.

2. T. bifidum, Gray. Erect, very slender, 1 ft. high, pale green and glaucous, the petioles and calyx more or less pilose-villous: stipules ovate-lanceolate, entire, setaceously acuminate: *leaflets linear-cuneate, the sides remotely toothed, apex bifid* and mucronulate: peduncles slender, exceeding the leaves: heads 6—15-flowered: calyx deeply 5-parted, the teeth subulate-setaceous, about equalling the minute pale rose-colored corolla: pod 1-seeded: seed rather narrowly obovate-oblong. Var. decipiens, Greene. Taller and stouter, the *leaflets cuneate-oblong* with closely serrulate margins aud *only a shallow notch at apex;* heads 15—30-flowered.—The type is frequent between San Jose and Vacaville, mostly within or to the eastward of the Mt. Diablo range. Only the variety is found about the Bay. April—June.

3. **T. ciliolatum**, Benth. Erect, 1—2 ft. high, glabrous: stipules narrow, acuminate; leaflets cuneate-oblong or obovate, ½—1 in. long, obtuse or retuse, serrulate: fl. purple, 3 lines long; *calyx-teeth* lanceolate, very acute, *rigidly ciliolate.*—Throughout the western part of the State, both seaward and in the interior. April—June.

4. **T. PROCUMBENS**, L. Ascending or suberect, slender, pubescent: leaflets cuneate-oblong, emarginate, denticulate, the terminal one on a longer stalklet: *heads ovate or oblong, very dense; fl. yellow;* banner deflexed over the other petals in age.—A small Old World clover, beginning to appear spontaneously with us.

5. **T. REPENS**, Rivinus, (1690). Perennial, diffuse, creeping, sending up erect long-stalked glabrous leaves and heads: leaflets obcordate, denticulate: *heads depressed-globose, at length umbellate: fl. white;* calyx-teeth unequal, lanceolate-subulate, shorter than the tube: pod about 4-seeded.—The *White Clover* of eastern and European meadows and pastures; a troublesome plant in lawns with us; sparingly naturalized.

+–+– *Flowers nearly or quite sessile, not reflexed; calyx-teeth-elongated, plumose, or at least hairy.*

++ *Perennial.*

6. **T. PRATENSE**, Tragus, (1552). Stoutish, ascending, 1 ft. high, pubescent: leaflets oval or obovate, often retuse, 1 in. long: *heads ovate,* 1 in. long, *sessile: corolla elongated-tubular, rose-purple.*—The *Red Clover* of eastern and Old World meadows; occasionally spontaneous with us.

++ ++ *Annuals.*

7. **T. Macræi**, Hook & Arn. Much branched, decumbent or almost prostrate, the slender branches 8—18 in. long, the herbage more or less villous- or pilose-pubescent: leaflets cuneate-oblong, obtuse, denticulate above the middle, 6—10 lines long: *heads nearly or quite sessile,* usually in a terminal pair, ovate, ¼—½ in. high; *calyx-teeth longer than the tube,* densely plumose-hairy, nearly equalling the small purplish corolla: pod 1-seeded. Var. **albopurpureum**, Greene. Often 1—1½ ft. high, ascending; *heads small, ovate-conical or sub-cylindrical, solitary at the ends of very long slender peduncles;* calyx-teeth slender, more delicately plumose, fully equalling the white-tipped purple corolla.—Common and variable.

8. **T. dichotomum,** Hook. & Arn. Erect or ascending, stoutish, 1—1½ ft. high, often flexuous and repeatedly dichotomous: pubescence longer than in the last, more spreading: leaflets cuneate-obovate or oblanceolate, the upper acute, ¾ in. long; sharply denticulate: *heads long-peduncled, ovate-conical ¾—1¼ in. high: calyx-teeth* setaceous, densely hairy, *equalling the red-purple corolla:* pod with close elevated striæ.—Plentiful on plains of interior, from Vacaville to Antioch.

9. **T. amœnum,** Greene. Commonly 2 ft. high, stout, simple or with few branches from the base, the heads 1—3, terminal and subterminal, herbage canescently villous: leaflets broadly obovate, retuse or obtuse, erose-denticulate, 1 in. long or more, 10 lines broad: *heads globose, in age oval 1½ in. high:* calyx-teeth linear-setaceous, plumose throughout, 3—4 lines long, much shorter than the *very showy corolla;* this *light rose-purple with dark centre.*—Vanden Station, Sacramento plains.

10. **T. columbinum,** Greene. Erect, nearly simple, 1 ft. high, somewhat silky-pubescent: leaflets 1 in. long, cuneate-oblong, obtuse, crenulate-denticulate: head ovate-conical, 1 in. high: *calyx-tube less than 1 line long; the filiform segments 5 lines,* soft and *silky-plumose throughout,* deeply concealing the minute purple corolla: pod striate, villous at apex.—Common about Vacaville; readily known by its pale dove-colored heads altogether soft and silky, exhibiting no flowers, but seemingly made up of the long, densely plumose calyx-teeth. May.

11. **T. olivaceum,** Greene. Simple or branched from the base, 1—1½ ft. high, glabrous except an appressed pubescence on the lower face of the leaves: petioles 1—2 in. long, with lanceolate acuminate entire stipules; leaflets as in the last, but somewhat serrulate: heads on long slender peduncles, hemispherical in flower, 1 in. or more broad and high; *calyx-tube 1 line long; the linear-setaceous teeth 5—6 lines, densely plumose toward the base only,* gradually less so above, *nearly naked at the rather rigidly setaceous tips;* corolla deep violet-purple, very small and concealed; pod striate, glabrous.—With the preceding, but more common; distinguished by its large olive-green heads.

12. **T. ARVENSE,** L. Related to the last two, but of different aspect; the numerous *branches lateral,* not basal; the leaves and heads short-stalked: heads oblong or cylindrical, ¾ in. long, or less: *calyx-teeth silky-plumose* throughout, *longer than the minute whitish corolla.*—The *Rabbit-foot* or *Mouse-ear Clover* of Europe, naturalized on the Atlantic coast, has been reported from Alameda Co.

* * *Heads subtended by a flat or concave (sometimes nearly obsolete) involucre.*

←*Corolla not inflated in age.*

↔*Involucre flat; heads a little one-sided.*

13. **T. Wormskjoldii,** *Lehm. Perennial,* spreading underground by slender root-stocks; stems decumbent, 3 in.—2 ft. long; herbage flaccid,

glabrous: leaflets obovate-oblong, obtuse, pectinate-denticulate, 1 in. long or more: heads hemispherical, 1 in. broad or more; involucre ½—¾ in. broad, laciniate-aristate: calyx-tube scarious, 10-striate, the alternate nerves less prominent, transverse veinlets 0; *teeth linear-subulate, much longer than the tube*, all entire, or 1 or more of them setaceously 2—3-parted: banner elliptical, deeply emarginate, pale purple; other petals darker.—On hills about San Francisco only a few inches high; in springy places, or along perennial streams, large and fistulous, forming dense masses, the leaflets often 4, and the calyx-teeth more or less cut into setaceous divisions. April—August.

14. **T. variegatum**, Nutt. *Annual*, glabrous, decumbent or prostrate, with very numerous slender branches: leaflets obcordate to obovate-oblong, minutely spinulose-serrate: upper stipules roundish, lacinately cleft: peduncles slender, longer than the leaves: laciniate involucre, shorter than the small (3—15-flowered) heads: calyx-tube about 15-nerved; the *teeth broadly subulate, tapering to a setaceous point*, longer than the tube, shorter than the corolla: fl. dull purple or whitish. Var. **melananthum**, Greene. More rigid, ascending, the branches often a foot long or more; heads larger: *calyx-teeth more triangular and only pungently acute* or acuminate, of a dark purple almost to the base; corolla deep purple. Var. **major**, Loja. Flaccid and procumbent, but very stout and fistulous, the branches often a yard long; leaflets oblong-cuneiform, 1 in. long or more; heads 1 in. broad more or less: calyx-teeth dark purple; petals purple with whitish tips.—Very common and variable. April, May.

15. **T. appendiculatum**, Loja. Glabrous, *flaccid, diffuse*: leaves long-petioled: leaflets cuneate-obovate or obcordate, serrulate-spinulose, mucronulate at apex: heads hemispherical, 1 in. or less in breadth: fl. purple: calyx-teeth lanceolate-linear, entire, longer than the tube: *keel of the corolla rostrate-attenuate*, longer than the wings.—Lake Merritt, Oakland, *F. K. Chesnut*.

16. **T. oliganthum**, Steud. Pale green, glabrous, *erect, slender*, with few ascending branches, 6—18 in. high: upper *leaflets linear, acute*, 1 in. long, spinulose-serrate: peduncles filiform, 2—3 in. long, exceeding the leaves: head small, 7—12-flowered; involucre reduced, laciniately divided: fl. pale purple and white: 2–3 lines long: calyx-teeth ovate-acuminate, pungent, entire, equal, shorter than the 10-striate tube. Var. **Sonomense**. Smaller, the *leaflets* broader, *oblong-cuneiform*, truncate, cuspidate: calyx-teeth subulate, aristate-pointed, equalling or even exceeding the tube. Var. **triflorum** (*T. triflorum*, Greene, Pitt. i. 5). Leaflets as in var. *Sonomense* nearly, but still broader and retuse: flowers fewer: calyx-teeth triangular, acuminate, only a third as long as the tube.—Type common in ravines and other shaded places. Var. *Sonomense* in Knights' Valley, Sonoma Co. Var. *triflorum* near Mt Diablo.

17. **T. tridentatum,** Lindl. Erect, 8—16 in. high, glabrous, neither viscid nor clammy: stipules setaceously laciniate, erect: *leaflets linear or lanceolate*, sharply serrate: heads 1 in. broad, the laciniate involucre much shorter than the flowers: fl. ½ in. long, bright purple with dark centre: calyx with 10-nerved tube, the *rigid segments broad at base, abruptly narrowed* to a subulate spinulose-tipped apex which is usually subtended by *a short stout tooth on each side.* Var. **scabrellum,** Greene. Slender, with long almost filiform peduncles and broad truncate cuspidate leaflets, and a sparse scabrous pubescence upon its stalklets and growing parts.—The type belongs to the seaboard, where it abounds in clayey soils, both on hills and plains. The var. *scabrellum* is from the plains of the San Joaquin. March—May.

18. **T. obtusiflorum,** Hook. Stout, erect, 1—3 ft. high, the herbage bright green, sparsely short-hairy under a lens; the inflorescence and growing parts *somewhat resinous-glandular:* stipules setaceously lacerate, broad and spreading, in age reflexed; leaflets elliptic-lanceolate, 1—1½ in. long, spinulose-serrate: heads more than 1 in. broad, on long stoutish peduncles: calyx-tube oblong-campanulate, ¼ in. long, with 10 prominent and as *many lesser nerves, these branching and forming reticulations above;* teeth subulate-spinose, entire: corolla ½ in. long, lilac-purple with dark centre.—Common on clayey hill-sides and stream banks in the open country along the base of the Mt. Diablo Range. May.

19. **T. roscidum,** Greene. Erect, with ascending branches, stout, 1—2 ft. high, stems flexuous, purple, leaves deep dull green, *soft-pubescent throughout and very clammy*, not at all resinous: stipules spreading or reflexed, setaceously fimbriate: leaflets 1 in. long, linear-lanceolate, pectinately setulose: heads as in the preceding (though not glandular), calyx the same; corolla white, with dark red-purple centre.—Plentiful in cañons, along streams, in Solano Co., etc.

++ ++ *Involucre concave; flowers developing equally all around.*

20. **T. microcephalum,** Pursh. Slender, much branched, decumbent or procumbent, *soft-pubescent:* leaflets obovate-cuneiform or obcordate, emarginate, denticulate; stipules ovate-acuminate, nearly entire; heads subglobose, very small, ∞-flowered, on slender peduncles; involucre many-cleft, segments entire: *calyx-teeth subulate*, broad, scarious, and sometimes toothed at base: fl. minute, pinkish: pod globose, 1-seeded. —Common. May.

21. **T. microdon,** Hook. & Arn. Larger than the last, not rarely 2 ft. high, *glabrous or nearly so*: involucre broader, deeply cup-shaped, equalling the head, its many lobes conspicuously toothed: *calyx-teeth* rigid, triangular, acute, *serrulate below.*—Abundant in many places.

←← *Corolla more or less inflated in age.*

22. **T. barbigerum,** Torr. Branches stout, with short internodes, nearly prostrate, 4—10 in. long; herbage deep green, *soft-pubescent:* petioles elongated; leaflets broadly obovate, obtuse, denticulate, ½ in. long or less: involucre as broad as the long-peduncled heads, 4—8 lines wide, shortly lobed and setaceously toothed: calyx-tube short, thin and at length scarious; *teeth setaceous-awned,* plumose, sometimes 2—3 parted, usually exceeding the small purple corolla: pod 2-seeded.— Frequent at Berkeley, San Francisco, etc.

23. **T. Grayi,** Loja. Erect, stout, with long internodes, 1—2 ft. high, sparingly branched, *villous* with long spreading hairs: leaflets 1 in. long, cuneate-oblong or elliptic-lanceolate, obtuse or acutish, sharply serrulate: heads long-peduncled, 1 in. broad; the involucre as broad: calyx-tube scarious, villous, 10-nerved; *teeth linear-subulate* from a triangular base, plumose, as long as the dark red-purple corolla.—Marin Co.

24. **T. fucatum,** Lindl. Branches stout and somewhat fistulous, often a foot long: leaflets 1 in. long, rhombic-obovate rather conspicuously spinulose-serrate or -dentate, in texture somewhat succulent: heads 1¼—2 in. broad, 13—20-flowered: fl. 1 in. long or more; calyx-tube campanulate, 1½ lines long; *none of the teeth as long as the tube,* all triangular, the *two upper short and acute, the three lower tapering to a setaceous point;* corolla cream-color, with a slight greenish tinge, fading pinkish, the keel-petals with a dark purple spot; legume rather long-stipitate.—In low meadow lands; most frequent near the Bay.

25. **T. flavulum,** Greene. Pale green and glaucescent, stoutish, often larger than the last but heads not half as large: leaflets ½—¾ inch long, broadly obovate, from pectinate-denticulate to entire: heads ½—1 inch broad, 5—12-flowered: flower seldom ½ inch long: *calyx-tube a line long, the shortest of the teeth decidedly longer, the 3 lower about twice as long,* all slender-subulate from a broad base: legume subsessile. —More common than the last; usually on higher ground.

26. **T. virescens,** Greene. Near the two preceding, but slender and half as large: leaflets inverse-deltoid, broadest at summit and truncate, sharply serrulate: slender peduncles twice the length of the leaves: heads less than 1 in. broad: *calyx-teeth all slender-subulate, the two upper shorter than the tube,* closely approximate, the *lower twice the length of the tube:* corolla greenish-yellow, 7—8 lines long.—Mountains and valleys of Marin and Sonoma counties.

27. **T. Gambelli,** Nutt. A span high, scarcely branching: long-peduncled heads usually only 3—5-flowered: upper calyx-teeth subulate, the *lower much larger, each cleft into 5—7 long setaceous segments.*—Hills at the eastern base of Mt. Diablo.

28. **T. amplectens,** Torr. & Gray. Light green and glabrous, small, slender, the branches 3—10 in. long: leaflets ½—¾ in. long, cuneate-obovate or -oblong, truncate or retuse, mucronately denticulate: peduncles slender; involucre half as broad as the heads, its *lobes broad, scarious-margined, obtuse,* sometimes cleft or toothed: *calyx cleft nearly to the base,* the subulate slenderly acuminate teeth very unequal, the larger rarely toothed or cleft: corolla ochroleucous, 2—3 lines long: pod membranaceous, translucent, finely reticulate with green veins, promptly dehiscent by one suture only, 4—6-seeded: seed small, transversely oval, emarginate at the hilum, coarsely tuberculate-rugose.— Not common; but found at Alameda, and in the Oakland Hills.

29. **T. hydrophilum.** Diffuse, glabrous, the branches flaccid though not very slender, 1—2 ft. long: stipules ovate, entire, subulate-pointed; *leaflets linear or oblong,* obtuse or truncate, repandly dentate or somewhat serrulate, 1 in. long: peduncles slender, little exceeding the leaves: heads 8—15-flowered: involucre of about 5 small ovate or oblong bracts: *calyx-teeth very long, subulate-aristiform:* corolla in age oblong, *slightly inflated and about equally so from end to end,* conspicuously striate: pod 2-seeded: seed transversely oblong, sinuous-rugose.—In low moist lands along the seaboard, preferring the vicinity of the salt marshes; but also around ponds among the hills, and even on subsaline plains of the lower Sacramento. A most distinct species every way, and one which, having its lowest leaves narrowest and its uppermost and later ones broadest, reverses that order of leaf-widening which is otherwise universal in Californian clovers. It is *T. diversifolium* of the Flora Franciscana; but Nuttall's species can not be identified by his meagre description, and it is hardly probable that he had this plant in view.

30. **T. Franciscanum.** Slender but wiry, the decumbent branches 5—10 in. long: *leaflets linear-cuneiform,* the very lowest entire, truncate and cuspidate, the others serrulate and acute; filiform peduncles far exceeding the leaves: segments of involucre oblong, obtuse; heads small, hemispherical or subglobose: *calyx-teeth short and subulate:* corolla red-purple, in age *inflated to the broadly ovate.* Var. **truncatum.** Larger and more flaccid, the leaflets ampler, nearly all linear and oblong-linear, truncate and scarcely toothed; heads larger: *corolla ochroleucous,* the keel tipped with dark purple, the whole in age *inflated almost to the turbinate or obpyramidal.*—Type common at San Francisco; the variety abundant in Napa and Solano counties, etc. The two forms comprise the *T. stenophyllum* of the Fl. Fr.; but that name is another of Nuttall's *nomina seminuda,* and his description is more suited to the variety next of the species.

31. **T. depauperatum,** Desv. Only a few inches high, branched from the base, flaccid, decumbent, glabrous: *leaflets* ½ in. long, *cuneate-oblong,* obtuse or emarginate, denticulate: head long-stalked, few-flowered:

involucre greatly reduced, *with truncate short lobes:* corolla larger than in the last, less inflated, red-purple: pod 1—2-seeded: seed little broader than long, rather angular, tuberculate-rugose.—Var. **stenophyllum.** Stems more slender and elongated: leaflets narrowly linear, the lowest emarginate, the upper acute: corolla smaller, ochroleucous.—Mostly in low places among the hills of the Coast Range.

32. **T. laciniatum,** Greene. Slender, flaccid, glabrous, ascending, 3—6 in. high: stipules ovate, acuminate, mostly entire: lower leaflets narrowly cuneiform, denticulate, *the upper broad,* truncate and 3-dentate at apex, *laciniately toothed or pinnatifid: involucre obsolete:* fl. 3—5, white with purple centre, inflated in age: pod 3—4-seeded: seed oval, with the strong corrugation running into a more or less distinctly favose coarse reticulation. Var. **angustatum.** *Leaves all linear,* truncate, entire.—Type from the vicinity of Byron Springs only; the variety there, and also in Sonoma Co., *Bioletti.*

11. MELILOTUS, *Morison* (SWEET CLOVER). Erect herbs with pin-, nately 3-foliolate leaves, the leaflets toothed, and small fragrant flowers in slender axillary racemes. Petals free from the diadelphous stamens, deciduous. Pod ovoid, small, scarcely dehiscent, 1—2-seeded.—Old World plants with sweet-scented herbage and very fragrant flowers. The following species are naturalized with us.

1. M. INDICA, Allioni. Annual, glabrous, 1—3 ft. high, bearing many racemes of minute yellow flowers.—Common in low grounds, chiefly near the salt marshes or along rivers. A good fodder plant.

2. M. ALBA, Lam. Stout, 3—6 ft. high: fl. larger, white, very fragrant. —Spontaneous in northern California; perhaps not within our limits.

12. MEDICA, *Tourn.* Herbs with pinnately 3-foliolate (rarely 5-foliolate) leaves and flowers 2, 3, or many, on axillary peduncles. Petals free from the diadelphous stamens, deciduous. Pod 1-several-seeded, falcate-incurved or coiled into a spiral.—Valuable forage plants, natives of Asia, brought to California, by way of Mexico or South America in early times; some of them now widely naturalized.

1. M. LEGITIMA, Clus. (1601). *Medicago sativa,* L. (ALFALFA). Perennial, erect, glabrous, 2—4 ft. high: leaflets cuneate-oblong or oblan-ceolate, toothed above: fl. ∞, racemose, violet: pod spirally coiled, unarmed.—Here and there spontaneous, but not very prevalent in the wild state, at least in our district.

2. M. LUPULINA (L.), Lam. Annual, slender, procumbent, 1—2 ft. long, soft-hairy: leaflets obovate, small: fl. minute, in small oblong heads, yellow: pod small, reniform or curved almost into a ring, black when ripe, 1-seeded.—Not rare in moist waste lands.

3. M. DENTICULATA (Willd.). (BUR CLOVER). Annual, much branched, decumbent, glabrous: leaflets obovate or obcordate, denticulate: fl. 2—3, yellow: pods coiled into 2 circles, their margins armed with hooked prickles.—Common everywhere: valuable as a forage plant, but the "burs" damaging to wool.

4. M. APICULATA (Willd.). Aspect of *M. denticulata*, but the pods unarmed, their margin beset on either side by a row of tubercles or murications, the whole surface reticulate.—In grain fields, etc.: not common.

5. M. ARABICA, Camerarius (1588). *Medicago Arabica*, Allioni (1785); *M. maculata*, Willd. (1801). Larger every way than *M. denticulata;* leaflets with a blackish purple irregular blotch in the middle: pods coiled into a spiral of 4 or 5 turns, thus becoming globular, not reticulate: the spines in 2 rows, divaricate, curved throughout —In moist shaded grounds.

13. **CYTISUS,** *Diosc.* (BROOM). Shrubs with green very leafy or nearly leafless often angular branches, palmately or pinnately 3-foliolate leaves (leaflets entire), and solitary or racemose yellow or white flowers. Calyx with campanulate tube and bilabiate limb. Petals broad; keel obtuse. Stamens monadelphous. Pod compressed, several-seeded.— Natives of the Old World; becoming spontaneous on our coast.

1. C. CANARIENSIS (L.), Greene. Much branched, 3—6 ft. high, soft-pubescent, the branches and branchlets very leafy: leaflets ¼—½ in. long: fl. yellow, in numerous terminal short racemes, fragrant; calyx with upper segment deeply, lower obsoletely 3-toothed at apex; banner not reflexed; keel deflexed, releasing the stamens.—Running wild on the grounds of the University at Berkeley.

2. C. SCOPARIUS (L.), Link. Size of the last, but sparingly leafy, the branches prominently angular: leaflets glabrous, often 1 only: fl. large, bright yellow, solitary or in pairs along the branchlets, in the leaf-axils, and apparently racemose: pod pilose along the margins.—Naturalized abundantly northward; more sparingly with us.

3. C. PROLIFERUS, L. f. Arborescent, branches terete and, with the young leaves, etc., silky-pubescent; leaflets 3, elliptic-lanceolate, 1 in. long or more: fl. white in lateral umbellate racemes: banner reflexed: keel shorter than the wings, enclosing the stamens: pod villous.—Native of Teneriffe; a valued forage shrub in some countries; escaped from cultivation at Berkeley. Jan., Feb.

14. **ULEX,** *Linn.* (FURZE, GORSE). Compact very thorny shrubs with simple prickle-pointed leaf-like organs, and scattered yellow flowers. Calyx of 2 nearly or quite distinct yellowish sepals. Banner nearly as

long as the other petals, not reflexed, scarcely even erect. Stamens
monadelphous. Pod few-seeded, little longer than the calyx.

1. U. EUROPÆUS, L. Three to six feet high, the numerous short
branchlets villous, ending in a stout spine: lower leaves sometimes
lanceolate, more commonly reduced to green spines ½ in. long: fl. ½ in.
long, yellow, solitary but often crowded on the branchlets; calyx villous.
—Spontaneous here and there about San Francisco. Feb.—June.

15. SPARTIUM, *Lobel.* (SPANISH BROOM). Branches stout, terete,
green and rush-like, glabrous, sparsely leafy with 1-foliolate leaves, or
leafless, bearing terminal loose racemes of large yellow flowers. Calyx
spathaceous, cleft to the base above, 5-toothed at apex. Banner roundish,
erect; keel acuminate. Stamens monadelphous. Pod compressed.

1. S. JUNCEUM, L. Native of southern Europe. Spontaneous near
San Francisco.

16. LUPINUS, *Catullus* (LUPINE). Leaves palmately 5—15-foliolate;
leaflets entire, sessile; stipules adnate, seldom conspicuous. Flowers
blue, pinkish or yellow, in terminal racemes, with bracts mostly caducous.
Calyx deeply bilabiate; upper lip notched, lower usually entire, occasionally 3-toothed or -cleft. Banner roundish; wings falcate-oblong,
commonly slightly united at tip in front of, and enclosing, the falcate
usually slender-pointed keel. Stamens monadelphous, dimorphous, 5
with longer and basifixed anthers, the alternate 5 with shorter and
versatile ones. Pods compressed, straight.

* *Pods several-seeded; cotyledons petiolate.*

+– *Annuals; flowers more or less verticillate.*

1. L. micranthus, Dougl. Rather *slender and weak*, branched from
the base, 6—18 in. high, pilose-pubescent, not at all succulent: leaflets
5—7, narrowly linear to linear-spatulate, ½—1½ in. long, on petioles
twice as long: raceme peduncled, verticils 3—5, often indistinct: pedicels
1½ lines long (in fruit 3 lines); upper calyx-lip with divergent lobes;
lower long, entire: *corolla 2 lines long*, blue, except the white and dotted
middle of the erect mucronulate banner, the white spot changing to light
blue; *wings narrow, appressed;* keel woolly-ciliate toward the apex; pod
5-seeded: seed quadrate-oval, whitish, with or without minute light
brown dots.—Common in sandy soils.

2. L. polycarpus, Greene. Erect, *stoutish, rather succulent,* 1—2 ft.
high, with firm ascending branches from midway of the stem, pubescent:
leaflets 7, narrowly oblanceolate, 1 in. long; glabrous above: racemes
with 4—7 very distinct verticils; pedicels 1 line long: upper calyx-lip
bifid, its ovate segments short, parallel; lower scarcely longer, 3-nerved,
slightly notched at apex: corolla 1½ lines long, deep blue; the obovate

retuse banner with a white spot; *wings* coherent at tip, *inflated, exposing the base of the broad short keel;* this ciliate below the apex: *pod rigid, slightly falcate, 7—9-seeded.*—Very common, preferring rich low meadow lands adjacent to the salt marshes; also occurring in a reduced form on the low plains of the interior.

3. **L. trifidus,** Torr. Slender, branched from the base, 6—10 in. high, pilose-canescent: racemes short (1—3 whorls): *upper calyx-lip deeply cleft,* segments divergent; *lower broad, deeply trifid:* corolla 2½ lines long, blue, the white spot on the banner permanent; keel deep, scarcely falcate, shortly and obtusely pointed, and with a few stiffish ciliolæ above the middle.—In sandy land, at San Francisco and Alameda.

4. **L. bicolor,** Lindl. Low, often diffuse, stoutish, 6—10 in. high, silky-pilose: leaflets 5—7, linear-spatulate, 1 in. long: upper calyx-lip bifid; lower twice as long, entire: *corolla 4—5 lines long,* blue and white, the white changing to red-purple; banner reflexed; keel falcate, acute, ciliate toward the apex: *pod small,* about 5-seeded.—Sandy soil about San Francisco, in a slender depressed very hairy form; also on gravelly crests of the Oakland Hills, where it is stouter, with ascending branches.

5. **L. pachylobus,** Greene. Stout, rigid, barely 1 ft. high, a few ascending branches from the base, hirsute throughout: petioles slender and long; leaflets 5—7, linear, ¾ in long: *racemes on stout peduncles, whorls 2—4:* fl. 3 lines long, subsessile, deep blue: calyx-lips broad, the upper very short, notched; lower entire and twice as long: *pod large,* (1¼ in long, 4—5 lines wide), very hirsute, 4—6-seeded.—Contra Costa Co., and northward, in the Coast ranges.

6. **L. nanus,** Dougl. Commonly 1 ft. often 2 ft. high, with many decumbent branches, *not succulent,* minutely and not densely villous-pubescent: leaflets oblanceolate, 1 in. long: *racemes* short-peduncled, *3—7 in. long,* of many rather indistinct whorls *of large deep purple fragrant flowers:* upper calyx-lip deeply cleft; lower 3-dentate: *corolla 6—7 lines long,* the orbicular retuse banner closely reflexed, the white middle part turning rose-red; wings lightly joined, forming an obliquely obovate inflated sac; falcate *keel with a long slightly ciliate beak.*—Common on plains, especially in sandy soil.

7. **L. carnosulus,** Greene. Erect, 1—2 ft. high, usually simple, *stout and succulent;* pubescence minute, appressed: leaflets oblanceolate, 1 in. long, obtuse, but with a small recurved mucronation: *raceme loose, distinctly verticillate:* upper calyx-lip deeply cleft; lower entire: corolla deep blue; *keel villous in the middle.*—Valleys of the Coast Range.

8. **L. affinis,** Agardh. Very stout and succulent, irregularly branching, 1—2 ft. high, the pubescence very sparse and short; stipules small, setaceous: leaflets 7, cuneate-obovate, obtuse or emarginate, 1—1½ in.

long, on stout petioles twice or thrice as long: racemes rather short-peduncled; whorls 3—7; bracts equalling the calyx; upper calyx-lip bifid; lower entire or 3-toothed: corolla 5—6 lines long, deep bluish purple; *keel broad, not strongly falcate, naked: ovary densely velvety;* pod glabrate, 1—2 in. long, 5—9-seeded.—Common in low, clayey soils.

+ + Perennials; fl. 6—7 lines long.

9. **L. polyphyllus,** Lindl. *Stem nearly simple,* very stout, somewhat fleshy, erect, 3—5 ft. high, pilose-pubescent, equably leafy up to the inflorescence: stipules adnate for half their length or more; petioles 6—12 in. long; *leaflets 11—15, lanceolate,* acute, hirsute beneath, glabrous above, *3—6 in. long:* raceme short-peduncled, dense, 1—2 ft. long: fl. subverticillate, long pedicelled, ½ in. long and as broad: calyx-lips of about equal length, the upper broader, both entire: wings bluish, banner red-purple: keel falcate, acuminate, naked: pod 1—1½ in. long, ¼ in. broad, 7—9-seeded.—In open marshy ground from near Point Bonita northward. May.

10. **L. formosus,** Greene. Stoutish and suberect, or more slender and decumbent, 2–3 ft. high, sparsely silky-pubescent: stipules long, linear-setaceous, persistent: *leaflets 7—9, linear-lanceolate, very acute,* 1—1½ in. long, equalling the petiole: raceme subsessile, more or less whorled, but rather dense: fl. 6—7 lines long, rich violet, the banner and wings equalling, the latter entirely enfolding, the less elongated *naked keel.* Var. **Bridgesii,** Greene. Stipules narrowly lanceolate, the whole plant *silvery-canescent,* and even villous; raceme distinctly pedunculate, the verticils more remote and distinct.—This, in various forms, is the common perennial lupine of fields and orchards among the Coast Range valleys, and on the plains beyond.

11. **L. sericeatus,** Kell. Stoutish, decumbent, ½—1 ft. high, very leafy, *canescent with a minute closely appressed silky pubescence:* stipules setaceously acuminate from an adnate base; leaflets 7, spatulate-oblong, obtusish, 1¼—2 in. long, on petioles as long: raceme short-peduncled: fl. large, in about 5 whorls, deep purple: calyx-lips large, the upper cleft, lower obscurely 3-toothed; *keel slender-pointed, lightly ciliolate.*— An elegant species apparently confined to a limited area in the mountains of Lake, Napa and Sonoma counties.

12. **L. latifolius,** Agardh. Stoutish, erect, branching, 2—4 ft. high, minutely appressed-pubescent, the stem not striate, dark green and shining, equably leafy, the basal leaves not long-stalked: stipules linear-setaceous; *leaflets 5—7,* broadly oblanceolate, *thin, mucronulate, pilose-ciliate on the margins* and the midvein beneath, 1—2½ in. long: racemes slender-peduncled, loose, the verticils often distinct; pedicels slender: calyx-teeth elongated, the upper notched slightly at the narrow apex;

fl. blue, *changing to dull brown*: keel ciliolate below the middle.—By streamlets and on wooded northward slopes of the Coast Range at low altitudes: common in the hills near Berkeley. May—August.

13. **L. littoralis,** Dougl. Stems clustered, decumbent or ascending, 1—2 ft. long, from yellow roots that are somewhat fleshy and fusiform; herbage canescently silky: leaflets 5—7, acute, 1 in. long, silky on both sides: fl. distinctly and rather remotely verticillate in a short-peduncled raceme: calyx-lips subequal, entire: banner red, shorter than the blue wings: keel ciliate: pod linear, hirsute: seeds linear, brown with black spots.—Near Point Reyes, on seashore sands.

+-+-+- *Suffrutescent or shrubby species.*

14. **L. albifrons,** Benth. Arborescent, the distinct trunk-like woody stem 1—3 ft. high, parted into spreading leafy branches, these ending in a rather long-peduncled loose raceme: leaflets 7—9, oblanceolate, 1 in. long or more, silvery-silky on both sides: fl. verticillate, large, deep blue: upper calyx-lip broad, cleft to the middle, or less deeply; lower entire; *petals subequal,* the broad banner with a whitish spot which soon changes to rose-purple; *keel ciliate:* pod 2 in. long, 5—9-seeded: seed oval, 2 lines long, brownish, encircled marginally by a dark line. Var. **collinus,** Greene. Smaller in all its parts and with no trunk-like stem, the branches decumbent from a short caudex.—Very common on clayey slopes and along ravines; the variety on rocky summits about the Presidio, and on the islands in the Bay. Feb.—April.

15. **L. jucundus,** Greene. Shrubby, 2—4 ft. high; the branches ascending, leafy, ending in a *long-peduncled rather loose raceme:* leaves silky-canescent; leaflets 7—9, narrowly oblanceolate, acute, not very unequal: fl. very distinctly whorled: calyx-lips subequal, the upper bifid: corolla ½ in. long, mainly dark violet, but with a yellow spot in the middle of the banner which soon turns to a dark tawny red, the very margin white changing to rose-red; *keel naked; banner notably smaller than the other petals.*—Vaca Mountains. *(L. tricolor,* Greene, not of garden catalogues).

16. **L. eminens,** Greene. Of almost arborescent form, 3—6 ft. high: branches stoutish ascending, very leafy, ending in a rather *short and dense short-peduncled raceme:* growing branches and both faces of the leaves silvery-silky: leaflets 7—9, very unequal, the longest 1¾ in., the smallest barely 1 in. long: fl. scarcely whorled in the raceme; upper calyx-lobe very broad, scarcely notched, the lower narrow, entire: corolla about ½ in. long, purple, the banner shorter than the other petals, changing from whitish to tawny; keel naked: *pods villous, rather short, almost erect,* about 4-seeded.—Description drawn from a plant of the Santa Inez Mts.; but the same appears to occur on Mt. Tamalpais.

17. **L. Chamissonis, Esch.** Commonly 3 ft. high, but never arborescent; the suffrutescent branches forming a more or less dense tuft and leafy throughout: the petioles short; *raceme elongated and dense, but scarcely peduncled:* fl. not very distinctly whorled, *of a lavender shade;* banner with a permanent yellowish spot.—Apparently confined to the sand dunes of the San Francisco peninsula and Point Reyes. April—July

18. **L. variicolor, Steud.** *Woody basal branches short, slender,* very tough, the decumbent, or often assurgent annual ones very leafy, 1 ft. long or less: pubescence of the leaves scanty, appressed, the stems often sparingly hirsute: leaflets 7—9, narrow, acute: *raceme short, the whorls often 3, 2 or 1 only:* fl. large; banner white or pale blue; wings blue; *keel ciliate* throughout its length: *pods large.*—Frequent on grassy northward slopes at the Presidio, San Francisco, and southward.

19. **L. arboreus, Sims.** From arborescent and 6—10 ft. high to suffrutescent and bushy; *slightly silky-pubescent:* leaflets about 9, narrowly lanceolate, ¾—1¾ in. long, acute, glabrate above: *raceme often 1 ft. long; fl.* whorled, *sulphur-yellow;* keel ciliate: pod 2—3 in. long, 8—12-seeded: seed oblong, dark.—Very common among the sand-hills, or in sandy soil, mostly near the sea or along the shores of the Bay.

20. **L. propinquus, Greene.** Near the last, but small and more bushy, usually 2—4 ft. high, *puberulent,* except the glabrous upper face of the leaflets: *racemes short* and short-peduncled, the *bracts squarrose,* very caducous: *petals violet,* the banner reddening in age; keel strongly ciliate: pods and seed nearly as in the preceding.—A seaboard species of wide range north and south. Point Reyes, etc.

* * Pods 2-seeded; cotyledons connate. Annuals with whorled flowers and persistent bracts.

21. **L. microcarpus, Sims.** *Branched from the base, or near it,* 1 ft. high or less, somewhat succulent, villous throughout: leaflets 9, cuneate-oblong, 1 in. long or more: racemes short-peduncled: bracts subulate-setaceous, equalling the calyx or shorter: fl. short pedicelled, purplish or flesh-color: *calyx densely hirsute;* upper lip short, subscarious, emarginate or cleft; lower obscurely 2—3-toothed.—Throughout the State, apparently in the interior only.

22. **L. densiflorus, Benth.** *Stem stout, simple below,* parted in the middle into numerous wide-spread branches, 2 ft. high, succulent, sparsely villous: racemes 6—10 in. long, long-peduncled: bracts setaceous from a broad base: fl. white or rose-color, the banner greenish-dotted: *calyx sparingly villous:* upper lip scarious, deeply cleft: lower long, toothed.—Very common, both on the seaboard and plains of the interior. A

yellow-flowered form, possibly distinct (*L. Menziesii*, Agh.) occurs in Napa Valley and near Antioch, and has sometimes been confused with the next. April, May.

23. **L. luteolus**, Kell. More slender, simple below, loosely branching above, 2 ft. high or more, *rigid, not succulent: racemes shorter and more dense;* bracts linear-setaceous: fl. rather small for the group (6 lines long), pale yellow, subsessile: upper lip of calyx ovate-lanceolate, entire; lower 3-toothed.—A mountain species, from Contra Costa and Sonoma counties; otherwise northern. June—Sept.

17. **XYLOTHERMIA**, *Greene.* A rigid much branched spinescent shrub, with small nearly sessile 1—3-foliolate exstipulate leaves, and large solitary almost sessile purple flowers. Calyx campanulate, repandly 4-toothed. Petals equal; banner orbicular, the sides reflexed; keel-petals oblong, obtuse, distinct. Stamens distinct. Pod linear, compressed, straight, several-seeded.

1. **X. montana** (Nutt.), Greene. Shrub 3—6 ft. high, the branches spreading widely: leaves crowded; leaflets ¼—¾ in. long, oblanceolate, acute, entire, somewhat silky when young: fl. near the ends of the stiff spinescent branchlets, on short 2-bracteolate peduncles, from pale rose- to deep red-purple, about ¾ in. long.—At middle elevations in the Coast Range; often forming dense thickets on hill-sides and summits.

18. **THERMOPSIS**, *Robert Brown* (FALSE LUPINE). Erect perennials with palmately 3-foliolate leaves, foliaceous stipules, and a terminal raceme of yellow flowers; the pedicels subtended by persistent bracts. Calyx campanulate, cleft to the middle, the two upper teeth often united. Banner roundish, shorter than the wings, the sides reflexed; keel nearly straight, obtuse, equalling the wings. Stamens distinct. Pod linear, flat, several-seeded.

1. **T. Californica**, Wats. Stipules broadly lanceolate, less than 1 in. long; leaflets obovate or oblanceolate, 1—2 in. long, silky-tomentose on both faces: pod 6—8-seeded. Var. **velutina**, Greene. Silvery-canescent with a dense velvety pubescence; leaflets more acute.—Type common on low hills toward the coast; the variety on Mt. Hamilton. May.

ORDER XXXIV. D R U P A C E Æ.

Shrubs or trees with bark exuding gum; bark, leaves and seeds more or less keenly bitter. Leaves alternate, simple, with small caducous stipules. Flowers perfect (except in *Osmaronia*), regular. Calyx tubular or campanulate, free from the ovary, the tube lined with a disk, deciduous; limb 5-lobed, imbricate in æstivation. Petals 5, perigynous. Stamens about 20, inserted within the petals on the disk of the calyx-tube. Pistil 1 (in *Osmaronia* 5); style simple; ovary 1-celled, 2-ovuled, becoming a drupe. Seed pendulous; cotyledons large, thick, fleshy; albumen 0.

1. PRUNUS, *Varro* (PLUM-TREE. PRUNE). Leaves convolute in the bud. Flowers in umbellate clusters from lateral buds, appearing before or with the leaves. Drupe ovoid, glabrous, glaucous; the thick sarcocarp pulpy, sweet or pleasantly acidulous, and with the distinctive flavor of plums; putamen bony, smooth, compressed, acutely edged on one margin, grooved on the other.

1. **P. subcordata,** Benth. Arborescent, 3—10 ft. high, much branched, more or less spinescent; nascent leaves and twigs finely pubescent, in age glabrate: leaves ovate, cuneate or subcordate at base, obtuse or acute, sharply serrulate, about 1 in. long, short-petioled: umbels 2—4-flowered; pedicels ¼—½ in. long, fl. white, ½ in. broad: drupe ¾ in. long, red, the pulp rather hard and unpalatable.—Hillsides and banks. Fl. March, April. Fr. August, Sept.

2. CERASUS, *Theophr.*, (CHERRY-TREE. CHOKE-CHERRY. ISLAY). Leaves conduplicate in the bud. Flowers corymbose or racemose from lateral buds which are often leaf-bearing. Drupe globose, glabrous, destitute of bloom; the sarcocarp sweet rather than acidulous (in our species), often keenly bitter, sometimes sour and astringent; putamen osseous or ligneous, smooth, mostly globose, not prominently margined.

* *Flowers corymbose, from lateral buds; drupe small, with bony putamen.*

1. **C. emarginata,** Dougl. Shrub 3—8 ft. high, branched from the base and clothed throughout with a smooth shining bark: leaves obovate, oblong or oblanceolate, obtuse, retuse or emarginate, on sterile twigs acutish, ¾—1½ in. long, finely crenate-serrulate, mostly uniglandular, and that, on the lower part of the blade, well above the junction with the petiole: fl. few, in a short corymb: fruit bright red, intensely bitter. —Hills of the Coast Range.

* * *Flowers racemose, from axillary leafless buds.*

3. **C. ilicifolia,** Nutt. Evergreen, often 12—18 ft. high, with well rounded head, the trunk clothed with a dark rough bark: leaves ovate or ovate-lanceolate, obtuse or acute, truncate or rounded at base, coarsely spinose-toothed, coriaceous, glabrous throughout, 1—2 in. long, short-petioled: racemes 1—2 in. long, leafless; fl. small: drupe ½ in. thick or more, slightly obcompressed, putamen scarcely ligneous; sarcocarp thin, sweetish, when ripe.—Oakland Hills, thence southward.

* * * *Flowers racemose at the ends of leafy branchlets; drupe small.*

4. **C. demissa,** Nutt. Deciduous, 3—12 ft. high: leaves ovate or oblong-ovate, acute or acuminate, rounded or cordate at base, sharply

serrate, more or less pubescent beneath, 2—4 in. long, with 1 or 2 glands on the petiole just below its summit: racemes 3—4 in. long, many-flowered: drupe globose, red or dark purple, astringent, putamen ligneous, globose.—Hills behind North Berkeley; also in San Francisco Co. Fl. April, fr. Sept.

3. OSMARONIA, *Greene.* (Oso Berry). Deciduous shrub, with flowers diœcious, in pendulous racemes terminating short leafy branch-lets. Calyx turbinate-campanulate, 5-lobed. Petals 5, broadly spatulate, erect in the pistillate flowers, spreading in the staminate. Stamens 15, in two rows, 10 inserted with the petals, 5 more deeply within the calyx-tube; filaments slender, short. Pistils 5; styles short, lateral, jointed at base. Drupes 1—4, ovoid, with thin pulp and osseous putamen. Seed solitary; cotyledons convolute.

1. **O. cerasiformis** (T. & G.), Greene. Stems 2—10 ft. high, the bark dark brown: leaves broadly oblanceolate, entire, obtuse or acutish, mucronulate, 2—3 in. long, short-petioled: racemes shorter than the leaves; bracts conspicuous: fl. white, very fragrant: drupes 6—8 lines long, slightly compressed, blue-black; pulp very thin, bitter.—Coast Range hills. Jan.—April.

Order XXXV. POMACEÆ.

Trees and shrubs with astringent but neither bitter nor poisonous properties; not gummiferous. Leaves alternate, simple or unequally pinnate, with caducous stipules. Flowers perfect, regular, racemosely or corymbosely clustered, white or reddish. Calyx-tube urceolate or campanulate, more or less coherent with the ovary, the usually short free portion lined with a staminiferous disk: limb 5-lobed, imbricate in æstivation. Petals 5, perigynous. Stamens mostly 20, inserted on the disk. Ovary of 2, 3 or 5 carpels, becoming a pome; styles as many as the carpels. Seeds usually 2 in each cell, collateral, ascending; cotyledons fleshy; albumen 0.

1. **AMELANCHIER,** *Lobel.* (Service-Berry). Shrubs with decid-uous oblong or rounded serrate or subentire leaves, and bracted racemose white flowers appearing with them in early spring; the bracts caducous. Calyx-tube broadly turbinate; segments as long as the tube, erect or reflexed in flower. Petals linear-oblong; plane. Stamens 20, shorter than the petals. Styles 3—5, coalescent at base or distinct; carpels as many, incompletely 2-celled, but only 1-seeded. Fruit small, berry-like, dark purple, more or less glaucous, the pulp sweet and edible.

1. **A. alnifolia,** Nutt. Arborescent, but seldom 10 ft. high: leaves nearly full grown at flowering time, but thin, dark green, oval or oblong-ovate, obtuse at both ends, coarsely serrate toward the apex, otherwise

entire, woolly-pubescent beneath, even in age: racemes ∞-flowered; bracts setaceous, long-woolly: calyx densely tomentose, the triangular lanceolate teeth closely reflexed: petals spatulate-linear, ¾ in. long, plane: stamens very short, not equalling the calyx-teeth.—Banks of streams, and, in dwarf form, on rocky hills.

2. HETEROMELES, *Rœmer* (CALIFORNIA HOLLY. CHRISTMAS BERRY). A small evergreen tree with simple coriaceous serrate leaves, and numerous small white flowers in terminal corymbose panicles. Calyx turbinate; limb 5-parted, the lobes at length inflexed over the carpels and becoming fleshy. Petals rounded, concave. Stamens 10; filaments dilated at base and slightly connate. Ovary 2—3-celled, 4—6 ovuled; styles and stigmas 2—3. Fruit ovoid, red, berry-like with dry mealy pulp; carpels free from the fleshy calyx-tube above the middle. Seeds 1—2 in each cell, erect; testa thin-cartilaginous.

1. **H. arbutifolia** (H. Ait. f), Rœmer. Usually 10—25 ft. high; nascent parts tomentulose: leaves dark green and shining, narrowly oblong to oblong-lanceolate, acute at both ends, sharply but not very closely serrate or dentate, 2—4 in. long: pome 3 lines long: seed one-half as long. —Very common along streams and on northward slopes. Fl. July, fr. Dec.

3. MALUS, *Tourn.* (APPLE-TREE. CRAB-APPLE). Small deciduous trees. Leaves simple, more or less serrate; Flowers rather large, reddish or white, corymbose at the ends of short lateral branchlets. Stamens 20. Styles 5, more or less united at base. Carpels 5; wholly covered by the adnate calyx-tube, chartaceous in fruit, 2-seeded. Pome large, globose, depressed at each end, the flesh acidulous, destitute of grit-cells.

1. **M. rivularis** (Dougl.), Rœm. Tree 15—25 ft. high: leaves ovate-lanceolate, acute or acuminate, 1—3 in. long, often slightly 3-lobed, sharply serrulate, more or less pubescent when young: corymb somewhat racemose; pedicels slender, 1 in. long: petals orbicular, 3–4 lines broad, white: pome red or yellow, short-cylindrical, ½ in. long or more.—The *Oregon Crab-apple* has been found as far southward as Sonoma Co.

2. **M. COMMUNIS,** DC., the common apple of the orchards, already of frequent occurrence by waysides, is becoming naturalized in California, as it already is in many parts of the world where it has been long cultivated.

ORDER XXXVI. R O S A C E Æ .

Herbs or shrubs often prickly, with alternate frequently compound leaves and mostly foliaceous commonly adnate stipules. Flowers perfect or unisexual, solitary, cymose, corymbose, or paniculate. Calyx free from the ovary, 4—5-cleft, the segments valvate (rarely imbricate) in æstivation. Petals perigynous, as many as the calyx-lobes and alternate with them, or 0. Stamens 5—∞, perigynous. Pistils 1—∞ ; ovary usually 1-celled and with 1 ovule, sometimes many-ovuled: ovules pendulous or ascending. Styles as many as the ovaries. Fruit an achene or an aggregation of drupelets, sometimes follicular and dehiscent by the ventral suture. Seeds with little or no albumen.

*　Unarmed shrubs: leaves simple.*

Pistils 3—5; mature carpels dry, dehiscent,
　　inflated, several-seeded,....................................OPULASTER　　1
　　not inflated, 1-seeded,....................................HOLODISCUS　　2
Pistil 1 { fruit a plumose-tailed achene,........................CERCOCARPUS　　3
　　　　 { fruit an achene without plumose tail,..............ADENOSTOMA　　4

*　*　Prickly shrubs.*

Fruit a mass of coherent drupelets,......................RUBUS　　10
Fruit a number of bony achenes, enclosed
　　in a red berry-like calyx-tube..........................ROSA　　11

*　*　*　Herbaceous plants; mostly perennial.*

Leaves unequally pinnate; pistil 1 only;
　　Fl. apetalous; mature calyx-tube beset with
　　retrorsely barbed prickles,..............................ACÆNA　　5
　　Mature calyx bur-like with hooked prickles..............AGRIMONIA　　6
Leaves variously compound or cleft; pistils several
　　or many; fruits achenes;
Small annuals, with minute green
　　apetalous flowers, and only 2 pistils,.................ALCHEMILLA　　7
Leaves palmate or pinnate; pistils ∞;
　　achenes on a dry usually hairy receptacle,.............POTENTILLA　　8
Leaves ternate; pistils ∞; achenes on a conical
　　fleshy receptacle,......................................FRAGARIA　　9

1. OPULASTER, *Medic.* (NINE-BARK). Shrubs unarmed with surculose shreddy-barked stems and simple more or less lobed and toothed deciduous leaves; stipules free and deciduous. Flowers white, in corymbs terminating lateral leafy branchlets. Calyx 5-lobed with campanulate tube. Petals 5, rounded. Stamens ∞, in several rows. Pistils 1—5, becoming as many inflated 2-valved several-seeded capsules which are alternate with the calyx-lobes when of the same number, slightly coherent toward the base. Seeds several, obovoid, with a shining crustaceous testa, and copius albumen.

1. O. capitatus (Pursh). Surculiform stems 10—20 ft. long, more or less tortuous and reclining or interlacing among the branches of small trees: leaves short-petioled, ovate, acute, more or less distinctly 3-lobed and coarsely toothed, 2—3 in. long, glabrous or stellate-pubescent: fl. in hemispherical corymbs: pedicels and calyx more or less tomentose: follicles usually 4, exceeding the calyx, 3—4 lines long, ultimately splitting into 2 valves.--Common along streams. April, May.

2. HOLODISCUS, *Maxim.* Unarmed deciduous shrubs with simple toothed or lobed extipulate leaves and terminal panicles of numerous small white flowers. Calyx deeply 5-cleft, nearly rotate. Petals 5, rounded, imbricate in bud. Stamens 20, inserted on an annular perigynous disk. Pistils 5, wholly distinct, becoming 1-seeded hairy carpels, alternate with the calyx-lobes, very tardily dehiscent by the dorsal suture only, or indehiscent.

1. **H. discolor** (Pursh), Maxim. Shrub 2—6 ft. high, the branches short, rigid, clothed with a gray more or less broken and shreddy bark: *leaves ovate, cuneately narrowed to a short winged petiole,* above the middle pinnately toothed or lobed, the lobes when present entire, deep green and nearly glabrous above, whitish-tomentose beneath; panicles erect on short erect or ascending branches; carpels more or less densely *hirsute throughout.*—On dry rocky slopes and summits. July.

2. **H. ariæfolius** (Smith). Shrub commonly 8—18 ft. high, with long spreading or recurved slender branches, these and the stem clothed with a smooth unbroken dark brown bark: leaves short-petioled, deltoid-ovate, 2—3 in. long, two-thirds as broad, *almost truncate at base, pinnately shallow-lobed* from base to apex, the lobes entire or toothed, green and glabrate above, slightly paler beneath with sparse villous appressed pubescence: panicle ample, 6—10 in. long, drooping in fl., erect in fr.: *carpels* compressed, *hirsute along both sutures,* the sides glabrous and covered with sessile globular resin-dots.—Woods of the Coast Range, at low elevations among the hills.

3. CERCOCARPUS, HBK. (MOUNTAIN MAHOGANY). Unarmed evergreen shrubs or trees with simple leaves, small stipules, and axillary solitary or fascicled apetalous flowers. Calyx salverform, the 5-lobed limb deciduous. Stamens ∞, in 2 or 3 rows on the limb of the calyx. Pistil 1; style terminal; stigma terminal; ovule solitary, ascending. Fruit a terete villous achene surmounted by a long villous twisted style. Seed linear; albumen 0.

1. **C. betulæfolius,** Nutt. Shrubby or arborescent, 6—15 ft. high, the stem with a gray thin flaky bark; branches spreading or recurved: leaves somewhat coriaceous, broadly obovate with more or less cuneate entire base, but coarsely serrate-toothed above the middle, conspicuously feather-veined, glabrous above, pubescent beneath, ½—2½ in. long: calyx-tube at length ½ in. long: tail of achene often 3 in. long.—Hills of the Coast Range.

4. ADENOSTOMA, *Hooker & Arnott* (CHAMISO) Unarmed evergreen shrubs with rigid linear entire sessile fascicled stipulate leaves, and small white flowers in closely panicled terminal racemes. Calyx obconical, 5-toothed, 10-striate, the orifice bearing 5 oblong glands. Petals 5, orbicular. Stamens 10—15, inserted in bundles alternate with the petals.

Pistil 1, simple; style laterally inserted and flexuous toward the base; ovary 1-celled, 1- or 2-ovuled, becoming an achene covered by the hardened persistent calyx-tube.

1. **A. fasciculatum,** Hook. & Arn. Shrub 2—20 ft. high, the virgate branches covered with leaf-fascicles: leaves linear-subulate, 2—5 lines long, pungently acute, glabrous, often resinous; stipules small, acute: fl. crowded, sessile; calyx 1 line long, bracted at base, the teeth much · shorter than the small petals: ovary obliquely truncate.—A most characteristic bush of the summits and elevated slopes of the Coast Range.

5. **ACÆNA,** *Mutis.* Perennial herbs, or the stems somewhat woody at the decumbent or creeping base. Leaves unequally pinnate; leaflets incised or pinnatifid. Flowers in terminal more or less spicate clusters. Calyx-tube oblong, contracted at the throat, persistent, at length armed with retrorsely barbed prickles; limb 3—7 parted, valvate, deciduous. Petals 0. Stamens 1—10. Pistils 1 or 2; ovary free from the calyx; style terminal; stigma capitate, multifid; ovule 1, suspended. Achene enclosed in the hardened calyx-tube.

1. **A. trifida,** Ruiz & Pavon. Stems 1 ft. high, leafy mostly at the creeping and woody base; herbage silky-villous: leaflets 9—13, oblong-ovate, 3—5 lines long, pinnately cleft into 3—7 segments: fl. small, greenish purple in an interrupted spike; filaments exserted: fr. ovate, 2 lines long, 3—4-angled; angles with 2—4 stout prickles, the intervals with shorter ones.—Grassy summits or northward slopes of hills along the sea coast. April—June.

6. **AGRIMONIA,** *Brunfels* (AGRIMONY). Perennials with odd-pinnate leaves and long slender terminal racemes of small yellow flowers. Calyx-tube urceolate; throat encircled by a border of hooked prickles; limb 5-lobed, at length connivent. Petals 5. Stamens 5—15, in 1 row. Pistils 2, distinct, free from the calyx; styles terminal; stigma dilated, 2-lobed: ovule pendulous. Achenes 1 or 2, enclosed in the bur-like calyx.

1. **A. Eupatoria,** L. Hirsute or glabrate, 2—4 ft. high, sparingly branched above: leaflets 5—7, usually 2—3 in. long with very small ones intervening, oblong-obovate, coarsely toothed, acute at each end; stipules large, semicordate, toothed or lobed: calyx in fruit ¼—⅓ in. long, the tube 10-sulcate above: achene 1, subglobose, 1 line thick.—Apparently widely disseminated in California, but seldom seen.

7. **ALCHEMILLA,** *Tragus.* Herbs of various habit; ours small annuals with leafy stems, and minute green flowers fascicled in the axils of the palmately lobed leaves. Calyx-tube urceolate; limb 4—5-cleft, with or without as many minute bractlets or intervening teeth. Stamens 1 or 2, minute. Pistils 1 or 2; style basal or lateral; ovule 1, ascending. Achene ovate, compressed.

1. **A. arvensis** (L.), Scop. Slender, 1—4 in. high, leafy, floriferous and hirsute-pubescent throughout, the calyx-tube densely hirsute: leaves 3-parted, the segments 2—3-cleft: calyx-tube much contracted under the 4-parted limb, bractlets minute. Var. **glabra**, Greene. Glabrous, even to the calyx-tube, which is broader than in the type, less constricted at the orifice, with larger bractlets.—Common along borders of thickets, or on open plains; the variety in the valley of the Sacramento.

8. **POTENTILLA**, *Brunfels* (FIVE-FINGER). Herbs with pinnately or palmately compound leaves, the leaflets usually toothed or cleft, and adnate stipules. Flowers axillary and solitary or in terminal cymes. Calyx from flat to campanulate, 5-cleft, valvate, with 5 alternating bractlets. Petals 5, rounded or elongated, yellow, red or white. Stamens 5—∞; filaments filiform or dilated. Pistils 1—∞; styles more or less lateral, deciduous. Achenes on a glabrous or hairy dry receptacle.

* *Stamens 10—30, uniform; filaments filiform, or dilated at base only: petals rounded.*

+ *Perennials; stamens more than 10.*

1. **P. Anserina**, L. Leaves often 1 ft. long; leaflets 7—21, with smaller ones interposed, oblong, sharply serrate, white-tomentose beneath, silky or glabrate above: *stems prostrate*, with long internodes, rooting at each joint and producing at each a tuft of leaves and one or more *long peduncled large yellow flowers:* petals $\frac{1}{4}$—$\frac{1}{2}$ in. long, exceeding the calyx: stamens 20—25: achenes 20—40: receptacle villous.—Along stream-banks, margins of ponds, or in springy places both along the seaboard and in the mountains.

2. **P. glandulosa**, Lindl. *Erect,* 1—2 ft. high, *glandular-pubescent and ill-scented:* leaves pinnate; leaflets 5—9, ovate or rhombic-ovate, coarsely and doubly serrate: cyme lax, leafy-bracted: fl. small; the pale yellow obovoid petals scarcely equalling the calyx: stamens 25, in 1 row, on the margin of the thickened disk: styles attached below the middle of the ovary.—Bushy hills.

+ + *Annuals or biennials; stamens 10.*

3. **P. millegrana**, Engelm., Wats. Tall, flaccid, soft-pubescent, leafy up to the inflorescence: leaves ternate, the radical on long slender petioles; leaflets cuneate-obovate, obtusely serrate at apex only; stipules ovate-lanceolate, entire: *cymes diffuse*; fl. very numerous; petals yellow; stamens about 10: achenes whitish.—Muddy banks of the lower San Joaquin.

4. **P. biennis**, Greene. Biennial, branched from the base, erect and rather stout, 1 ft. high or more, the stems purple, leafy, the whole herbage pubescent and minutely glandular: stipules oblong-lanceolate,

obtuse or acute, the lowest entire, the upper more or less toothed or lobed: leaflets 3 (rarely 5), cuneate-flabelliform, irregularly incised, the broad teeth or lobes mucronulate: *cymes mostly contracted and dense:* petals small, yellow, spatulate-oblong, scarcely equalling the calyx: stamens about 10: achenes minute, whitish.—Moist places in the mountains.

* * *Perennials; petals obovate to linear; stamens 10, alternately long and short, the filaments petaloid-dilated.*

, ← *Cymes lax, dichotomous; bractlets large, often exceeding the calyx-lobes.*

5. **P. frondosa,** Greene, Erect or decumbent, 1½—3 ft. high, leafy throughout, *viscidly hirsute and heavy-scented:* radical leaves with 7—9, cauline with 5—7 *leaflets;* these *1—2 in.* long, oval or oblong, *doubly incised, thin and finely rugose;* stipules ovate-lanceolate, coarsely incised: cyme widely spreading, loose and leafy: calyx short-campanulate, the large spreading *bractlets exceeding the segments,* trifid at apex: stamens very unequal: petals ligulate, erect or little spreading, white.—Near Martinez, *Frank Swett.*

6. **P. Californica** (Ch. & Schl.), Greene. Size and habit of the last, but stem less leafy, leaves mostly radical: *herbage glandular-pubescent, very fragrant; leaflets* 11—21, the uppermost more or less confluent, the lower distinct but approximate, ¼—¾ *in.* long, *broadly cuneiform,* toothed or deeply incised at the rounded apex: calyx ½ in. high, short-campanulate; bractlets exceeding the calyx-lobes, usually 3-toothed at the broad apex, the middle tooth longest: petals white, spatulate, spreading or suberect. Var. **elata,** Greene. More slender than the type, equally fragrant: leaflets deeply and incisely once or twice cleft: bractlets of the calyx like the segments triangular-lanceolate, entire.— The type is common on wooded slopes about San Francisco and Oakland. The variety is of Napa Co., and northward.

7. **P. Kelloggii,** Greene. Stems stout, ascending, or almost prostrate, 1—2 ft. long; *herbage glandless, scentless, canescent with a sho· t dense silky pubescence:* leaflets 11—15, obovate, coarsely toothed, ½—¾ in. long: calyx-tube cupulate; lobe, lanceolate, ¼ in. long, equalled by the oblong entire bractlets: petals pure white, spatulate oblong, ¼ in. long.—In sandy soil at Alameda, Lake Merced, etc.

← ← *Cymes more condensed; bractlets smaller than the calyx-lobes; leaflets in many pairs, deeply incised or lobed.*

8. **P. tenuiloba,** Greene. Stems 1 ft. high; *herbage canescently villous:* leaflets ¼—½ in. long, cuneate-obovate, deeply parted into 4—8 linear lobes, or the uppermost narrower, few-lobed or linear and entire: cymes compact: calyx 2 lines long; lobes linear, surpassed by the *oblong-spatulate white petals.*—Sonoma Co., *Bigelow,* and southward.

9. **P. Micheneri,** Greene. Stems 6—8 in. high, tufted, reddish, and, with the *younger leaves, somewhat villous, the mature herbage glabrous*: leaflets in about 15 pairs, small and crowded, 5—7-parted into oblong obtuse entire segments: young cymes compact; *petals cuneate-obcordate,* exceeding the calyx, white: all 10 stamens with oblong-petaloid white filaments, the alternate ones smaller.—Southern flanks of Mt. Tamalpais.

9. FRAGARIA, *Brunfels* (STRAWBERRY). Perennial stoloniferous herbs with 3-foliolate leaves; the leaflets coarsely toothed: scapes cymosely ∞-flowered. Flowers as in *Potentilla,* but the numerous achenes borne on an enlarged pulpy edible receptacle. Petals white.

* *Leaves light-green, of thin texture; achenes superficial.*

1. **F. Californica,** Ch. & Schl. Often 10 in. high, commonly smaller: leaflets cuneate-obovate, rounded, sparingly villous on both sides: scapes and petioles slender: fl. ½ in. broad; calyx-teeth and often the petals also more or less toothed: fr. small, globose.—Common along the sea-board; preferring wooded or bushy slopes.

* * *Leaves dark-green, of firmer texture; each achene inserted in a small depression of the receptacle.*

2. **F. Chilensis,** Ehrh. Diœcious: scapes and petioles short; leaflets cuneate-obovate, nearly glabrous and somewhat shining above, villous beneath: fl. 1 in. broad.—Sandy banks and grassy slopes near the sea.

10. RUBUS, *Vergil.* Shrubs with stems unarmed or prickly, erect, reclining or prostrate. Leaves simple and lobed, or compound; stipules adnate. Flowers white or red, solitary, corymbose or panicled. Calyx persistent, 5-lobed, without bractlets. Petals (5) and stamens (∞) perigynous. Pistils 2—∞, crowded on an elevated receptacle, ripening into a coherent body of small drupes, so forming the aggregate fruit called a raspberry or blackberry.

* *Fruit concave beneath, parting freely from the receptacle.*

← *Unarmed; leaves ample, palmately lobed.*

1. **R. parviflorus,** Nutt. Erect, 3—8 ft. high; the bark of the main stem becoming brown and shreddy; branchlets and pedicels hirsute and more or less glandular-hispid: leaves membranous, 4—12 in. broad, irregularly serrate, the 3—5 lobes acute or acuminate: fl. few, in loose terminal clusters, white or pinkish, 1—2 in. broad: carpels ∞, tomentose; fr. hemispherical, scarlet when ripe, "sweet and pleasantly flavored." Var. **velutinus,** Greene. Leaves smaller, of much firmer texture, densely velvety-pubescent, evenly serrate: fr. dry, insipid.—The type is found only in the mountains of the interior or easterly parts of the State. The variety belongs to the seaboard, where it is common. Fl. March; fr. June.

⊢ ⊢ Stems prickly; leaves 3-foliolate.

2. **R. spectabilis,** Pursh. Stoutish, 5—10 ft. high, sparingly armed
with stout straight prickles: leaves occasionally simple: leaflets ovate,
acute or acuminate, doubly serrate, often more or less lobed, the veins
beneath and the stalks and stalklets sparingly villous: *fl. 1—3, large,
red: fr. large, ovoid, red or yellow,* glabrous. Var. **Menziesii,** Wats.
Foliage somewhat tomentose and silky.—Mendocino Co., *Bolander,*
northward, in moist woods. The variety is of the San Francisco district,
growing on wooded banks of streams, mostly near the sea. April—June.

* * *Fruit persistent on the elongated receptacle.*

3. **R. vitifolius,** Ch. & Schl. Stems woody, very prickly and glaucous,
weak and trailing or suberect, 5—20 ft. long: leaves pinnately 3—5-folio-
late; leaflets ovate to oblong, coarsely toothed, glabrous or more or less
pubescent or tomentose: stipules oblanceolate to linear: fl. imperfect;
staminate large, with elongated petals; pistillate small, with petals short
and relatively broad: *fr. oblong. black and sweet.*—Very common on
banks of streams throughout the Coast Range and in the interior. Fl.
Jan.—April; fr. May, June.

11. **ROSA,** *Varro* (WILD ROSE). Prickly shrubs with unequally
pinnate leaves, adnate stipules and solitary or corymbose large flowers.
Calyx-tube globose or urceolate: limb 5-parted; bractlets 0. Petals 5,
rounded, spreading. Stamens ∞, on a thickened margin of the silky
disk which lines the calyx-tube. Pistils ∞; ovaries free and distinct;
styles subterminal; ovules solitary, pendulous. Fruit of osseous achenes
enclosed in the fleshy-enlarged red berry-like calyx-tube.

* *Calyx-lobes deciduous from the fruit.*

1. **R. gymnocarpa,** Nutt. Slender, 1—4 ft. high, armed with scattered slender and weak straight prickles: leaflets 5—9, rather remote,
glabrous, oval, sharply doubly serrate, ½—1 in. long: fl. 1, 2 or 3, barely
1 in. broad: calyx-lobes ovate, with few or no appendages: fr. 3—5 lines
long, oval or oblong, nearly or quite closed at summit: seeds few, smooth.
—Common in shady places, near streams and on bushy northward
slopes. March—May.

* * *Calyx-lobes persistent.*

2. **R. Sonomensis,** Greene. Slender, 1 ft. high, with many very
leafy branches *well armed with straight prickles:* stipules short, almost
truncate, narrow, the margin closely glandular-ciliolate, at length revolute: leaflets 5, remote, broadly ovate or nearly orbicular, truncate or
somewhat cordate at the slightly inequilateral base, ¼—½ in. long, the
margin evenly and coarsely serrate, the serratures minutely glandular-
denticulate, both surfaces glabrous: *fl. many, small, in dense terminal*

corymbs: calyx-tube round-pyriform, *glandular-hispid;* lobes ovate-lanceolate, acuminate, without foliaceous tip or appendages, erect in fruit.—At the Petrified Forest in Sonoma Co., also on Mt. Tamalpais.

3. **R. Californica,** Ch. & Schl. Erect, branching, 3—8 ft. high; *prickles few, stout, usually recurved,* mostly infrastipular in pairs: foliage deep green, of firm texture, more or less glandular and tomentose; stipules entire: leaflets 5—7, ovate or oblong, acute or obtuse, the serratures mostly simple, spreading rather than falcate-incurved: corymb few- or many-flowered; pedicels pubescent and glandular; calyx-lobes foliaceous-tipped: fruit globose, 4—6 lines thick, the persistent lobes erect.—The common wild rose of middle parts of the State.

ORDER XXXVII. **CALYCANTHEÆ.**

A small order, placed here on account of the analogy subsisting between it and some Rosaceæ in point of floral structure; but probably in no wise related to that order. It is represented in our district by one species of

1. **BUTNERIA,** *Du Hamel* (SWEET-SCENTED SHRUB). Fragrant shrubs with opposite entire exstipulate leaves, and solitary terminal large red or purple flowers. Sepals ∞, in many ranks, inserted on a persistent obconical tube; the outer successively shorter and bract-like, the inner longer and colored like the petals; all deciduous. Petals ∞, on the mouth of the tube, the inner shorter. Stamens ∞, inserted on the upper part of the tube within; filaments short, persistent. Pistils ∞, distinct, inserted on the base and sides of the calyx-tube; styles terminal. Achenes enclosed in the dry thin fibro-ligneous calyx-tube. Seed erect; albumen 0; cotyledons foliaceous, convolute.

1. **B. occidentalis** (H. & A.), Greene. Shrub 6—12 ft. high: leaves dark-green, ovate to oblong-lanceolate, scabrous, 3—6 in. long: peduncles 1—3 in. long; petals and larger sepals linear-spatulate, 1 in. long or more; inner petals incurved: sterile filaments linear subulate, densely villous: fruiting calyx ovate, 1¼ in. long: achenes villous, 4 lines long.—Common along streams in the lower mountains. Flowers of a dull dark red. May—August.

ORDER XXXVIII. **SAXIFRAGEÆ.**

Herbs or undershrubs (*Ribes* shrubby) with simple alternate usually exstipulate leaves, the petiole often stipulaceously dilated at base. Stems mostly simple below, commonly leafless and scape-like. Inflorescence mostly either cymose, racemose or paniculate. Calyx of about 5 sepals, often more or less coherent below and united to the base of the ovary. Petals as many or 0. Stamens 5 or 10, perigynous or hypogynous. Ovary of about 2 carpels more or less cohering below, commonly distinct and diverging at apex; style often wanting and stigmas sessile on the tips of the lobes of the ovary. Fruit capsular or follicular (in *Ribes* baccate). Seeds many, small, albuminous.

* *Herbaceous perennials.*

Stamens 10;

* * *Stems more or less woody.*

1. SAXIFRAGA, *Pliny* (SAXIFRAGE). Short-stemmed or stemless herbaceous plants with simple leaves, their petioles commonly sheathing at base. Flowers in cymose thyrsoid or panicled clusters. Sepals distinct, or at base conjoined to each other and the base of the ovary. Petals entire, imbricate in bud. Stamens 10, inserted with or below the petals, on the base of the calyx, or between it and a fleshy disk. Carpels 2, usually partly united, dehiscent by the inside of the divergent beaks. Seeds with thin coat and no wing or appendage

1. S. Mertensiana, Bong. Scape and leaves from a scaly-bulbous base, glandular-pubescent, ½—1 ft. high: *leaves thin and pale, round-cordate,* crenately or incisely many-lobed, ¾—1½ in. broad, on long petioles which are scarious-dilated at base: cymose panicle loose, the branches often flowering at apex and bearing granular bulblets down the sides: petals 1—2 lines long, oval or oblong, white, with a pair of oval green spots near the base: *filaments somewhat petaloid-dilated.—* Sonoma Co., and northward.

2. S. Californica, Greene. *Leaves few, rather thick,* reddish-veined, sparsely glandular-villous, *oval, oblong or elliptical,* 1—2 in. long, on broad petioles of ½—1 in.; margin coarsely crenate to repand-denticulate, rarely either sharply dentate or nearly entire: scape 6—18 in. high, loosely cymose-paniculate: calyx nearly free from the ovary, the sepals reflexed: petals oblong, thrice the length of the sepals, white or rose-tinted: *filaments subulate,* inserted under the edge of an elevated perigynous disk which equals the summit of the ovary.—Plentiful on cool northward slopes. March—May.

2. THEROFON, *Raf.* Perennial herbs with erect leafy stems, and corymbose or paniculate cymes of white flowers; leaves round-reniform, palmately lobed or toothed, the teeth with callous-glandular tips; the petioles stipularly dilated at base. Calyx 5-lobed; lobes valvate, but early open in the bud; the tube more or less adherent to the ovary. Petals 5, entire, imbricate or convolute in bud. Stamens 5, short, alternate with the petals. Capsule 2-celled, dehiscent down the beaks. Seeds minutely granulate or papillose.

1. **T. elatum,** (Nutt.). Slender, 1—2 ft. high, glabrous or glandular-pubescent, the bases of the petioles bearing brown bristly hairs: leaves thin-membranaceous, 5—7-lobed, 1—3 in. broad: calyx-lobes lanceolate-triangular; tube oval and urceolate in fruit: petals cuneate-oblong, obtuse, persistent, in age recurved: seeds elongated-oblong, acute at one end, dark brown, rather densely tuberculate.—Shady banks and rocky margins of streams. May—August.

3. **TELLIMA,** *Robert Brown.* Perennial herbs, with leaves chiefly radical, round-cordate, toothed or palmately divided, their petioles stipulaceously dilated at base. Flowers in a simple terminal raceme. Calyx campanulate or turbinate, 5-lobed, free from the ovary, or adherent to it at base or even to above the middle; the short triangular lobes valvate in bud. Petals 5, laciniate-pinnatifid, or 3—7-lobed, or entire, distant, sometimes involute in bud. Stamens 10, short, included. Ovary short, 1-celled, with 2 or 3 parietal placentæ; styles 2 or 3, very short; stigmas capitate. Capsule conical, opening between the short beaks. Seeds very numerous, with a close coat.

* *Corolla regular, the petals greenish, sessile by a broad base, laciniately pinnatifid; styles and placentæ 2.*

1. **T. grandiflora** (Pursh), Dougl. Stoutish, 1—2 ft. high, from rather coarse-tufted rootstocks; herbage rough-hirsute: leaves round-cordate, more or less lobed, 2—4 in. broad: calyx ¼—½ in. long, inflated-campanulate: petals deeply tinged with red: seeds light brown, oval, strongly rugose-pitted.—Wooded hills, or sometimes in open ground, from Santa Cruz northward. May, June.

* * *Corolla slightly irregular, the petals white or pinkish, entire or lobed or toothed, short-unguiculate; styles and placentæ 3.*

+*Calyx turbinate, the tube more or less adherent to the ovary.*

2. **T. affinis** (Gray), Boland. Stems one or several from a slender horizontal or ascending tuberiferous rootstock, commonly ½ ft. high, scabrous-hirsute: radical leaves very few, round-reniform, slightly lobed, 1 in. broad; cauline relatively broader, 3-lobed to the middle, the lobes coarsely toothed: calyx 2½ lines long; pedicels rather longer: lower petals 4 or 5 lines long, 3-toothed, the upper narrower and a trifle shorter, entire: styles short, not exserted from the calyx: seeds oblong, dark brown, faintly striate-pitted or almost smooth.—Frequent on shady hillsides. April—June.

++ *Calyx campanulate, only the base adherent.*

3. **T. heterophylla** (T. & G.), Hook. & Arn. Slender, 1 ft. high, scabrous-hirsute: *lowest leaves* ¾—1 in. broad, *with 5 shallow rounded lobes:* cauline more deeply 3-lobed or -parted: pedicels very short, the broad truncate-based calyx appearing almost sessile: *petals* (at least the lower 3) *obtusely 3-lobed:* styles glabrous: seeds muriculate.—Common in the Coast Range.

4. **T. Bolanderi** (Gray), Boland. Near the last but larger, often 2 ft. high, more hirsute: radical leaves 1½—2½ in. broad; cauline more divided; *petals* 3—4 lines long, obovate or oval, *the upper entire, the lower often with a lateral tooth on each side:* seeds muricate-scabrous.—Southern slope of Mt. Diablo, *Brewer.*

4. TIARELLA, Linn. Perennial herbs with simple or 3-foliate alternate more or less distinctly stipulate leaves, and a terminal panicle or raceme of small white flowers. Calyx 5-parted, the lobes valvate. Petals 5, entire, unguiculate. Stamens 10, inserted with the petals into the base of the calyx; antlers with 2 parallel cells. Ovary 1-celled, compressed, the two valves early separating and becoming unequal, one becoming lanceolate-elongated, the other remaining short. Seeds few at the base of each placenta.

1. **T. unifoliata,** Hook. Pubescent, 6—15 in. high; leaves thin, ovate-cordate, rounded or triangular, 3—5-lobed, the lobes crenate-toothed, the radical ones long-petioled, the cauline few, small, short-petioled: panicle narrow and raceme-like: petals almost filiform.—From San Mateo Co., *Kellogg,* northward, in woods.

5. HEUCHERA, *Linn.* (ALUM-ROOT). Perennial herbs, with leaves and flowering stems from a short branching caudex, the former long-petioled, palmately veined, roundish cordate, slightly lobed. Stems somewhat scapiform, bearing few alternate reduced leaves and a panicle or thyrse of cymose-dichotomous clusters of small white flowers. Calyx campanulate, 5-lobed, the tube adherent to the ovary below; lobes obtuse, imbricate in bud. Petals 5, small, entire. Stamens 5, alternate with the petals; anthers 2-celled. Capsule 1-celled, with 2 parietal placentæ, 2-beaked, dehiscent between the beaks. Seeds horizontal, oval, muriculate or hispidulous.

1. **H. micrantha,** Dougl. Leaves thin, 1—3 in. broad, ovate-cordate, 5—9-lobed, hairy on the veins beneath: *stem villous, bearing a few small leaves and a loose panicle* often 1½ ft. long: *calyx campanulate,* 1—2 lines long, acute at base, shorter than the slender pedicels, puberulent: narrowly spatulate petals and slender filaments white, well exserted.—Common in shady ravines. May—July.

2. **H. pilosissima,** Fisch & Mey. *Hirsute with rusty and viscid spreading hairs:* leaves 1—3 in. broad, round-cordate, obtusely lobed

and crenate: *stem* 1—2½ ft. high, *naked or few-leaved, rather densely and thyrsoidly paniculate: calyx* densely hairy, *subglobose*, the tube rounded, the lobes incurved: filaments and narrowly spatulate petals little exserted. Var. **Hartwegi,** Wats. Stems 2—3 ft. high: panicle more open; the whole plant, and especially the calyx, less hairy.—In the Coast Range, and apparently not common; at all events seldom seen.

6. **PARNASSIA,** *Tourn.* Glabrous stemless perennials, with entire petioled exstipulate leaves and simple 1-flowered scapes. Calyx 5-parted; the base free from or adnate to the base of the ovary. Petals 5, oval or oblong, imbricate in bud, white, with conspicuous green veins, widely expanding, tardily deciduous. Stamens 5, alternating with the petals, and with as many clusters of short gland-tipped sterile filaments. Ovary ovate, 1-celled, with 3 or 4 parietal placentæ; stigmas as many, closely sessile each directly over its corresponding placenta. Capsule 3 –4-valved from the apex, the valves placentiferous in the middle.

1. **P. Californica,** Greene. Radical leaves ovate or ovate-oblong, 1—2 in. long, tapering from the broad and sometimes slightly rounded base to a long or short petiole: scapes 1—2 ft. high, the very small sessile but not clasping leaf borne much above the middle: petals oval or obovate, sessile, entire, ¾ in. long: sterile filaments about 20 in each set, united to the middle, each tipped with a conspicuous antheroid protuberance.—In the mountains above New Almaden, *J. Burtt Davy.*

7. **WHIPPLEA,** *Torrey.* Slender diffuse hairy undershrub, with short-petioled leaves, and terminal naked-peduncled clusters of small white flowers. Calyx white like the petals, 5-cleft: tube adnate to lower part of the ovary. Petals 5, ovate or oblong, narrowed at base. Stamens usually 10; filaments subulate. Ovary 3—5-celled, with a single ovule in each cell; styles distinct; stigmas introrse. Capsule septicidally parting into distinct cartilaginous 1-seeded portions which open ventrally only. Seeds oblong, with a short obtuse appendage at each end.

1. **W. modesta,** Torr. Stems 1 ft. long or more: leaves thin, ovate or oval, somewhat toothed or entire, 1 in. long or less: fl. 2 lines broad or less: calyx tube nearly hemispherical: capsule globular; styles deciduous from it.—Borders of thickets, or in deep woods, in the Coast Range.

8. **RIBES,** *Fuchs.* Shrubs, with alternate palmately lobed often resinous-glandular or viscid leaves; the stipules when present adnate to the petiole. Flowers racemose (rarely solitary) on short leafy shoot-from lateral buds; pedicels subtended by a bract and usually bibracteolate about midway. Calyx-tube adnate to the globose ovary and more or less produced above it, 5-lobed (4-lobed in n. 9), the lobes commonly spreading or reflexed, usually colored. Petals 5, mostly smaller than the calyx-lobes, inserted in or near the sinuses. Stamens 5, alternate

with the petals. Ovary 1-celled; placentæ 2, parietal; styles 2, more or
less united; stigmas terminal. Fruit a berry, crowned with the withered
remains of the flower.

     * *Unarmed: leaves convolute in bud; calyx-tube elongated.*

1. **R. tenuiflorum**, Lindl. Shrub 5—10 ft. high, nearly glabrous,
glandless: leaves light green, 3—5-lobed at apex, not at all cordate:
racemes ∞-flowered; bracts green and conspicuous: fl. bright yellow,
scentless; calyx salverform, the tube ½ in. long or more, thrice longer
than the oval lobes: berry glabrous, amber-colored and translucent,
acidulous when ripe.—Wild Cat Creek, *Behr;* also near Niles.

     * * *Unarmed; leaves plaited in the bud; calyx-tube broader.*

2. **R. glutinosum**, Benth. Often 6—15 ft. high: *leaves thin*, 3—5 in.
broad, glutinous when young, glabrous or more or less pubescent in
age, *not rugulose;* petioles very abruptly dilated at base and obscurely
ciliolate: racemes long-peduncled, pendulous, very many-flowered: calyx
with 2 caducous bracteoles at base, cleft scarcely to the middle, the *tube
cylindrical,* the whole from pale pink to rose-color: *berry* large, globose,
blue with a dense bloom, and glandular-hispid; pulp black, dry, insipid.
Var. **melanocarpum.** Ripe *berries black,* without any trace of bloom.—
Very common along streams among the hills. The variety at Berkeley,
and in Santa Clara Co. Fl. Jan.—April; fr. August, Sept.

3. **R. malvaceum**, Smith. More rigid and compact, 3—6 ft. high:
leaves thick, 1—2 in. broad, *strongly rugulose* and somewhat scabrous
above, more or less densely white-tomentose beneath; the slight stipular
dilatation of the petiole only obscurely ciliolate: racemes short-peduncled,
dense; pedicels and ovaries whitish-tomentose: *calyx-tube* subcylindrical,
abruptly dilated and broadest just above the ovary; segments short,
spreading, the whole rose-color: petals white, roundish or subreniform:
berry oval, ⅓ in. long, purple, glaucous; pulp soft and sweet.—On dry
open hills of the Coast Range, from Bolinas Ridge, *Drew,* and Vaca
Mts., *Jepson,* southward. Fl. Nov.—March; fr. May.

     * * * *Thorny; leaves plaited; flowers few.*

     ⊢ *Fl. 5-merous; calyx-lobes reflexed.*

4. **R. divaricatum**, Dougl. Nearly glabrous: stems clustered, the
widely spreading branches 5—12 ft. long: leaves roundish, 3—5-lobed;
the lobes incisely toothed: peduncles elongated, slender, drooping,
3—9-flowered; pedicels with a small broad bract at base: fl. ⅓ in. long;
calyx green without, dark livid purple within, the oblong-linear lobes
exceeding the campanulate tube; *petals white, fan-shaped, plane, the
margins convolutely overlapping:* filiform villous filaments and deeply
cleft style long-exserted: berry small, glabrous, black, agreeable.—
Along streams and on northward slopes.

5. **R. Victoris,** Greene. Shrub 5 ft. high; branches very prickly: leaves and growing branchlets pubescent and viscid: pedicels short, deflexed, with 1 or 2 persistent bracts and as many short-pedicellate greenish flowers ½ —¾ in. long: calyx-tube short-campanulate, much exceeded by the greenish lobes: *petals* 1½ lines long, white, *thinnish* involute, *acute, more or less toothed* at apex; filaments stoutish, little surpassing the petals; anthers large, subsagittate, mucronate; berry glandular-hispid.—By streams in the Coast Range north of San Francisco, and in the Vaca Mts.

6. **R. Californicum,** Hook. & Arn. Shrub 2—4 ft. high, with *very rigid and flexuous glabrous branches: leaves small,* 3—5-lobed and incised, sparsely glandular-puberulent when young, *not at all viscid or heavy-scented,* in maturity glabrous: peduncles very short, 1—3-flowered; the very short pedicels each with a small round-ovate bract beneath: calyx-tube very short, the reflexed lurid-purple segments thrice as long; *petals* white, *thick,* strongly involute, *truncate and erose-toothed at summit;* filaments stout, thrice the length of the petals, the anthers ovate-oblong, mucronate, reddish; ovary glandular-hispid: *berry* large, *prickly.* —On bleak hills. Fl. Feb., March.

7. **R. subvestitum,** Hook. & Arn. Tall leafy open and rather handsome shrub 5—10 ft. high; branches usually more or less setose-hispid: *leaves* more or less glandular-pubescent, *very viscid and heavy-scented:* peduncles 1—3-flowered; pedicels elongated: calyx-tube broadly campanulate, 1½ lines long, the red-purple reflexed segments nearly twice as long: *petals* white-waxy, *truncate, entire,* strongly involute; filaments well-exserted: ovary densely glandular-hairy: berry large, as densely clothed with short stiff gland-tipped hairs; pulp soft, sweet.—Very common in the Coast Ranges from at least Sonoma Co. to Monterey. March, April.

8. **R. Menziesii,** Pursh. Size and habit of the last: leaves more than 1 in. broad and of greater length, deeply 3-cleft, the lobes coarsely incised, usually *soft-pubescent beneath, seldom or never viscid:* peduncles slender, pendulous, 1—2-pedicellate above the middle, the bracts small, persistent: fl. ¾ in. long; *calyx* of a rich red-purple, *pubescent exteriorly,* the tubular-funnelform tube about half as long as the ligulate reflexed segments: *petals* large, thickish, *truncate,* involute, cream-color or whitish: filaments subulate, not exserted, only the large linear-oblong mucronate white anthers borne beyond the petals: ovary densely echinate: *fruit very prickly.*—Marin Co. April, May.

← ← *Fl. 4 merous; calyx-lobes erect.*

9. **R. speciosum,** Pursh. Shrub 6—10 ft. high, with long leafy red-bristly branches: subaxillary spines 3, united at base: *leaves subcoriaceous,*

dark green, very smooth and shining above, rounded and 3-lobed; lobes short, crenately-toothed: peduncles pendulous, 2—5-flowered: *fl. bright red, often 2 in. long* from the base of the ovary to the tips of the long-exserted stamens; calyx cylindraceous, the 4 (rarely 5) lobes erect; anthers oval, small; ovary bristly: berry small, rather dry, densely prickly.—Frequent along the seaboard, southward; not certainly known as within our limits.--March—May.

Order XXXIX. CRASSULACEÆ.

Succulent herbs with exstipulate leaves. Flowers symmetrical, cymosely arranged. Sepals 3—20, more or less united at base. Petals as many, inserted in the bottom of the calyx, distinct or cohering below to form a gamopetalous corolla. Stamens as many or twice as many as the petals, when of the same number alternate with them; filaments subulate. Ovaries as many as the petals, opposite to them, each with or without a hypogynous scale at base. Fruit follicular. Seeds attached to the margins of the suture, small, albuminous.

1. **TILLÆA,** *Micheli.* Small and slender fleshy glabrous annuals. Leaves opposite, entire. Flowers minute, axillary, white or pinkish. Sepals and petals 3—5, distinct or united at base. Stamens as many. Carpels distinct; styles short-subulate. Follicles 1—several-seeded. Seeds striate lengthwise.

    * *Fl. clustered; petals acuminate; carpels 1—2-seeded.*

1. **T. minima,** Meiers. Simple or with few or many ascending branches, 1—3 in. high: herbage very light green when young, in age reddish: internodes short: leaves ovate or oblong, obtuse, 1 line long, connate: fl. in short axillary panicles, mostly subsessile, occasionally some with long pedicels: sepals 4, ½ line long, acute, nearly or quite equalled by the linear-lanceolate acuminate petals: carpels acute, not longer than the petals.—Very common in clayey soils in the hilly districts. March—May.

    * * *Fl. solitary; petals oval or oblong; carpels several-seeded.*

2. **T. Drummondii,** Torr. & Gray. Stems *very slender, dichotomous, diffuse,* rooting at some of the lower nodes, 1 in. long or more: leaves oblong-linear, slightly connate: pedicels at length equalling or exceeding the leaves: petals red, fully equalling the obtuse carpels, and twice or thrice the length of the calyx-lobes.—Common in moist low places in wheat fields near Suisun. May.

3. **T. Bolanderi,** Greene. Stems *stoutish, simple,* 2—5 in. long, the lower portion with long internodes and rooting at the nodes; leaves linear or linear-oblong, acutish, subterete, slightly connate; fl. short-

pedicellate, the pedicel in fruit elongated and surpassing the leaves: petals oblong, acutish, equalling the carpels, more than twice the length of the ovate calyx-segments.—Frequent on muddy shores about San Francisco. May.

2. SEDUM, *Columna* (STONE-CROP). Glabrous perennials or annuals. Flowers in cymes, mostly secund. Sepals 4 or 5, united at base. Petals as many, distinct. Stamens twice as many. Carpels distinct, or rarely connate at base, few- or many-seeded.

* *Perennial.*

1. **S. spathulifolium,** Hook. Glaucous and often pulverulent: stems 4—6 in. high, ascending from a branched and rooting caudex: leaves flat, obovate or spatulate, obtuse, 6—10 lines long: fl. 3 lines long: petals yellow, lanceolate, acute, twice longer than the ovate acute sepals.— Rocky places on the northward slopes of hills and mountains from San Francisco and Berkeley northward.

* * *Annuals.*

2. **S. radiatum,** Wats. Stems 3—6 in. high, decumbent at base: leaves oblong or oblong-ovate, obtuse or acutish, somewhat clasping by the narrower base, ¼—½ in. long: fl. sessile; sepals short, triangular; petals yellow, narrowly lanceolate, acuminate, 3 lines long: carpels broad, abruptly divergent from the united bases.—Mt. Hamilton. Strictly annual, though propagating by deciduous buds formed in the axils of the lowest leaves.

3. **S. pumilum,** Benth. Slender, erect, 1—3 in. high: leaves 1—2 lines long, ovate-oblong: fl. sessile in sparingly branched cymes; calyx-lobes minute, triangular; petals yellow, linear, acute, 1½ lines long: follicles short, 1-seeded, the seed erect, filling the cavity.—Hills of Napa Co. and eastward.

3. COTYLEDON, *Nicander.* Succulent herbs coarser than *Sedum* and larger, but quite like them in all other respects save that the petals are more or less united into a tube, and the follicles erect or suberect rather than spreading; the inflorescence in ours compound-cymose.

1. **C. cæspitosa,** Haw. Nearly or quite acaulescent, *dull green*: leaves ovate-oblong to oblong-lanceolate, acute, the larger 1½—3 in. long: flowering branches ½—1 ft. high, with broadly triangular-ovate clasping bracts: pedicels short and stout, subtended by broad bracts: sepals ovate, 2 lines long or less: petals yellow, broadly lanceolate, acute, 4—5 lines long: carpels ovate-oblong, about 3 lines long—From near San Francisco northward.

2. **C. farinosa** (Lindl.), Baker. Short-caulescent, *more or less white-farinose:* leaves rather flaccid, ascending, lanceolate, acuminate, the

larger ones 2—4 in. long, acute: flowering branches 6-10 in. high, with scattered broadly ovate to lanceolate clasping bracts: bracts ovate-lanceolate: pedicels 1—3 lines long: sepals broadly lanceolate, ¼ in. long: petals yellow, oblong-lanceolate, mostly acuminate, 4—6 lines long: carpels ovate-oblong, ¼ in. long.—Near Sonoma.

Order XL. FICOIDEÆ.

Very succulent herbs or shrubs. Leaves plane, triquetrous or terete, without stipules. Calyx-lobes usually 5, unequal, foliaceous. Petals very many and linear or 0. Stamens 5—∞, with slender filaments, inserted on the calyx-tube. Styles 4—20. Fruit 4—20-celled, dehiscent stellately across the summit, or circumscissile, or indehiscent. Seeds usually numerous and minute.

1. MESEMBRYANTHEMUM, *Breyne.* Flowers large, terminal. Calyx-tube adnate to the ovary. Petals and stamens very numerous. Fruit structurally capsular, but in ours juicy and baccate.

1. **M. æquilaterale,** Haw. Perennial, glabrous, glaucescent, the stout prostrate stems several feet long, the short flowering branches ascending; leaves opposite, very fleshy, triquetrous with linear sides, 1—3 in. long: fl. solitary, subsessile, 1½ in. broad, bright rose-purple: calyx-tube turbinate, ½ in. long or more; the larger lobes as long: stigmas 6—10: fr. large, fragrant, edible.— On banks and cliffs near the sea; also Australian and Chilian.

2. SESUVIUM, *Linn.* (Sea Purslane). Flowers small, axillary and terminal. Calyx-tube free from the ovary; lobes 5, apiculate on the back near the top, scarious-margined, often purplish within. Petals 0. Stamens 5—∞, inserted at the top of the calyx-tube. Styles 3—5. Capsule ovate-oblong, 3—5-celled, circumscissile, ∞-seeded.

1. **S. Portulacastrum,** L. Stems prostrate, 1 ft. long or more: leaves linear- to oblong-lanceolate, ½—1½ in. long, acute or obtuse: fl. sessile or pedicellate: calyx 3—5 lines long, the lobes purple: stamens numerous.—Lower San Joaquin, *Bioletti.*

3. TETRAGONIA, *Linn.* Perennial, with alternate plane fleshy leaves and axillary greenish apetalous flowers. Calyx 4-cleft, adherent to the ovary, 4—8-horned; the lobes yellowish within. Stamens several. Styles 3—8; ovary 3—8-celled. Fruit osseous, nut-like, indehiscent, 3—8-celled, the cells 1-seeded.

1. **T. expansa,** Murr. Leaves petiolate, rhombic-ovate, acute or acuminate, entire, more or less crystalline-papillose, 1—2 in. long: fl. sessile, 1-3 in each axil: fr. 4-horned, about ⅓ in. long, scarcely as broad.—Beaches of San Francisco Bay, in Marin and Alameda counties.

Order XLI. EPILOBIACEÆ.

Herbs, often with hard shrubby-looking stems shedding a thin papery outer bark. Leaves simple, usually alternate, entire, toothed or pinnatifid. Flowers axillary to the leaves, or in bracted or naked racemes or spikes, rarely panicled, usually 4-merous. Calyx-tube partly or wholly adherent to the ovary; lobes valvate in bud. Petals borne on the throat of the calyx-tubes or at the sinuses of the lobes, convolute in bud. Stamens 2—8. Style single; stigma capitate or 4-lobed; ovary 2 or 4-celled. Seeds naked or appendaged; albumen none.

Free portion of the calyx-tube deciduous from the ovary;

Seeds with a coma { fl. small, purplish or white..............EPILOBIUM 1

{ fl. large, scarlet.........................ZAUSCHNERIA 2

Seeds { fl. yellow.......... ....ŒNOTHERA 3

not { fl. purple or { Calyx-lobes reflexed { petals sessile.GODETIA 4

comose { " unguiculate.CLARKIA 5

{ rose-color { Calyx-lobes ascending............... BOISDUVALIA 6

Only the segments of the calyx free from the ovary,

these persistent { fl. apetalous...................... ....ISNARDIA 7

{ petals yellow.................... ....JUSSLÆA 8

1. EPILOBIUM, *C. Gesner* (WILLOW-HERB). Tube of calyx little prolonged beyond the ovary; limb deeply 4-cleft, campanulate or funnelform, or 4-parted to the base with the lobes spreading, deciduous. Petals 4, spreading or erect, often emarginate or bifid, purplish or white. Stamens 8, the 4 alternate ones shorter; anthers elliptical or roundish, fixed near the middle. Stigma oblong, clavate, or with 4 spreading or revolute lobes. Capsule mostly linear, 4-sided, 4-celled, 4-valved. Seeds numerous, ascending; the summit bearing a tuft of long white hairs.

*   *Annuals, with terete stems; leaves alternate (except the lowest).*

1. **E. minutum,** Lindl. *Diffusely branched from the base,* the mostly decumbent branches ½—1 ft. long, puberulent: leaves ovate-lanceolate or lanceolate, entire or repand-denticulate, ½—¾ in. long: fl. solitary in all the axils; petals obcordate, white or with a tinge of rose: 4 long stamens equalling the style; stigma clavate, the lobes at length expanded and fimbriate: capsule pedicellate, about 1 in. long: seeds rather few, smooth.—In the Coast Range, on dry hills. April—June.

2. **E. paniculatum,** Nutt. Erect, *slenderly paniculate-branched above,* 1—10 ft. high, from wholly glabrous to minutely and densely glandular-pubescent: leaves narrowly lanceolate or linear, obscurely serrulate. 1—2 in. long, with smaller ones fascicled in the axils, the floral reduced to subulate bracts: corolla cruciform; the rose-colored petals quadrate-oblong, abruptly and often deeply notched, rose-purple and veiny, 1—2 lines long, rotate-spreading: capsule pedicellate, 1 in. long, attenuate at each end, often arcuate: seeds minutely papillose.—Dry ground; common. July—Nov.

* * *Perennials; leaves mostly opposite (except the upper).*

3. **E. Franciscanum,** Barbey. Very stout, 2—4 ft. high, *pubescent with soft short glandular hairs:* stem reddish, subterete, but with delicate sharp angles running down from the leaf-bases: lower leaves opposite, with short but distinct petiole, blade 2—4 in. long, oblong lanceolate, rounded at base, serrulate: racemes dense, leafy-bracted, the red-purple or pale flowers appearing somewhat corymbose: *petals* ¼ in. long or more, *deeply emarginate:* capsule 2 in. long: seed obovoid-oblong, acutely pointed at base, the hyaline papillæ forming close longitudinal lines.— Plentiful in springy places, along streamlets and shores of ponds about San Francisco. June—Dec.

4. **E. Watsonii,** Barbey. Size of the preceding, but not stout, the terete stems with less marked lines, *somewhat hoary with a soft pubescence:* leaves oblong-lanceolate, rather obtuse, denticulate, rounded to short-winged petioles: fl. not crowded, suberect in the axils of the more reduced and acute upper leaves, rose-red; *petals elongated-obcordate:* seeds more coarsely granulate-striate.—On Russian River, Sonoma Co.

5. **E. holosericeum,** Trel. Loosely branched, at least the upper leaves and branches *canescent with sub-appressed hairs;* leaves oblong-lanceolate, obtuse or sometimes acute, remotely serrulate, attenuate, or abruptly contracted and then cuneately narrowed, to short petioles: fl. small, scattered on the elongated branches, pale: mature capsules on peduncles equalling the floral leaves: seeds short-beaked, very minutely papillose-striate.—Solano Co., along streams.

6. **E. adenocaulon,** Hausskn., var. **occidentale,** Trel. Tall, with paniculate ascending branches and long internodes; *branches, inflorescence and capsules glandular-pubescent;* leaves ovate- or triangular-lanceolate, ascending, abruptly rounded to short winged petioles, prominently denticulate, the floral small, acute at both ends: fl. small: capsule slender, short-pedicellate; seed elongated, obovoid, minutely striate.—Common by streams and about springy places.

7. **E. Californicum,** Hausskn. Tall, slender, more sparingly branched, glabrous below; *pubescence of the buds, pods, etc., of coarse ascending, not glandular hairs:* leaves lanceolate, acutish, rather remotely serrulate, short-petiolate: fl. scattered: fruiting peduncles slender, almost equalling the floral leaves: capsules nearly glabrous: seeds almost beakless.— Apparently along the seaboard only, and less common than the last.

2. ZAUSCHNERIA, *Presl.* Perennial herbs spreading by subterranean shoots. Flowers racemose along the leafy branches, large, scarlet. Calyx-tube globose-inflated just above the ovary, thence becoming narrow-funnelform, 4-lobed, within bearing 8 small scales, 4 erect and 4 deflexed. Petals 4, little exceeding the calyx-lobes, obcordate or deeply cleft. Stamens 8, the 4 alternate ones shorter; anthers linear-oblong, attached by the middle. Stigma peltate or capitate, 4-lobed. Capsule slender-fusiform, 4-angled, 4-valved, ∞-seeded. Seeds comose.

1. **Z. Californica,** Presl. Decumbent, 1—3 ft. high, *canescent with a minute but dense tomentose pubescence: leaves linear-lanceolate,* ¾—1½ in. long, entire or denticulate, thickish, *seldom at all feather-veined:* fl. 1¾ in. long; calyx-tube narrow-funnelform, twice the length of the linear-lanceolate segments, these surpassed by the deeply cleft petals: capsule nearly glabrous, distinctly pedicelled: seeds oblong-obovate.— In the Coast and Mt. Diablo Ranges, from Lake Co. southward, on dry open ground. July—Nov.

2. **Z. latifolia,** Greene. Decumbent, seldom 1 ft. high, commonly *nearly glabrous: leaves from broadly ovate to ovate-lanceolate,* ½—1 in. long, very acute, more or less serrate-toothed, thin, *conspicuously feather-veined:* fl. 1 in. long; calyx-tube narrowly cylindrical for about 2 lines above the globose base, thence widening abruptly to a funnelform throat, the whole not longer than the petals: capsule subsessile, glabrous. —Eastern base of the Mt. Diablo Range. June—Nov.

3. **ŒNOTHERA,** *Linn.* (EVENING PRIMROSE). Herbs exceedingly diverse in habit. Leaves alternate. Flowers yellow, white or purplish, axillary, spicate or racemose. Calyx-tube prolonged above the ovary, mostly deciduous. Petals 4, mostly vespertine as to time of opening, and evanescent, usually obcordate, or flabelliform. Stamens 8, equal, or those opposite to the petals shorter; anthers various. Ovary 4-celled, ∞-ovuled; style filiform; stigma 4-lobed or capitate. Capsule from membranaceous to woody, more or less perfectly 4-valved and dehiscent, or indehiscent. Seeds in 1 or 2 rows in each cell, horizontal or ascending, naked, often angled.

* *Tall annuals or biennials; calyx-tube elongated, deciduous from the ovary; fl. in a leafy spike, vespertine; stigma-lobes linear.*

1. **Œ. biennis,** L. Erect, 3—5 feet high, the older parts, and especially the capsules, hirsute: leaves rather thin, lanceolate, denticulate: calyx-tube 1¼ in. long; segments ¾ in., their tips very short, not contiguous: petals 1 in. long, light yellow: filaments subulate, short, the long anthers exserted: style short; stigma-lobes green, not widely spreading: capsule 1¼ in. long, tapering from below the middle to apex, scarcely angular, the valves separating at apex only.—Along the Sacramento River in Solano Co., *Œ. Jepsonii,* of Fl. Fr., but probably not specifically distinct.

2. **Œ. GRANDIFLORA,** Ait. Erect, 3—5 ft. high; stem and inflorescence scabrous and sparsely hirsute; the ovate-lanceolate denticulate leaves minutely and sparsely pubescent: calyx-tube 1—2 in. long, the segments

almost as long, their slender tips elongated: petals obcordate, 1½—2 in. long, yellow, turning to deeper yellow: filaments filiform, declined: style shorter than the petals; linear stigma-lobes ¼ in. long, yellow: capsule obtusely quadrangular, slightly tapering from near the base: seeds sharply angled.—Common in cultivation, and sparingly naturalized about Oakland, Alameda, etc. Differing from Œ. *biennis* by its large almost scentless flowers, declined stamens, etc.

* * *Acaulescent; fl. diurnal, yellow, erect in bud; calyx-tube filiform above the ovary; stigmas capitate.*

+— *Perennials; calyx-tube persistent; capsules not winged.*

3. **Œ. ovata,** Nutt. Sparingly pubescent: leaves mostly oblong-lanceolate, entire or denticulate, often somewhat undulate, occasionally pinnatifid, 3—8 in. long: calyx-tube 1—4 in. long: petals ½—¾ in. long: capsules partly subterranean, chartaceous, 1 in. long, tapering above, scarcely dehiscent: seeds ovoid-oblong, smooth.—Very common in open grounds. Feb.—May.

+—+— *Annual; calyx-tube deciduous; capsules winged.*

4. **Œ. graciliflora,** Hook. & Arn. Herbage green and pilose: leaves linear, entire or obscurely denticulate: calyx-tube not longer than the leaves; segments short: petals 3—5 lines long, obcordate, turning greenish: capsule hard coriaceous, ½ in. long or less, angled at base, 4-winged above, the wings obliquely truncate and hairy; seeds smooth. —Hillsides and plains in Contra Costa Co.

* * * *Caulescent; calyx-tube obconic or short-funnelform; stigma capitate; capsules sessile, mostly contorted.*

+— *Maritime plants, with short primary axis bearing crowded elongated narrow leaves, and radiating decumbent or prostrate shrubby-looking flowering branches; fl. diurnal; capsules contorted.*

5. **Œ. spiralis,** Hook. Radiating branches stout, procumbent, 1—3 ft. long: *leaves from spatulate to ovate-cordate,* 1—3 in. long, *entire or dentate,* more or less hirsute: petals 4—6 lines long: anthers linear-oblong, fixed in the middle: capsule acutely quadrangular, hirsute: seeds ovate, acute at base, compressed, dark brown.—Plentiful on the sand hills of San Francisco, flowering almost throughout the year.

6. **Œ. micrantha,** Hornem. Size and habit of the last but more slender and hirsute, the small calyx densely hairy: *leaves from narrowly oblanceolate to linear-oblong,* 2—4 in. long, acutish, *more or less undulate:* petals 1—2 lines long, entire or emarginate: capsule 4-angled, contorted, rather slender, gradually attenuate upwards, sparsely hirsutulous.—At San Francisco, near the Presidio, etc., and southward along the coast.

+ + *Plants not maritime, erect at base and with ascending branches; capsules narrow.*

++ *Radical leaves narrow and petiolate, the cauline broad, sessile; capsules sharply angled, much contorted.*

7. **Œ. hirtella**, Greene. Stoutish, erect, simple, or with a few ascending branches from the base, 6—10 in. high, short-hirsute: radical leaves oblanceolate, denticulate, 1½ in. long; cauline ovate, sessile, ½ in. long, coarsely toothed and more or less undulate or crisped: petals 1 line long or more: capsules hirsute, narrow, attenuate upwards, once or twice coiled: seeds pale, smooth, more or less regularly rhombic-ovate. —Common in the hill country away from the sea, from Sonoma Co. and Solano southward.

++ ++ *Without radical leaves; branches many, slender, leafy; capsules narrowly linear, slightly or not at all contorted.*

8. **Œ. strigulosa** (F. & M.), Torr. & Gray. Erect-spreading, ½—1 ft. high, all but the older parts clothed with short white hairs: leaves ½ in. long, linear-lanceolate, acutish, denticulate, subsessile: *petals broadly obovate*, 1½ *lines long*, yellow, turning deep red: *anthers roundish*, basifixed: capsule about ¾ in. long, sessile, straight or arcuate, not contorted, scarcely attenuate at apex.—Sandy soil; common.

9. **Œ. campestris**, Greene. Branched from the base and bushy, 6—10 in. high and as broad, more or less hirsute-pubescent throughout: leaves linear-lanceolate, 1 in. long, dentate: *petals very broadly cuneate-obovate*, 4--5 *lines long*, turning brick-red: *anthers linear-oblong*, ¾ line long, fixed toward the middle and versatile: pods more than 1 in. long, narrowly linear, slightly incurved. Var. **cruciata**, Greene. Petals half as large, narrowly obovate or oblong, often emarginate.—Common on the plains from Antioch southward.

4. **GODETIA**, *Spach.* Erect simple or branching annuals. Leaves alternate, entire or denticulate. Flowers mostly purple, showy, in leafy spikes or racemes. Calyx-tube above the ovary obconic or short-funnel-form, deciduous. Petals 4, broad, sessile, entire, emarginate or cleft, diurnal and lasting for two days or more. Stamens 8, unequal, the filaments opposite the petals shortest: anthers perfect, elongated, attached by the base, erect or recurved. Ovary 4-celled, ∞-ovuled; style short; stigma-lobes short, linear or roundish. Capsule ovate to linear, 4-sided, coriaceous. Seeds ascending or horizontal, in 1 or 2 rows, obliquely angled, the upper part tuberculate-margined.

* *Flowers in a strict dense spike; capsule ovate or oblong.*

← *Tips of the calyx-lobes not free in the bud; sides of capsule not 2-costate: seeds in 2 rows in each cell.*

1. **G. purpurea** (Curtis), Wats. Stem erect, 6—15 in. high, puberulent: leaves oblong or lanceolate-oblong, obtusish, entire, glaucescent:

calyx-tube funnelform, as long as the segments: petals broadly obovate,
½ in. long or more, crenulate, deep purple: stamens much shorter than
the petals: stigma lobes broad and short, dark purple: capsule ovate-
oblong, ½—¾ in. long, hairy, the sides nearly flat, with a strong
midvein.—Dry hills of the Mt. Diablo Range.

+— +— *Tips of the calyx-lobes slightly free in the bud; capsule 2-costate on
at least two of the sides; seeds in 1 row in each cell.*

2. **G. lepida**, Lindl. Branching above, pubescent with short
appressed hairs: *leaves ovate-lanceolate, entire*, slightly pubescent: calyx-
tube obconical, very short, greatly surpassed by the segments: *petals
rounded and emarginate at apex*, pale purple, with a dark red cuneate
spot at summit: stigma purple, cruciform: capsule ovate-oblong, sessile,
closely ribbed and sulcate, white-villous.—Mt. Diablo Range.

3. **G. micropetala**, Greene. Slender, simple, 1—3 ft. high, puberulent: leaves 1 in. long, narrowly lanceolate, entire, sessile: spike rather
short: calyx-tube scarcely 2 lines long; segments 4 lines, the slender
elongated tips twisted in the bud: *petals linear-lanceolate, only 3 lines
long*, entire or erose: stigma purple, the lobes broad and short: capsule
sessile, ¾ in. long, linear-oblong, abruptly pointed, hirsute, the alternate
sides bicostate.—Contra Costa Co., near Walnut Creek, Martinez, etc.

4. **G. Arnottii**, (T. & G.), Walp. Nearly glabrous, *slightly glaucous*,
1—2 ft. high, *densely flowered at the leafy summit:* leaves mostly opposite,
except the floral, oblong-lanceolate, obscurely denticulate: calyx-tube
short: corolla deep purple: *stigma purple, the lobes oval:* capsule cylin-
drical-conic, bicostate on the sides.—Common in the Sacramento Valley.

5. **G. albescens**, Lindl. Rigid, pubescent, the branches very short,
crowded at the summit: leaves glabrous, glaucous, lanceolate, entire:
fl. sessile, densely crowded among the upper leaves on the short branchlets: calyx-tube funnelform, as long as the segments: petals obcordate,
½ in. long, pale purple, with a small darker spot in the centre: *stigma-
lobes narrow, greenish: capsule* oblong, 8-sulcate, *acuminate, villous:*
seeds roundish, scabrous.—Solano Co. and southward.

* * *Flowers in loose spikes or racemes; capsules mostly linear;
seeds in 1 row.*

+— *Racemes erect in bud; calyx-lobes distinct and reflexed in flower;
capsules sessile.*

6. **G. quadrivulnera** (Dougl.), Spach. Very slender, 1—2 ft. high,
puberulent: leaves linear or linear-lanceolate, entire or slightly denticulate: calyx-tube obconic, 2—3 lines long: petals purplish with a dark
spot at summit, 3—6 lines long: *stigma-lobes short, purple: capsules 5—10
lines long*, attenuate at apex, bicostate at the alternate angles——Common
toward the coast everywhere.

7. **G. viminea** (Dougl.), Spach. Glabrous, 1—3 ft. high: leaves linear-lanceolate, entire, narrowed at base, 1—2 in. long; puberulent: calyx-tube 2—4 lines long: petals purplish with a dark spot at summit, ¾—1¼ in. long: stamens short, subequal: *stigma-lobes linear-oblong,* purple: *capsules* 1—1½ *in. long,* pubescent, slightly bicostate on the sides.—From middle parts of the State northward.

8. **G. tenella** (Cav.), Wats. Puberulent, slender, erect, ½—1½ ft. high; leaves linear, acute or obtuse, mostly entire, more or less narrowed at base, ½—2 in. long: calyx-tube obconic, 1—3 lines long: petals 3—5 lines, deep purple: stigma-lobes purple: *capsule linear, attenuate at apex,* 8—14 lines long, *quadrangular,* the sides not costate but the midvein usually prominent.—Common.

+ + *Racemes nodding in the bud; calyx-lobes united and turned to one side under the open corolla; capsule pedicellate.*

9. **G. amœna** (Lehm.), Lilja. Slender, 1—2 ft. high, puberulent: leaves lanceolate or oblanceolate, entire or denticulate: *calyx-tube obconic,* 2—4 lines long: petals 8—15 lines long, white, pink or purple, with a dark purple spot near the base: filaments rather stout; anthers deep crimson, the vacant upper end white or yellowish: stigma-lobes linear: *capsule* 1—1½ *in. long, narrowed at each end;* pedicel 2—6 lines long.—Along the coast; the typical white-flowered form seldom seen.

10. **G. rubicunda,** Lindl. Near the preceding, but often 4 ft. high: *calyx-tube longer, funnelform:* petals purple, with an orange spot at base: anthers orange-red, the empty end bright yellow: *capsule sessile, scarcely attenuate at apex.*—Of more northerly general range than the preceding.

11. **G. hispidula,** Wats. Simple, a few inches high, often 1-flowered: *pubescence hispidulous:* leaves narrowly linear, 1—2 in. long: calyx-tube 2—3 lines: petals purple, ½—1 in. long: filaments slender: stigma-lobes linear: *capsules* ⅓—¾ in. long, *attenuate at apex, below abruptly contracted* to a short pedicel.—A little known species of the valleys of the Sacramento and San Joaquin.

12. **G. biloba** (Durand), Wats. Slender, 1—2 ft. high, sparingly branching, nearly glabrous: leaves linear to narrowly lanceolate, 1—2 in. long, obscurely denticulate, the lower on long and slender petioles: *petals* light-purple ½—¾ in. long, *cuneate-obovate, deeply 2-lobed:* capsule puberulent, ½—¾ in. long, attenuate at apex, narrowed at base into a short pedicel.—Plentiful in the Briones Hills of the Mt. Diablo Range in Contra Costa Co. not far from Martinez; otherwise only in the Sierra.

5. **CLARKIA,** *Pursh.* Erect sparingly branched annuals, with racemose or spicate purple flowers nodding in the bud. Calyx-tube more or less prolonged above the ovary, deciduous. Petals 4, unguiculate, often

lobed or cleft. Stamens normally 8, but those opposite the petals often sterile or rudimentary, or sometimes wanting; anthers oblong or linear, fixed by the base. Ovary 4-celled; style elongated; stigma with 4 broad spreading lobes. Capsule linear, attenuate above, coriaceous, straight or somewhat curved, 4-angled, 4-celled, 4-valved to the middle. Seeds angled or margined.

     * *Calyx-tube elongated and almost filiform; stamens 4 only.*

1. **C. concinna** (F. & M.), Greene. Simple, or with a few subcorymbose branches, 1—2 ft. high, glabrous or puberulent: leaves ovate, entire: calyx-tube almost filiform, 1 in. long: corolla regular; petals ½—¾ in. long, cuneate-obovate, 3-lobed, the *middle lobe broadest, little longer than the others:* filaments subulate; anthers recurved after dehiscence, somewhat villous: stigma subpeltate, the lobes short, rounded: capsule subcylindrical, in maturity obscurely quadrangular, acutish.— Coast Range at considerable elevations. May, June.

2. **C. grandiflora** (F. & M.), Greene. Near the preceding, but diffusely branching from the base: *corolla larger, irregular*, the 3 upper petals approximate, ascending, the lower one remote from these and declined, the middle lobe of each attenuate to a claw and far surpassing the others.—Very common in the Mt. Diablo Range, perhaps identical with the preceding.

3. **C. Breweri** (Gray), Greene. Glabrous, 1—2 ft. high: leaves lanceolate, entire, short-petioled: calyx-tube slender, more than 1 in. long, abruptly dilated at base: corolla irregular; petals round-obcordate, with a *linear-spatulate middle lobe* proceeding from the deep, or rather shallow sinus and *far exceeding the others:* filaments clavate: anthers densely white-villous along the sutures, erect after dehiscence as before: capsule sessile, 1 in. long, curving away from the stem: seeds large, tuberculate, conspicuously winged.—On Mt. Hamilton.

     * * *Calyx-tube obconic; petals never 3-lobed; stamens 8, all perfect.*

4. **C. elegans**, Dougl. Glabrous or puberulent, reddish and glaucous, erect, 1—6 ft. high, simple or somewhat branching, stout and rigid: leaves broadly ovate to linear, repand-dentate: petals entire, the *rhomboidal limb about equalling the linear claw:* filaments with a densely hairy scale on each side at base: capsule ½—¾ in. long, stout, sessile, 4-angled, somewhat curved, often hairy.—On open or half-shaded hillsides, very common.

5. **C. rhomboidea**, Dougl. Puberulent or glabrous, 1—3 ft. high, rather slender: leaves thin, entire, oblong-lanceolate to -ovate, 1—2 in. long: *petals with rhomboidal limb and short broad claw which is often broadly toothed:* filaments with hairy scales at base: capsules pedicellate, 8—12 lines long, 4-angled, glabrous, curved near the base.—Of wider range than the last; equally common.

6. BOISDUVALIA, *Spach.* Annuals, rigid and leafy, rather low (except the first species); the leaves alternate, sessile. Flowers small, purple, in leafy-bracted spikes. Calyx-tube funnelform above the ovary, deciduous; lobes not reflexed in flower. Petals 4, obovate-cuneiform, sessile, 2-lobed. Stamens 8, all perfect, unequal; filaments slender, naked at base; anthers oblong, fixed near the base. Ovary 4-celled, several-ovuled; stigma-lobes short, somewhat cuneate. Capsule membranaceous, ovate-oblong to linear, nearly terete, acute, dehiscent to the base. Seeds in 1 row in the cell.

1. **B. densiflora** (Lindl.), Wats. Stoutish, sparingly branching, 1—5 ft. high, *soft-pubescent* throughout: lower leaves lanceolate, acuminate, serrate-toothed; the floral broader, entire: flowers in rather dense terminal spikes: calyx 1½—3 lines long, half as long as the purple petals: capsules ovate-oblong, glabrous or villous, 2—4 lines long; cells 3—6-seeded, the partition separating from the valves and adhering to the placenta: seeds nearly a line long. Var. **imbricata,** Greene. Less canescent than the type, the whole plant larger and coarser; spikes thick and dense, the capsules concealed under the very broad acute closely imbricated bracts.—Abundant in low grounds. July—Oct.

2. **B. stricta** (Gray), Greene. Canescent *with a short stiff spreading pubescence;* plant slender, seldom 1 ft. high: leaves lanceolate or linear, narrow at base, entire or denticulate, the floral not differing from the others except as being smaller: *flowers in a loose simple spike,* minute: capsules linear-acuminate, 4—6 lines long; cells 6—8-seeded: seeds ½ line long or less, ovate.—Frequent from Santa Clara Co. northward.

3. **B. cleistogama,** Curran. *Pale and glaucescent,* glabrous or hispidulous; 4—10 in. high, rather slender: *leaves ovate-lanceolate,* ½—1½ in. long, *remotely serrate:* fl. small, rose-red, the earliest ones cleistogamous: capsule rather coriaceous: seeds numerous.—Common on the lower Sacramento plains; also in Sonoma Co. .

7. ISNARDIA, *Linn.* Herbs (ours creeping and aquatic or riparian) with entire opposite leaves, and axillary commonly apetalous 4-merous flowers. Calyx-tube prismatic, not produced beyond the ovary; lobes 4, persistent. Stamens as many or twice as many. Ovary broad at apex and usually flattened, or crowned with a conical style-base; stigma capitate, 4-grooved. Capsule 4-celled, dehiscent by lateral slits or terminal pores. Seeds very many, minute.

1. **I. palustris,** L. Glabrous; stems 4—10 in. long: leaves all opposite, oval or ovate, acute, ½—1 in. long, tapering to a short petiole: fl. sessile, 1 in each axil: petals rarely present, minute, reddish: capsule oblong, 2 lines long or less, somewhat 4-angled.—On muddy shores in the Sacramento and San Joaquin valleys.

8. JUSSIÆA, *Linn.* Habit of the preceding. Leaves attenuate. Flowers 5-merous; petals yellow. Calyx-tube not angular, elongated, not produced beyond the ovary; lobes conspicuous, persistent. Capsule in ours indehiscent. Seeds very numerous.

1. **J. diffusa,** Forsk. Perennial, the stout floating stems 1 ft. to 2 yards long: herbage altogether glabrous: leaves obovate to obovate-oblong and even lanceolate, obtuse or acute, 1—2½ in. long, on petioles of ½—1 in.; stipules gland-like or slightly scale-like: fl. 6—8 lines broad, deep yellow; the petals obtuse: fr. 1 in. long, spongy; the pedicel ½ in. or more.—Plentiful, forming extensive floating masses, covering the surface of sluggish waters of the lower Sacramento and San Joaquin.

ORDER XLII. **HALORAGEÆ.**

Plants herbaceous and mostly aquatic, with small inconspicuous usually apetalous flowers sessile in the axils of leaves or bracts. Calyx, in fertile flowers, adnate to the ovary, its limb short or absolute. Fruit indehiscent and nut-like, 1—4-celled, with a single seed suspended in each cell. Cotyledons small and short. Albumen copious.

1. **LIMNOPEUCE,** *Vaillant.* Erect stoutish but low perennial aquatics. Stem simple, short-jointed, with linear entire leaves in whorls of 8 or 12. Calyx-tube globular; the limb entire. Petals 0. Stamen 1; filament subulate. Ovary 1-celled; style becoming filiform and elongated, stigmatic throughout. Fruit oblong-ovoid, nut-like, 1-seeded.

1. **L. vulgaris,** Vaill. Stem ½—1 ft. high; herbage glabrous: leaves ½—1 in. long, acute: calyx ½ line long: style and stamen rather conspicuous: nutlet nearly 1 line long.—In shallow ponds and pools, and about springy places, but not often met with.

2. **MYRIOPHYLLUM,** *Matthiolus* (WATER-MILFOIL). Aquatic perennials. Leaves verticillate, the submersed ones pinnately divided into capillary or filiform segments; the emersed ones pectinate, or toothed, or entire. Flowers spicate or axillary, commonly unisexual; the staminate with a very short calyx-tube, and 2—4-lobed limb or none. Petals 2—4. Stamens 4—8. Calyx of pistillate fl. with a more or less deeply 4-grooved tube and 4 minute lobes or none. Styles 4, short, often plumose and recurved. Fruit quadrangular, when ripe splitting into 4 one-seeded carpels.

1. **M. spicatum,** L. Stems often many feet long, growing in deep waters, branching above: *flowers in emersed short-peduncled verticillate spikes* 2—3 in. long; bracts reduced and inconspicuous; submersed leaves in whorls of 4 or 5: petals 4, deciduous: stamens 8: carpels rounded on the back, with a deep wide groove between them.—Mountain Lake, San Francisco. July.

2. **M. hippuroides**, Nutt. Stems 4—8 in. long, growing in mud or shallow water, the *emersed branches erect, simple, leafy, flowering throughout*: submersed leaves in whorls of 4 and 6, with 6—8 pairs of capillary pinnæ; emersed ones often alternate, linear-lanceolate, serrate or dentate, or the uppermost entire; the lowest often pinnatifid: petals often pinkish and somewhat persistent: stamens 4: carpels carinate and somewhat roughened; deep grooves between them.—In Marin Co., also on the lower San Joaquin. June—Sept.

3. **CALLITRICHE**, *Columna*. Small and slender, growing in water or on moist shaded ground. Leaves opposite, linear, spatulate or obovate. Flowers solitary in the axils, subtended by a pair of falcate or lunate membranous bracts, mostly consisting of a single stamen and pistil. Filaments elongated: anthers reniform, the cells ultimately confluent. Styles 2, filiform, papillose. Fruit sessile or peduncled, 4-celled, more or less carinate or winged on the margins, 4-lobed, the lobes united in pairs, forming 2 discs with a groove between them, at maturity parting into 4 compressed carpels, each 1-seeded.

1. **C. marginata**, Torr. Usually terrestrial and very small; when aquatic the submersed leaves linear, 1-nerved, passing gradually into the emersed, which are oblanceolate or spatulate, 3-nerved: styles elongated, reflexed, deciduous: mature *fruit on slender pedicels, often buried in the mud, deeply emarginate at both ends*, broader than high, the margins of the thick carpels widely divergent and narrowly winged.— Low grounds, among growing grain, etc., from San Mateo and Alameda counties northward. June.

2. **C. palustris**, L. Usually aquatic, with linear retuse or bifid submersed leaves, and spatulate or obovate emersed ones, these rounded or truncate or retuse at apex, narrowed into a margined petiole, and profusely dotted with stellate scales: *fr. sessile, oblong, with a small apical notch*, narrow-winged above, deeply grooved between the lobes.— In sluggish or stagnant shallow pools. June.

ORDER XLIII. CERATOPHYLLÆ.

Represented by a single species of the genus

1. **CERATOPHYLLUM**, *Linn*. (HORNWORT). Aquatic herbs, with rigid verticillate leaves, these usually pinnatifid and the segments toothed. Flowers clustered in the leaf-axils, involucrate, unisexual. Involucre multifid. Calyx and corolla wanting. Stamens 14—20. Ovary ovate, 1-celled; style filiform, incurved. Fruit a small nutlet: the seed pendulous. Albumen 0; cotyledons 4, verticillate, 2 larger than the others; plumule conspicuous, compound.

1. **C. demersum**, L. Stem 1—2 ft. long, nearly glabrous; internodes short; leaves in whorls of 6 or 8; the linear segments acute, aculeate-

toothed: achene 2 lines long or more, elliptical, somewhat compressed, short-stipitate, with a short spine or tubercle on each side near the base, not margined; style equalling the achene.—At San Francisco, according to *Chamisso*.

Order XLIV. S A L'I C A R I Æ.

Herbs with entire leaves, and axillary or spicate mostly 5-merous purplish flowers. Calyx tubular, enclosing the ovary but free from it; the petals and definite stamens borne on the throat of it. Style 1. Capsule mostly 1-celled by the vanishing of the thin partitions. Seeds numerous, small, on a central placenta, exalbuminous.

1. **LYTHRUM,** *Linn.* Calyx cylindrical, 10—12-angled or -striate, 10—12-toothed; the teeth alternately long and erect and shorter and incurved. Petals 5 or 6, inserted on the throat of the calyx-tube alternately with the erect teeth. Stamens from the middle or the base of the calyx-tube, as many or twice as many as the petals. Style filiform; stigma capitate.

* *Petals minute, pale.*

1. **L. Hyssopifolia,** L. *Annual, simple or branching, erect, 4—10 in. high:* herbage pale, glabrous: lowest leaves opposite: fl. subsessile in the axils of the alternate leaves, very small, whitish or pale-purple.—Not rare in the Coast Range, from Humboldt Co. southward throughout the State; also, in a large form, in the interior, near Stockton, etc.

2. **L. adsurgens,** Greene. *Stoloniferous perennial, the 5-angled branches 1—3 ft. long, decumbent or assurgent;* herbage pallid, glabrous, slightly succulent: calyx 2½ lines long, 12-striate, the striæ at length widening below: petals pale purple.—Plant very near the preceding in all points except its great size and perennial stoloniferous habit. Common in wet places near the Bay, at West Berkeley, etc.

* * *Petals larger, bright red-purple.*

3. **L. Californicum,** Torr. & Gray. Stoloniferous perennial, the roots spreading near the surface of the ground: stem erect, 2—3 ft. high, simple below, paniculately branching above: lower leaves lanceolate; upper and floral linear: striæ of the calyx not wing-margined; teeth very short.—In marshy land; also along streams, and in springy places.

2. **AMMANNIA,** *Houston.* Glabrous opposite-leaved annuals; the flowers 2 or more in each axil. Calyx subglobose, more or less distinctly 4-angled, 4-toothed, usually with horn-shaped appendages alternating with the teeth. Petals 4, purplish, small and deciduous, sometimes wanting. Stamens 4—8. Capsule globular.

1. **A. coccinea,** Rottb. Erect, stoutish, ½—2 ft. high, with few spreading branches: stem 4-angled: *leaves* linear-lanceolate, 1—3 in.

long, *with a broad auricled base*: fl. 1—5 in each axil, mostly sessile: calyx 1½ lines long, in fruit becoming 2 lines broad: petals small, bright purple: *capsule bursting irregularly.*—Common along the lower Sacramento and San Joaquin.

2 **A. humilis,** Michx. Smaller; *leaves* linear-oblanceolate, *not auricled* at base but tapering, sometimes short-petiolate: fl. 1—3 in each axil: calyx globular, the accessory teeth as long as the lobes or shorter: petals small, purplish: *capsule globular, dehiscent septicidally.*—Habitat of the preceding, but less frequent.

ORDER XLV. **L O A S E Æ.**

Rigid herbs clothed with stinging or jointed and barbed hairs. Bark of stems often white and deciduous. Leaves without stipules. Calyx-tube adnate to the 1-celled ovary. Stamens often very numerous, and some of the outer petaloid. We have but the following genus.

1. **MENTZELIA,** *Plumier.* Annuals and biennials. Leaves alternate, mostly coarsely toothed or pinnatifid. Flowers solitary or cymose, large or very small, yellow. Calyx-tube cylindrical, ovoid or turbinate; the 5-lobed limb persistent. Petals 5 or 10. Stamens ∞, inserted on the throat of the calyx; filaments free, or in clusters opposite the petals, filiform, or the outer more or less dilated and without anthers. Ovary truncate at summit, 1-celled; ovules horizontal or pendulous, in 1 or 2 rows on the 3 parietal placentæ. Capsule mostly cylindrical, opening irregularly at the summit. Seeds angled or compressed.

    * *Annuals, small-flowered; petals 5 only; stamens rather few.*

1. **M. affinis,** Greene. Stoutish, often 2 ft. high, simple and leafy below, widely branching above; leaves lanceolate, deeply sinuate-pinnatifid: fl. scattered, ½ in. broad; calyx-lobes attenuate-subulate, 2—3 lines long: capsule 1 in. long, almost linear, hispid with short stiff hairs which have a pustulate base: seeds prismatic, with grooved angles.—Plains of the San Joaquin, and far southward.

2. **M. micrantha** (H. & A.), Torr & Gray. More slender, the *inflorescence in age compactly dichotomous:* leaves ovate, from entire to serrate-toothed: *fl. very small;* 5 of the filaments petaloid and emarginate: capsules 3—6 lines long, few-seeded: seed prismatic, twice as long as broad, the base often oblique, angles with very shallow groove, sides faintly tuberculate.—Santa Clara Co., in the mountains. May—July.

3. **M. Lindleyi,** Torr. & Gray. Slender, simple or bushy-branched, 1—3 ft. high; leaves ovate to narrowly lanceolate, 2—3 in. long, from pectinate-pinnatifid to coarsely toothed: fl. axillary and terminal: calyx-lobes rather broadly lanceolate, ½—¾ in. long: fl. vespertine: petals obovate, abruptly acuminate or cuspidate, 1 in. long or more, golden

yellow: filaments many, very slender, unequal, the longest almost equalling the petals; anthers minute, oval: capsule 1 in. long or more: seeds angular, tuberculate.—Common in the Mt. Diablo Range.

* * *Flowers large; petals about 10; stamens very numerous.*

4. **M. lævicaulis** (Dougl.), Torr. & Gray. Biennial, stout, branched above, 2—3 ft. high; stem white, scarcely roughened: leaves lanceolate, sinuate-toothed, 2—8 in. long: fl. sessile on short branches, 3—4 in. broad, light yellow, diurnal: calyx-tube naked; segments 1 in. long or more: petals oblanceolate, acute, almost equalled by the numerous stamens: capsule 1¼ in. long, 3—4 lines in diameter: seeds minutely tuberculate, 1½ lines broad.—In the mountain districts at low altitudes, and on the plains near the foothills. July—October.

ORDER XLVI. ARISTOLOCHIACEÆ.

Shrubs or perennial herbs, with alternate entire mostly cordate or reniform exstipulate leaves, and solitary apetalous perfect flowers. Perianth lurid-purple or greenish, with a valvate regularly or irregularly 3-lobed limb: the tube more or less adnate to a 6-celled ovary, which becomes a 6-valved capsule, or a berry. Stamens 6—12, on the ovary, more or less adnate to the style; anthers extrorse. Styles usually 6, united at the base. Seeds in 1 or 2 rows in each cell.

1. **ARISTOLOCHIA,** *Diosc.* (PIPE-VINE). Perianth very irregular; tube inflated above the ovary, deciduous from it. Anthers 6, sessile and adnate to the short simple style. Stigma 3—6-lobed or -angled. Fruit capsular, 6-angled, 6-valved, septicidally dehiscent.

1. **A. Californica,** Torr. A deciduous shrubby climber, 6—10 ft. high, pubescent with short silky hairs: leaves ovate-cordate, obtuse or acutish, 2—4 in. long, short-petioled: peduncles slender, 1—2 in. long, with a small cordate or obovate bract in the middle: calyx-tube broadly saccate and doubled upon itself, 1—1½ in. long from the base to the top of the curvature, ½ in. broad, little contracted at the throat; limb bilabiate, the upper lip of 2 broad obtuse lobes, with a disk-like thickening on the inner side: anthers contiguous in pairs under each of the 3 broad stigma-lobes: ovary linear-clavate, pubescent: capsule spongy-coriaceous, obovate, attenuate to a slender base, 1½ in. long, 6-winged; spongy.—Banks of streams; not very common. April.

2. **ASARUM,** *Diosc.* (WILD GINGER). Perianth regular, campanulate; limb 3-cleft, persistent, the tips of the segments infolded in the bud. Stamens 12, nearly free from the styles, the alternate ones shorter; connective continued beyond the anthers, pointed. Styles 6, more or less united. Capsule globose, fleshy, irregularly dehiscent. Seeds large, thick, in 2 rows in each cell.

1. **A. caudatum,** Lindl. Acaulescent, with creeping aromatic root-stocks, the branches of these bearing 2 long-petioled leaves and a pedunculate flower: leaves cordate-reniform, somewhat cucullate, acutish or obtuse, 2—4 in. long, sparingly pubescent above: peduncles slender, 6—15 lines long: ovary 4 lines broad: calyx-lobes oblong, with long-attenuate apex, 1—2½ in. long: filaments stout, the free apex of the connective shorter than the anther: styles united, equalling the stamens: seeds 1½ lines long, ovate.—Woods of the Coast Range. April.

ORDER XLVII. CUCURBITACEÆ.

Herbs, tendril-bearing, trailing or climbing, commonly scabrous and succulent. Flowers axillary to the alternate leaves, unisexual. Calyx-tube coherent with the ovary; limb of 5 lobes or teeth. Corolla with petals more or less united. Stamens 5, more or less united; anthers 2-celled, or one of them 1-celled. Ovary 2—3-celled; stigma 3—5-lobed. Fruit fleshy. Seeds large, usually compressed, exalbuminous; cotyledons fleshy.

1. **CUCURBITA,** *Pliny* (SQUASH. PUMPKIN). Flowers solitary. Caylx-tube campanulate, 5-lobed. Corolla campanulate, 5-cleft to the middle or lower, the lobes recurved. Sterile fl. with stamens at the base of the corolla; filaments distinct; anthers more or less united, flexuous. Fertile fl. with 3 rudimentary stamens; ovary oblong, with 3 placentæ and many horizontal ovules; style short; stigmas 3, 2-lobed Fruit indehiscent; in our species with a hard shell-like rind.

1. **C. fœtidissima,** HBK. Root large, fusiform: stems long, trailing: leaves scabrous, triangular-cordate, acute, the slight lobes rounded or angled, mucronate-denticulate; petiole shorter than the blade: tendrils 3—5-cleft: fl. 3—4 in. long, yellow; corolla-lobes obtuse, mucronate: calyx-tube ½ in. long, the linear lobes as long: fr. globose, 2—3 in. thick, smooth, yellow, on a peduncle 1—2 in. long; shell filled with a fibrous bitter pulp: seed thin, obovate, 4—5 lines long, obtusely margined.—From San Joaquin Co. southward, on low plains.

2. **MICRAMPELIS,** *Raf.* (BIG-ROOT). Membranous-leaved trailing or climbing herbs, with simple tendrils, and small white or greenish flowers, the fertile solitary, the sterile racemose or panicled from the same axil. Calyx-tube broadly campanulate; teeth small or obsolete. Corolla rotate or campanulate, deeply 5—7-lobed, with elongated segments. Sterile fl. with stamens at base; filaments short, united: anthers distinct or more or less coherent. Fertile fl. pedicellate, with or without abortive stamens. Ovary globose or oblong, bristly, 2—4-celled; cells 1—4-ovuled: style short; stigma 2—3-parted or -lobed. Fruit prickly, fibrous and watery-pulpy, dehiscing irregularly near the apex. Seeds

large, ovoid or more rounded, more or less compressed, encircled by a mere marginal line; hilum linear, acute; cotyledons thick, remaining within the integuments after germination. Our species perennials with very large fleshy fusiform roots.

* *Leaves rather longer than broad; corollas rotate.*

1. **M. fabacea** (Naud.), Greene. Glabrous, or the younger parts with scattered short curved hairs: stem 10—30 ft. long: leaves 2—6 in. broad, of round-ovate general outline, more or less deeply and angularly 5—7-lobed; lobes abruptly acute, mucronate, the sinuses obtuse: sterile fl. 15—30 in slender racemes, the pedicels 1—2 lines long; corolla 3—4 lines broad, of a dull or greenish white: fertile fl. 5—6 lines broad, without abortive stamens: ovary globose, densely echinate, 2-celled, 4-ovuled: *fr. globose, 2 in. long, densely covered with stout pungent spines* ½—1 in. long: *seeds 4*, obovoid, 10 lines long, 6 lines broad. Var. **agrestis**, Greene. Stems 2—4 ft. long, prostrate or merely trailing; leaves and *fruits much smaller*, the latter *armed only sparsely with very short spines.*—The type is common all along the seaboard, growing in thickets and climbing high over shrubs and small trees. The variety is a weed in grain fields along the eastern foothills of Mt. Diablo. Fl. Jan.—April.

* * *Leaves broader than long; corollas campanulate.*

2. **M. Marah** (Wats.), Greene. Stems 3—30 ft. long: leaves reniform or round-cordate, 3—6 in. broad, pedately lobed: racemes of sterile flowers ½—1 ft. long; corolla ½ in. long or more, campanulate, clear white: fertile fl. with abortive stamens; pedicel slender, 2—6 lines long: ovary oblong-ovate, acuminate, more or less clothed with soft spines, 2—3-celled; ovules 1—4 in each cell, ascending or horizontal, attached to the outer side of the cell: *fr. ovate-oblong, 4 in. long, attenuate at each end*, more or less muricate with short weak spines: *seeds horizontally placed*, somewhat elliptical or nearly orbicular, *compressed*, 1 in; long and about half as thick.—Shady banks, or open northward slopes, trailing or high-climbing; common about Mt. Tamalpais, also in Alameda and Contra Costa counties.

3. **M. Watsonii** (Cogn.), Greene. Slender, not succulent, glaucous: stems 6—8 ft. long: leaves broad-reniform, 5—7-lobed, 2—4 in. broad, the lobes broader above, sinuate-toothed or -lobed: sterile racemes few-flowered; fl. small, white: fertile fl. without abortive stamens, on slender pedicels 1—2 in. long; ovary glabrous or muricate: *fr. nearly globose, 1 in. thick or more*, naked or with a few weak spines near the base, *2-celled, 2—4-seeded: seed nearly globose*, ½ in. thick, attached to the outer side of the cell, marginless.—Vaca Mts March—May.

Order XLVIII. ARALIACEÆ.

Herbs, shrubs or trees, with mostly hollow stems, and alternate lobed or compound leaves. Flowers small, in simple panicled or racemosely arranged umbels. Calyx joined to the ovary, entire or toothed. Petals 5, deciduous. Stamens 5, inserted around the border of the calyx outside of an epigynous disk. Ovary more than 2-celled; styles as many as the cells, sometimes connate. Fruit berry-like. Seeds pendulous; embryo minute; albumen fleshy.

1. **ARALIA,** *Vaillant* (SPIKENARD). Our species a very coarse perennial herb, with ternately compound leaves and large serrate leaflets. Calyx 5-toothed or entire. Disk depressed or 0. Fruit laterally compressed, becoming 3—5-angled, fleshy externally; endocarp chartaceous.

1. **A. Californica,** Wats. Unarmed, 6—10 ft. high: leaflets cordate-ovate, 4—10 in. long, abruptly acuminate, simply or doubly serrate: umbels in loose terminal and axillary compound or simple racemose panicles 1—2 ft. long, each umbel subtended by several linear bractlets: fl. 2 lines long; disk and style-base obsolete; styles united for half their length: fr. 2 lines long.--In shaded and moist ravines.

2. **HEDERA,** *Pliny* (IVY). Shrubby, climbing by aerial roots. Leaves coriaceous, evergreen, simple, lobed. Flowers in a terminal panicle of umbels. Calyx 5-toothed. Styles united into a single very short one. Berry smooth, black; seeds 2—5.

1. **H. HELIX,** Gerarde (1633). Leaves ovate, angularly 3—5-lobed, those of the sterile and young shoots more deeply so than those of the flowering branches; these bushy, erect, projecting a foot or more from the climbing main stem: umbels globose: fl. yellowish-green.--The *English Ivy,* common on trees in parks, and on buildings, and well adapted to our climate, fruits freely here, and will often be met with wild.

Order XLIX. UMBELLIFERÆ.

Herbs with mostly hollow, striate, angled or fluted stems, and usually compound leaves which are prevailingly alternate; the petiole dilated or even sheathing at base. Flowers small, in simple or compound umbels (sometimes sessile and therefore capitate). Calyx almost wholly adnate to the 2-celled ovary. Petals 5, mostly valvate in bud, usually inflexed at apex in flower. Stamens 5, alternate with the petals; anthers ovate, subdidymous. Styles 2, simple, more or less dilated at base into a *stylopodium*. Fruit of 2 closely approximated and often ribbed, sometimes winged, always 1-seeded carpels; the intervals between the ribs usually occupied by one or more oil-tubes or *vittæ*. The face by which the two carpels meet or partly cohere is called the *commissure*. A slender prolongation of the axis between these faces is called a *carpophore*, which, in maturity, is apt to split into 2 branches, with a carpel suspended from each.

* *Umbels simple, or imperfectly or irregularly compound.*

Low, flaccid; leaves simple;
    Stems creeping; peduncles erect, scapiform...............HYDROCOTYLE   1
    Stems trailing; peduncles short, not scape-like...........BOWLESIA   2
Rigid, branching; leaves elongated, firm, setaceously
    toothed or lobed...................................ERYNGIUM   3
Umbels irregularly compound; fruit without
    ribs, subglobose, prickly or tuberculate...............SANICULA   4

* * *Umbels regularly compound; leaves compound, often finely dissected.*

Fruit somewhat flattened laterally, not broadly winged;
    Oblong or rounded; oil-tubes 2 or 3 in the intervals........VELÆA   5
    Broadly ovate; ribs obtuse; oil-tubes none................CONIUM   6
    Ovate or oblong; ribs corky; oil-tubes 2 or more.........SIUM   7
    Broad-ovate or rounded; ribs broad; oil-tubes solitary....CICUTA   8
    Styles elongated; oil-tube 1 in each interval; seed
        flat on the face....................................ŒNANTHE   9
Styles not elongated; oil-tube 1 to the interval: seed terete.......APIUM   10
Fruit very small; ribs obscure: oil-tubes solitary.........APIASTRUM   11
Fruit ovate or oblong; ribs filiform; oil-tubes 1 to
    each interval.......................................CARUM   12
Carpels 5-angled; ribs slender; oil-tubes
    several to the interval.............................PIMPINELLA   13
Calyx with turgid border and no teeth....................FŒNICULUM   14
Ribs of carpel beset with hooked prickles;
    Seed nearly flat on the face.......................DAUCUS   23
    Seed deeply channeled on the face..................CAUCALIS   24
                  attenuate at base.........MYRRHIS   20
Carpels much elongated { short-beaked.................CHÆROPHYLLUM 21
                long-beaked ....SCANDIX   22
Fruit not compressed; ribs corky........................ŒNANTHE   9
Fruit slightly compressed dorsally; some of the
    ribs narrowly winged...............................SELINUM   15
Fruit strongly compressed dorsally and winged;
    Lateral wings broad; dorsal less prominent.............ANGELICA   16
    Lateral wings thick, corky: dorsal obscure...........LEPTOTÆNIA   17
    Lateral wings thin, coherent until maturity;
        Oil-tubes running the whole length of the carpel......PEUCEDANUM   18
        Oil-tubes obclavate, running from the apex to
            below the middle of the carpel....................SPHONDYLIUM   19

1. HYDROCOTYLE, *Tourn.* (MARSH PENNYWORT). Low glabrous herbs, growing in or near water, with creeping stems. Leaves rounded, toothed or lobed, sometimes peltate; stipules scale-like. Flowers inconspicuous, in simple umbels, or in whorls one above another, on a scapiform erect peduncle. Calyx-teeth obsolete. Petals entire, acute. Fruit flattened laterally, suborbicular, acutely margined, and with 2 or more less prominent ribs or nerves on each side; oil-tubes 0; carpels coherent.

1. **H. prolifera,** Kell. Herbage light green and flaccid: *leaves about 1 in. broad, peltate,* emarginate at base, *simply crenate,* on petioles 1—4 in. long: peduncles equalling or exceeding the leaves: fl. in 1—4 whorls, each 4—12-flowered, with many bractlets; pedicels 1—6 lines long: fr. 1 line wide, emarginate at base; ribs 2 on each side, prominent.—Said to occur near San Francisco; common in the Suisun marshes. June—Aug.

2. **H. ranunculoides,** L. Herbage dark green, fleshy: *leaves* 1—2 in. broad, *round reniform, 3—7-cleft,* the lobes crenate; petioles 2—10 in. long; peduncles much shorter (½—3 in.), reflexed in fruit: fl. 5—10 in a capitate umbel: fr. very shortly pedicellate, 1—1½ lines broad, with thickened but scarcely angled margins, rather obscurely nerved on each side, longer than the pedicels.—In shallow ponds, margin of lakes, etc., along the seaboard.

2. BOWLESIA, *Ruiz & Pavon.* Slender very flaccid herbs, with sparse stellate pubescence, and opposite simple leaves with scarious lacerate stipules. Flowers minute, white, in simple few-flowered umbels on axillary peduncles. Calyx-teeth rather prominent. Petals elliptical, obtusish. Fruit broadly ovate, with narrow commissure, turgid, becoming depressed on the back, without ribs or oil-tubes.

1. **B. lobata,** R. & P. Annual, the slender stems more or less dichotomous, 2 in. to 1 ft. long: leaves round-reniform or cordate, ½—1½ in. broad, shorter than the slender petioles, deeply 5-lobed; lobes acutish, entire or few-toothed: umbels short-peduncled, 1—4-flowered: fr. 1 line long, sessile or nearly so, pubescent, the inflated calyx not adherent to the carpels.—Among rocks, under trees, etc., on hillsides. April, May.

3. ERYNGIUM, *Nicander* (BUTTON SNAKEROOT). Perennials with rigid coriaceous spinosely toothed or divided leaves, and white or blue flowers sessile in dense heads which are encircled by a series of bracts forming an involucre; each flower also subtended by a rigid bract. Calyx-teeth manifest, rigid, persistent. Fruit ovoid or obovoid, scarcely compressed, covered with hyaline scales or vesicles; ribs obsolete; oil-tubes 0; carpels and seeds semiterete.

1. **E. armatum,** C. & R. Diffusely branching, 1 ft. high or more: radical leaves oblanceolate, serrately or spinosely dentate or incised, attenuate to a margined petiole; cauline narrower, sessile: heads peduncled, globose, ½ in. thick; *bracts of involucre* triangular-lanceolate, entire, thick-margined, 1 in. long and *much exceeding the head; bractlets similar and as prominent:* fr. with lanceolate-acuminate calyx-lobes longer than the styles.—Common in low ground.

2. **E. Vaseyi,** C. & R. Smaller, branching above: leaves oblanceolate, irregularly spinose-serrate, attenuate at base: *involucral bracts*

narrow, rigid, spinescent at tip and spinose-toothed, 1 in. long or less; bractlets similar: fr. with lanceolate acuminate-cuspidate calyx-lobes exceeding the short styles.—With the last.

3. **E. petiolatum,** Hook. Erect, 1—5 ft. high, branching above: *radical leaves oblanceolate,* irregularly spinose-serrate, *narrowed to an elongated fistulous petiole,* or the very lowest reduced to a long terete petiole; cauline mostly sessile: heads peduncled, globose, ½ in. high; involucral bracts linear-lanceolate, spinosely tipped and toothed, often 1 in. long; *bractlets* lanceolate, cuspidate-tipped, little exceeding the flowers, *scarious-winged below:* fr. with calyx-lobes like the bractlets but smaller, shorter than the long styles.—Perhaps not in our district.

4. **E. articulatum,** Hook. More or less branching, erect, decumbent or rarely prostrate: radical and lower leaves consisting of a long articulated petiole with or without a small lanceolate entire or laciniate blade; cauline sessile: bracts of involucre ½ in. long, exceeding the heads, linear, cuspidate, spinosely toothed; *bractlets tricuspidate,* little exceeding the flowers, *the central cusp largest:* calyx-lobes lanceolate, cuspidate, little exceeding the styles. Var. **microcephalum,** C. & R. Very small and slender: bracts ovate-acuminate, little surpassing the heads, these only 2—3 lines long; calyx-lobes short-mucronate.—In swamps and wet meadows.

5. **E. Harknessii,** Curran. Slender, not rigid, dichotomously branching, 3—4 ft. high: leaves much as in the last, but *blade of the lowest with perfectly entire and unarmed margin;* cauline petiolate, sparingly soft-spinulose on the margin: heads round-ovate, ¾ in. high, blue: *bracts of the involucre* longer than the head but *deflexed:* calyx-segments subulate, pungently mucronate, equalling the long styles.—In the Suisun Marsh. August—October.

4. **SANICULA,** *Brunfels* (SANICLE). Glabrous perennials (n. 1 biennial), with chiefly radical leaves, these mostly palmately divided and sometimes subdivided. Flowers unisexual, in irregularly compound few-rayed umbels; these involucrate with sessile leaf-like usually toothed bracts; the bracts of the involucels usually small and entire. Calyx-teeth persistent. Fruit subglobose or obovoid, densely uncinate-prickly or tuberculate; ribs obsolete; oil-tubes many.

 * Mature fruit pedicelled; leaves palmately lobed or divided.*

1. **S. Menziesii,** Hook. & Arn. Biennial: stem solitary, erect, branching loosely above, 2—5 ft. high: *leaves* 2—3 in. broad, *of rounded outline,* but with deep broad lobes and cordate base, the shining *surface delicately rugose;* the 3—5 lobes sharply toothed, the teeth setaceously tipped; involucre small, of 2 or 3 narrow leaflets; the involucels of 6—8 lanceo-

late entire bracts a line long: fr. obovate, a line long or more, *covered with hooked prickles.*—Abundant in open woods, and along streams in shade of thickets. May, June.

2. **S. arctopoides,** Hook. & Arn. The whole *herbage of a greenish yellow*, and with an offensive odor: main stem simple, very short; the many scape-like *flowering branches at first depressed*, later becoming elongated and divergent, 3—6 in. long, each bearing an umbel of 1—3 elongated rays: leaves deeply 3-parted, the lanceolate segments once or twice laciniately cleft: involucre of 1 or 2 leaflets; heads large, ½ in. broad, encircled by 8 or 10 oblanceolate mostly entire bracts which are yellow and resemble the rays of a composite: fr. 1½ lines long, *naked at base, strongly armed above.*—Bleak hills near the sea, at San Francisco, etc. Feb.—April.

* * *Mature fruit sessile; leaves palmately divided (except in n. 4).*

3. **S. nudicaulis,** Hook. & Arn. Stems several, slender, erect, 1 ft. high or more: leaves long petioled, of cordate outline, 3-parted; *divisions laciniately once or twice pinnatifid,* the segments with widely spreading acute often spinosely pointed teeth: fl. yellow, in many small heads disposed in compound umbels terminating sparingly leafy branches: fr. naked at base, uncinate-bristly above.—Wooded hills, along borders of thickets, etc., towards the sea. March—May.

4. **S. maritima,** Kell. Stoutish, 1 ft. high, rather fleshy: *radical leaves long-petioled, the lowest oblong-cordate, not lobed,* but crenate-dentate; some of the later more or less deeply 3-lobed, 2—4 in. long: involucre of large leaf-like lobed or parted bracts: umbel of about 3 elongated rays: fl. yellow, the sterile ones short-pedicellate: *fr. nearly naked below,* prickly above, 2 lines long.—In lowlands adjacent to salt marshes near Alameda, San Francisco, etc. March—May.

* * * *Fruit sessile; leaves pinnately divided and subdivided.*

5. **S. bipinnatifida,** Dougl. Stoutish, slightly fleshy, 1—2 ft. high, *herbage of a peculiarly dark green:* leaves mostly radical, but an opposite pair on the stem near the base, with 1—3 above these, all pinnately 3—7-parted, the divisions incisely toothed or lobed, decurrent on the toothed rachis, the teeth acutely or somewhat setaceously pointed: umbel of 3 or 4 greatly elongated rays: *fl. very dark purplish red:* fr. 1½ lines long, prickly.—Very common on hillsides and open grounds generally. March—May.

6. **S. bipinnata,** Hook. & Arn. Erect and rather slender, from a somewhat fusiform-tuberous and perhaps only biennial root, 1—2 ft. high: *leaves not fleshy, the segments* or leaflets remote, *not decumbent,* narrowly obovate, cuneate, mucronate-dentate: umbel compound: fl. yellow: fr. naked at base, echinate above.—Foothills of the inner Coast ranges. Feb.—April.

7. **S. saxatilis,** Greene. *Stems many, depressed,* 1 ft. long, *from a fleshy napiform root:* leaves ternately pinnate, the ultimate segments broad, coarsely toothed: branches repeatedly dichotomous, with pedicellate heads in all the forks: small involucels of very unequal foliaceous entire or toothed bractlets: fr. strongly tuberculate, the *tubercles of the upper part ending in a broadly subulate incurved point.*—Summit of Mt. Diablo, growing among loose rocks.

8. **S. tuberosa,** Torr. Very slender, the solitary erect freely branching stem 6—18 in. high, from a *small roundish* not deeply seated *tuberous root: leaves small, finely twice or thrice pinnate,* the ultimate segments small: umbels 1—4-rayed, small: fl. yellow, the sterile ones long pedicelled: *fr. broader than long, tuberculate.*—Rocky hills, in sterile clayey soil. March—May.

5. **VELÆA,** *De Candolle.* Glabrous or pubescent. Roots thick, elongated, yellow, fragrant. Leaves mostly radical, pinnately or ternately compound. Involucre sometimes wanting. Involucels conspicuous. Flowers yellow. Fruit somewhat flattened laterally, with prominent equal filiform ribs, and thin pericarp. Oil-tubes conspicuous, 3—6 in the intervals, 4—10 on the commissural side.

1. **V. Hartwegi** (Gray), C. & R. Subacaulescent, light green, the petioles and *veins somewhat scabrous:* leaves biternate and quinate; leaflets obovate or oval-oblong, 1—2 in. long, mostly confluent, coarsely and deeply mucronate-serrate: peduncles 1—2 ft. high; umbel 16—20-rayed, usually without involucre, but the umbellets subtended by *linear-oblong reflexed bractlets;* rays 2½—4 in. long; pedicels short: fr. nearly orbicular, smooth, 3—4 lines long, 2½—3 lines broad, sharply ribbed. Near San Francisco.

2 **V. Kelloggii** (Gray), C. & R. More slender than the last, mostly puberulent: leaves triternate; leaflets ovate, ½—¾ in. long; usually 3-lobed: umbel 8—16-rayed, mostly without involucre, the *involucels of small linear bractlets*; rays 1—3 in. long: fr. 1—2 lines long, nearly as broad, retuse at base, the ribs filiform.—Hills of the Coast Range, in wooded or open ground.

6. **CONIUM,** *Linn.* (POISON HEMLOCK). Tall glabrous biennial, with large ternately-dissected thin leaves, and compound umbels of small white flowers terminating the paniculate branches. Calyx-teeth obsolete. Fruit broadly ovate, laterally compressed; carpels with 5 prominent obtuse often undulate or crenulate ribs, and no oil-tubes.

1. C. MACULATUM, L. Root fusiform: stem stout, fistulous, 3—7 ft. high, glaucescent, spotted with purple: leaves a foot long or more, two-thirds as broad: segments ½ in. long, pinnatifid, the lobes acute: umbels 12—20-rayed: rays 1—1½ in. long: fr. 1½ lines long, shorter than the pedicels.—Waste grounds, in shady places.

7. SIUM, *Diosc.* (WATER PARSNIP). Glabrous perennial aquatics, with angled stems, pinnate leaves with leaflets pinnatifid or serrate, and white flowers; the involucres and involucels of several bracts. Calyx-teeth minute. Fruit oblong, ovate or nearly globose; ribs prominent or obscure; oil-tubes few or many in the intervals.

* *Fruit with corky ribs; oil-tubes between them.*

1. **S. heterophyllum,** Greene. Stem stoutish and brittle, strongly angular and somewhat flexuous, 3 ft. high, from a cluster of fleshy fibrous roots, these thickened below the middle: lowest leaves simple, 2—10 in. long, rhombic-lanceolate, serrate or laciniate, on a stout fistulous petiole which is still longer and usually submerged; the later radical 3-lobed or -parted, thus passing to the cauline which are pinnate, mostly with only 2 or 3 pairs of leaflets, these broadly lanceolate, acute, serrate: bracts of involucre broadly lanceolate, acute at each end: fr. 1½ lines long, broadly ovoid; oil-tubes broad, solitary between the ribs, 2 on the face: seed angular.—Common in brackish swamps, under the influence of tide-water, at Suisun, Stockton, etc.

* * *Fruit with angled corky covering; oil-tubes beneath this.*

2. **S. erectum,** Huds. Stems angular, 1—3 ft. high, from a stolon-iferous crown, usually erect, corymbosely branching above: leaflets about 6 pairs, ovate oblong to linear, 1½—2 in. long, often laciniate at base, the upper ones usually more or less deeply incised: peduncles 1—2 in. long: rays 1 in. or less: involucre and involucels of 6—8 linear entire lanceolate bracts: fr. ⅔ line long, less compressed than in the above: oil-tubes small in twos and threes, concealed beneath the corky covering (confluent ribs).—San Mateo Co.

8. CICUTA, *Besler.* (WATER HEMLOCK). Glabrous tall branching perennials of marshes and stream banks. Rootstocks short and erect, or horizontal and rooting from beneath. Leaves pinnately or ternately compound. Umbels of white flowers many-rayed; involucre small or 0; involucels of several small bractlets. Calyx-teeth small, acute. Stylo-podium depressed. Fruit broadly ovate or rounded, slightly compressed laterally, but the commissure narrow; ribs broad, obtuse, corky; oil-tubes solitary in the intervals. Seed subterete.

1. **C. Bolanderi,** Wats. *Roots* numerous, very coarse, 4—7 in. long, *whorled around the base of a short-conical strictly erect axis:* stem stout, erect, 4—9 ft. high, purplish below and very glaucous, paniculate from below the middle: radical leaves on petioles 2 ft. long or more, the blade twice or thrice pinnate: leaflets narrowly lanceolate-acuminate, 2—4 in. long, closely and sharply serrate, the setaceous tips of the teeth some-what spreading.—Marshes about Suisun Bay; also in similar situations (always within reach of tide-water) near Napa.

2. **C. Californica**, Gray. *Rootstock horizontal, freely branching*, the branches ¼—1 ft. long, the *older portion slender* (⅙ in. thick or more) *with long internodes. upper end abruptly clavate-enlarged and short-jointed:* stem erect, 3—6 ft. high: lowest leaves bipinnate, the upper simply pinnate; leaflets ovate-lanceolate: involucre nearly obsolete: seed sometimes with 2 oil-tubes iu the intervals.—In eddies and along the margins of swift-flowing mountain streams of the Coast Range only, from near Santa Cruz and Mt. Hamilton to the Oakland Hills.

9. **ŒNANTHE,** *Diosc.* Aquatic perennials, with glabrous decompound leaves and involucrate umbels. Calyx-teeth prominent, acute. Stylopodium short-conical: styles elongated in age. Fruit oblong, not compressed, with broad commissure, rounded corky ribs, and oil-tubes solitary in the intervals. Seed compressed dorsally, flat on the face.

1. **Œ. Californica** (H. & A.), Wats. Rootstocks erect or ascending, 1—2 in. long, ¾ in. thick, solid: stem solitary, decumbent or procumbent, rooting at the lower joints, erect above and with one or more umbelliferous branches: leaves ternate and bipinnate (or the upper ones simply pinnate), the pinnæ nearly sessile: leaflets approximate, ovate, acutish, toothed, at base often lobed, ½—1 in. long: fr. 1½ lines long, oblong, obtuse at each end, tipped with the long spreading styles; ribs and commissure corky: oil-tubes at the angles.—Very common, forming dense masses covering shallow pools, back of the salt marshes and among the hills. April—Nov.

10. **APIUM,** *Brunfels.* Glabrous biennial, with pinnately or ternately compound leaves, and nearly naked umbels of small whitish flowers. Calyx-teeth obsolete. Stylopodium depressed or 0. Fruit ovate or broader; the carpels straight, obtusely ribbed; oil-tubes solitary in the intervals. Seed nearly terete.

1. **A. GRAVEOLENS, L.** (CELERY). Biennial, with fibrous roots: stem erect, 2—3 ft. high, branching freely: leaves pinnate; leaflets in 1 or 2 pairs, cuneate-obovate or rhomboidal, sparingly toothed, 1—2 in. long, those of the uppermost leaves 3 only, oblanceolate, nearly entire: umbels sessile or short-peduncled; rays 6—12, slender, 1 in. long: fr. ⅖ line long.—Common in marshy grounds throughout the Bay region.

11. **APIASTRUM,** *Nutt.* A small and rather delicate branching annual, with leaves dissected into linear segments. Umbels sessile in the forks, or opposite the leaves, naked, few-rayed. Calyx-teeth obsolete. Petals ovate, concave, obtuse. Stylopodium depressed; styles short. Fruit cordate, laterally compressed, the commissure narrow; ripe carpels incurved, with 5 often obscure rugulose ribs; oil-tubes broad and solitary in the intervals, with a narrow one under each rib.

1. **A. augustifolium,** Nutt. A few inches to nearly a foot high; branches more or less dichotomous: leaves 1—2 in. long, biternately or

triternately dissected into almost filiform segments: rays of umbel very
unequal: fr. ½ line long, somewhat broader, the 5 primary ribs occasionally supplemented by 4 less prominent intervening ones.—Common on
bushy hills. April—June.

12. CARUM, *Turner.* Glabrous erect rather slender herbs, our species
perennial, with tuberous or fusiform or coarse-fibrous usually fascicled
roots, pinnately ternate leaves with few linear leaflets, and involucrate
umbels of white flowers. Calyx-teeth small. Fruit ovate to linear-
oblong; pericarp thin, with obtuse often filiform ribs; oil-tubes solitary
in the intervals.

1. **C. Kelloggii,** Gray. Stems several, 3—6 ft. high, from a *strong
tuft of coarse hard fibrous roots:* lower leaves ternate, the pinnate divisions
with linear segments 1—3 in. long or more: involucre and involucels
prominent, somewhat scarious: calyx-teeth subulate, conspicuous: fr.
oblong, 1½—2½ lines long; stylopodium prominent, styles as long:
seed sulcate beneath the large oil-tubes.—On open plains and hillsides.

2. **C. Gairdneri** (T. & G.), Gray. Stem solitary, 1—4 ft. high, from
a *fascicle of fusiform tuberous roots:* leaves mostly simply pinnate, with
3—7 linear or almost filiform leaflets 2—6 in. long, the lowest rarely
themselves pinnately divided, the uppermost cauline usually simple:
involucre of few bracts or 0: involucels of linear-acuminate bractlets:
fr. ovate, ½—1 line long, with long styles: seed terete.—Dry hills.

13. PIMPINELLA, *Brunfels.* Perennials with decompound foliage
and nearly naked umbels. Calyx-teeth obsolete. Fruit ovate, laterally
compressed but with broad commissure; carpels 5-angled, with distant
usually slender ribs, and several oil-tubes in the intervals. Seed somewhat flattened dorsally, with plane or slightly convex face.

1. **P. apiodora,** Gray. Stoutish, erect, glabrous, 2—3 ft. high, sweet-
scented: leaves mostly radical, 2—3-ternate: leaflets cuneate-ovate,
laciniately pinnatifid and toothed, 1 in. long: umbels long-peduncled,
6—15-rayed; rays 1—2 in. long, hispidulous-puberulent: fl. white or
pinkish: fr. broadly ovate (not known in its mature state), 1½ lines
long: oil-tubes 4—6 in the intervals, 8 or more on the face.—Hills of
Marin Co., near Sausalito.

14. FŒNICULUM, *Pliny* (FENNEL). Perennial, erect and tall, with
dark green striate stem, and equally dark sweet-scented and -flavored
leaves dissected into countless linear-setaceous leaflets. Flowers yellow,
in umbels destitute of bracts and bractlets. Calyx with turgid border
and no teeth. Fruit oblong; carpels 5-ribbed; oil tubes solitary in the
intervals, 2 on the face.

1. F. VULGARE, Gerarde. Cultivated from ancient times, and formerly
in high repute as a medicinal and culinary herb; naturalized in many

parts both of the Old World and the New, and common in central and
southern California, attaining the height of 3—6 ft.; readily known by
its dark green finely dissected foliage and large umbels of greenish-
yellow small flowers. May—Sept.

15. SELINUM, *Theophr.* Caulescent, branching perennials. Calyx-
teeth obsolete. Fruit with prominent crenulate disk. Carpel decidedly
winged; oil-tubes usually only one in each interval, 2—4 on the face.

1. **S. Pacificum,** Wats. Leaves ternate-bipinnate; segments ovate,
acutish, 1 in. long, laciniately toothed and lobed; peduncles stout, the
umbel about 15-rayed; bracts of involucre 1 in. long, equalling the rays,
lobed and toothed; involucels of several linear entire or 3-toothed bractlets: fr. oblong, 3—4 lines long; wings narrow; oil-tubes conspicuous,
rarely 2 in the intervals: seed channelled under the dorsal oil-tubes.—
Near Sausalito, and in the Mission Hills.

16. ANGELICA, *Braunschweig.* Perennials, stout and tall. Segments of the large pinnately or ternately compound leaves broad,
toothed; petioles dilated. Umbels many-rayed, nearly or quite naked.
Flowers white or purple. Calyx-teeth minute or oboslete. Fruit ovate
or oblong, strongly flattened dorsally, with a very broad commissure,
margined by a broad somewhat scarious wing; dorsal ribs prominent,
more narrowly winged; oil-tubes 1—3 in the intervals.

1. **A. tomentosa,** Wats. *Hoary-tomentose,* or the stem in age glabrate:
leaves quinately bipinnate; *leaflets firm,* ovate, acute, very oblique at
base, 2—4 in. long, the lower sometimes lobed, serrate with unequal
acute teeth: umbels naked, often dense: rays 1—3 in. long: fr. 3 lines
long, broadly elliptical, the lateral wings thin, the dorsal acutish: seed
thin, plane on the face, channeled on the back by the impressed dorsal
oil-tubes.—Banks of streamlets among the hills.

2. **A. Californica,** Jepson. Stem 4 ft. high, glabrous, only the *leaves*
and ends of the rays *puberulent: leaflets* broadly ovate, 2 in. long, *thinnish,*
the lower often lobed or divided at base, all irregularly serrate and the
serratures mucronate: peduncles with broadly dilated bracts about in
the middle: rays 40--50, unequal, 1—6 in. long: fr. oblong, about 4¼
lines long, the dorsal and intermediate ribs winged: oil-tubes 3 in the
intervals, 2 on the face.—Vaca Mountains.

17. LEPTOTÆNIA, *Nutt.* Glabrous subacaulescent perennials, with
thick often very large fusiform roots, pinnately decompound leaves.
Fruit strongly compressed dorsally, oblong or elliptical, with thick corky
lateral wings, the dorsal and intermediate ribs filiform or obscure; oil-
tubes 3—6 in the intervals, 4—6 on the face, mostly small, sometimes
obsolete.

1. **L. dissecta,** Nutt. Leafy at base, 1—3 ft. high: leaves broad, 1 ft. long, ternate and thrice pinnate; segments ovate or oblong, ½—1 in. long, laciniate-pinnatifid and toothed, puberulent on the veins beneath and along the margins: umbel 8—20-rayed, involucrate with few linear bracts, the bractlets of the involucels more numerous: fl. yellow or purplish: fr. sessile or nearly so, 5—9 lines long, about 3 lines broad: seed face plane.—On dry hills.

18. PEUCEDANUM, *Theophr.* Perennials of diverse habit, ours mostly low and subacaulescent, with fusiform root. Leaves ternately or pinnately dissected. Involucre 0: involucels usually present. Flowers white or yellow. Calyx-teeth obsolete or manifest. Fruit strongly flattened dorsally, oblong to suborbicular, glabrous or tomentose; carpel with dorsal and intermediate ribs filiform and approximate, the lateral ones developed into a broad thin wing which until maturity is coherent with that of its companion carpel, forming a broad scarious wing to the fruit as a whole. Oil-tubes 1—8 in the intervals, 2—10 on the face.

* *Stout; leaves finely dissected; fruit-wings broad; fl. white.*

1. **P. eurycarpum,** C. & R. Root tuberous-enlarged: stem 1—2 ft. high, branching, pubescent: leaves subdivided into countless small linear cuspidate segments: umbel 3—12-rayed, with involucels of lanceolate acuminate often united bractlets; rays ½—4 in. long; pedicels 1—5 lines: *fr. glabrous*, 5—9 lines long, *broadly elliptical*, the wings as broad as the body or broader, the ribs filiform; oil-tubes large, solitary in the intervals, 2 on the face.—Plains and hills of the interior.

2. **P. dasycarpum,** Torr. & Gray. Subacaulescent from a fusiform root, tomentose-pubescent: leaves small, with countless short linear segments: peduncles stout, ½—1 ft. high; umbel 6—12-rayed; involucels of linear-lanceolate more or less tomentose bractlets; rays 1—3 in., pedicels 3—5 lines long: *fr. nearly orbicular*, 4—7 lines long, *nearly glabrous or coarsely pubescent*, the thin scarious wings broader than the body: oil-tubes large, usually solitary in the intervals, 4 on the face: seed deeply sulcate under the oil-tubes.—In the interior.

3. **P. tomentosum,** Benth. Subacaulescent, more or less densely villous-tomentose and purplish: leaves cut into very small filiform or very narrow segments: peduncles 1 ft. high or more: umbel of 4—8 equal rays 1—3 in. long; involucels of linear-lanceolate or ovate-acuminate bractlets: calyx-teeth manifest: *fr. ovate to orbicular*, 5—9 lines long, *densely tomentose;* wings rather thick, from somewhat narrower to even broader than the body, the prominent ribs concealed by the tomentum: oil-tubes mostly 3 in the intervals, 4 on the face.—Common on bushy hills and open plains.

156

UMBELLIFERÆ.

* * *More slender, leaves much dissected; fl. yellow.*

4. **P. utriculatum**, Nutt. Rather slender, usually erect and branching, 1 ft. high or more, glabrous or puberulent: petioles short, their margins greatly dilated and forming a membranous saccate cavity; ultimate segments of the decompound leaves narrowly linear, ½ in. long or less: umbel 5—20 rayed, with involucels of dilated obovate often toothed petiolulate bractlets: fr. glabrous, broadly elliptical, 2—5 lines long; *wings thin, as broad as the body;* oil-tubes large and solitary in the intervals, 4—6 on the face.—On plains and open hills.

5. **P. caruifolium**, Torr. & Gray. Herbage and general aspect of the last, but acaulescent or nearly so; petioles without bladdery dilatation; leaf-segments ½—2 in. long; bractlets of involucels often lanceolate: fr. 3—4 lines long: *wings narrow and thickish; ribs obsolete:* oil-tubes indistinct, 2 or 3 in the intervals, none on the face.—Plains and hills.

* * * *Leaves not finely dissected, fl. yellow.*

6. **P. robustum**, Jepson. Acaulescent, *glabrous, glaucous,* 2 ft. high: leaves pinnately ternate; *leaflets broadly ovate or oblong, sessile, sparingly toothed or serrate at summit,* otherwise entire: scapes 1—3, very stout, greatly dilated under the rays, these 15—21, unequal, also dilated at summit; involucre and involucels 0: fr. 2½—3 lines wide, 4½—5 lines long, the wing half as broad as the body; oil-tubes solitary in the intervals, 6 on the commissural face.—Plains of the Sacramento, in Solano Co. May, June.

7. **P. Hassei**, C. & R. Caulescent, stout, 1—2 ft. high, glabrous, glaucescent: leaves biternate, in long petioles; *leaflets broadly ovate, with cuneate base, irregularly lobed, coarsely mucronate-toothed,* 2—3 in. long: umbel long-peduncled, 8—10-rayed, with involucre and involucels of oblanceolate, or linear, or linear-setaceous bracts and bractlets: fr. very large, glabrous, with very broad wings; oil-tubes solitary in the intervals.—Vaca Mountains, Solano Co.

8. P. SATIVUM (L.), Wats. Biennial, branching, 2—4 ft. high: *stem leafy, angular or fluted;* herbage nearly glabrous, of a somewhat yellowish green: leaflets of the pinnate leaves large, ovate or oblong, incisely toothed: involucre and involucels small or 0: fr. oval, 2—3 lines long, broadly winged, prominently ribbed; oil-tubes solitary in the intervals. —The *Parsnip* of farms and gardens, native of Europe; spontaneous here and there by waysides and in waste lands.

19. **SPHONDYLIUM**, *Tourn.* (Cow Parsnip). Perennial or biennial, with stout hollow fluted stem, ample lobed or compound leaves, and very large umbels of white flowers. Calyx-teeth small or obsolete. Fruit round-obovate, very much flattened dorsally, somewhat pubescent.

Carpel with dorsal ribs filiform, the margin winged; wings coherent when young: oil-tubes solitary in the intervals, obclavate, extending from the apex downward to or below the middle of the carpel.

1. **S. lanatum** (Michx.) Stem 3—8 ft. high: leaves ternate, 1—2 ft. long, the stout petioles and veins hirsute beneath, the base of the petiole much dilated; leaflets 4—10 in. long, rounded and subcordate, the lobes somewhat palmately arranged, acuminate, toothed: rays many, 3—6 in. long: fl. large, white, irregular, the outer petals being larger: fr. broadly obovate, 4—6 lines long, slightly pubescent.—In wet open ground, or in moist thickets. March, May.

20. MYRRHIS, *Morison* (SWEET CICELY). Perennials with thick aromatic roots, rather slender stems not tall, ternately-compound mostly radical leaves: involucres and involucels reduced or obsolete. Flowers white. Calyx-teeth obsolete. Fruit linear to linear-oblong, more or less attenuate at base, acute at summit, glabrous or bristly along the ribs. Carpel nearly pentagonal in section, flattened dorsally if at all. Oil-tubes obsolete in mature fruit. Seed-face slightly concave to deeply sulcate.

1. **M. occidentalis** (Nutt.), Benth. & Hook. Stoutish, puberulent or pubescent: leaflets oblong, 1½—4 in. long, acute, coarsely serrate, rarely incised: umbel 5—12 rayed, naked or with 1 or 2 bracts; rays 1—5 in. long, mostly erect; pedicels 1—3 lines: *fr. 7—12 lines long, 1½ lines wide, obtuse at base,* glabrous, with prominent acute ribs; the mostly conical stylopodium together with the style ½—1 line long.—Dry woods.

2. **M. nuda** (Torr.), Greene. Slender, 2—3 ft. high, more or less pilose-pubescent: leaves twice ternate; leaflets 1—2 in. long, ovate, acute or obtusish, rather deeply cleft and toothed: umbel long-peduncled, 3—5-rayed, naked or with small caducous bracts or bractlets; pedicels ¼—¾ in. long: *fr. slender, 3—7 lines long, with slenderly attenuate base:* carpels acutely ribbed; stylopodium very short.—Common in shady woods.

21. CHÆROPHYLLUM, *Columna.* Rather slender annuals with ternately compound leaves, and small white flowers in almost naked umbels. Calyx-teeth obsolete. Fruit lanceolate, or ovate-oblong and beaked at summit, the beak not as long as the body; ribs of carpel equal; oil-tubes present.

1. **C. ANTHRISCUS** (L.), Lam. Weak and often half reclining; small umbels opposite the leaves, about 3-rayed: fr. about 2 lines long including the short beak, roughened with short rigid incurved bristles.—In sandy soil at Alameda, etc.

22. SCANDIX, *Theophr.* Annual, with pinnately decompound leaves cut into countless slender segments. Flower and fruit much as in *Chærophyllum,* except that the beak of the carpel far exceeds the body.

1. S. Pecten veneris, Dod. Erect, 1 ft. high more or less leafy throughout, but radical leaves ample, of oblong outline, cut into many short ligulate acuminate lobes: bractlets of involucels many: fr. ½—3 in. long including the beak which is the conspicuous part of it, the body and the margins of the beak with tubercles ending in short prickles.—A weed in fields and by waysides.

23. DAUCUS, *Galen.* More or less hispid annuals and biennials, with pinnately decompound leaves, involucres and involucels of lobed or divided bracts, and white flowers. Outer rays of umbel longest, in fruit connivent over the inner, giving a concave top to the umbel. Calyx 5-toothed. Fruit ovate or oblong; carpels semiterete or dorsally flattened; primary ribs filiform and bristly, the secondary more prominent, winged with a row of more or less united barbed prickles. Oil-tubes solitary under the secondary ribs. Seeds nearly flat on the face.

1. D. pusillus, Michx. Annual, erect, or the branches short and decumbent, ½—2 ft. high, retrorsely hispid; leaves bipinnate, the segments pinnatifid, with short narrowly linear lobes; rays 2—6 lines long, nearly equal; involucre bipinnatifid, equalling the umbel; involucels equalling the greenish white flowers: fr. 1½—2 lines long, short-pedicellate, the prickles usually equalling or exceeding the width of the body: seed slightly concave on the face.—Nearly all parts of the State; on bluffs and hills near the sea, often depressed and condensed.

2. D. Carota, L. Biennial, stout, 2—3 ft. high, hispid: involucre of many pinnatifid bracts equalling the large umbel; bractlets scarious, with an herbaceous midrib: fl. white, but the central one of each umbellet abortive and dark purple: fr. oblong-ovoid, the spines as long as its diameter: fruiting umbel deeply concave, resembling a bird's nest.—The *Carrot* of the gardens; already becoming a wayside weed.

24. CAUCALIS, *Theophr.* Scarcely distinct from *Daucus,* but fruit more compressed laterally; the seed face deeply-channelled.

1. C. nodosa, Huds. Branching at base, the long branches reclining, leafy throughout and retrorsely hispid: leaves pinnate, with pinnatifid divisions: umbels small, naked, subsessile opposite the leaves: carpels unequal, the larger one a line long; surface tuberculate and prickly, the prickles barbed or incurved at summit.—Obscure weed, from Europe.

2. C. microcarpa, Hook. & Arn. Erect, slender, 6—15 in. high, nearly glabrous: leaves much dissected, hispidulous: umbels terminal and at the ends of the branches, subtended by two or more foliaceous dissected bracts, 3—6 rayed; rays slender, 1—3 in. long; umbellets few-flowered, the pedicels unequal; involucels of short entire bractlets: fr. oblong-ovoid, 2 lines long, armed with uncinate prickles.—Very common.

Order L. CORNEÆ.

Trees, shrubs or undershrubs, with opposite exstipulate leaves, and naked or involucrate cymose or capitate inflorescence. Calyx-tube coherent with the ovary; limb 4-lobed or obsolete. Petals 4, epigynous, valvate in bud. Stamens 4, alternate with the petals; anthers 2-celled. Style filiform; stigma simple. Fruit drupaceous, 1—2-seeded. Seed pendulous; embryo minute; albumen fleshy.

1. **CORNUS,** *Pliny* (DOGWOOD). Deciduous shrubs, or low semi-herbaceous plants. Drupe globose, ovoid or oblong; putamen 2-celled, 2-seeded.

* Flowers white, not involucrate, cymose.

1. **C. glabrata,** Benth. Shrub 5—12 ft. high, with gray bark, and nearly or quite glabrous twigs and foliage: leaves oblong to narrowly ovate, acute at each end, or acuminate at apex, 1—2 in. long, green alike on both faces; petioles short, slender: fl. in many small open flat-topped cymes: *fr. globose, white; stone little compressed,* not furrowed, broader than high, breadth 2 lines or more.—Coast and Mt. Diablo Ranges.

2. **C. Greenei,** C. & E. Size and habit of the last: twigs and inflorescence appressed-pubescent: leaves ovate, obovate or oval, acutish or rounded at base, acute or acuminate at apex, appressed-pubescent or glabrate above, beneath scarcely lighter but with a sparse appressed pubescence of stiffish hairs of which some are straight, others curved: fl. large, in loose paniculate cymes: calyx-teeth triangular: styles with enlarged greenish tips: *fr. dark blue; stone globular, not furrowed,* slightly ridged.—Wooden Valley, Napa Co., *Jepson.*

3. **C. pubescens,** Nutt. Shrub 6—15 ft. high, with smooth reddish branches: leaves ovate, acute, 2—4 in. long, paler and more or less pubescent beneath: fl. in convex cymes: *fr. white, subglobose,* 2 lines broad; *stone somewhat flattened,* mostly oblique, *with a more or less prominently furrowed edge,* the sides more or less prominently ridged. Var. **Californica,** C. & R. Pubescence said to be loose and spreading; leaves more rounded and broader; stone smaller, etc.—Throughout the State, the variety chiefly; the type being of more northerly habitat.

* * Flowers greenish, sessile on a thick convex receptacle, subtended by 4—6 large white petaloid bracts.

4. **C. Nuttallii,** Audubon. Tree 15—70 ft. high, with ascending or widely spreading branches and smooth bark: leaves 3—5 in. long, obovate, acute at each end, pubescent: bracts of involucre usually 6, obovate to oblong, 1¼—3 in. long, abruptly acute to acuminate, white, often tinged with red: head ½—1 in. broad, very dense: fr. 5—6 lines long, scarlet.—Coast Range from Monterey northward. May—July.

Order LI. GARRYACEÆ.

Consists of the genus

1. GARRYA, *Douglas.* Evergreen shrubs with greenish bark, and opposite entire coriaceous leaves. Flowers diœcious, in axillary pendulous aments, solitary or in threes between the decussately connate bracts. Petals 0. Calyx of sterile flowers 4-parted, with linear valvate segments. Stamens 4; filaments distinct; disk and rudimentary ovary 0. Calyx of fertile flower with a shortly 2-lobed or obsolete limb; disk and rudimentary stamens 0; ovary 1-celled, with 2 pendulous ovules; styles 2, stigmatic on the inner side, persistent. Fruit capsular, circumscissile at about the middle, or indehiscent. Seeds coated, with an acidulous or bitter red pulp which is never in contact with the wall of the pericarp, the inside of this being glabrous and polished.

1. **G. elliptica,** Dougl. Stems clustered, 5—15 ft. high: *leaves* ½—3 in. long, *dark green, elliptical,* rounded or acute and mucronate at apex, truncate or rounded at base, *the margin undulate,* glabrous above, tomentose beneath: aments solitary or several; the sterile 4—10 in. long, their silky bracts truncate or acute; calyx-segments cohering at tip: fertile aments stouter, 2—6 in. long; bracts acute or acuminate: ovary sessile, densely silky-tomentose: fr. globose, 4 lines thick.—In rich shady places along streams. Fl. Feb.; fr. Sept.

2. **G. Fremonti,** Torr. Shrub 5—10 ft. high, glabrate: *leaves light green, ovate or oblong, not undulate,* 1½—2½ in. long: aments 2—3 in. long, with acute somewhat silky bracts: ovaries nearly glabrous; fr. globose, 2 lines or more in thickness, short-pedicellate.—From Mt. Hamilton northward, on dry slopes and summits.

Division III, SYMPETALÆ PERIGYNÆ.

Petals united below, and, with the stamens, inserted on the calyx, usually near its summit, the tube being more or less adherent to the ovary (free from it in *Daphnoideæ*).

Order LII. DAPHNOIDEÆ.

We have but one member, of the genus

1. **DIRCA,** *Linn.* (LEATHERWOOD). Branching deciduous shrubs, with smooth and very tenacious brown bark: the wood also very tough and flexible. Flowers in fascicles of about 3, appearing before the leaves, but from the same buds, and these of yellowish or whitish very silky caducous scales, which appear as an involucre to the flowers. Perianth corolla-like, tubular, but slightly oblique, yellowish, nodding, 4-lobed. Stamens 8, inserted at base of the perianth-tube, exserted; filaments filiform; anthers small, oblong. Ovary sessile, 1-celled; style longer than the stamens. Perianth deciduous from the growing ovary, this becoming a somewhat drupaceous small fruit.

1. **D. occidentalis,** Gray. Shrub 4—7 ft. high: bud scales densely white-villous: leaves oval with rounded base, 1—3 in. long: perianth canary-yellow, subsessile, 3—4 lines long, rather deeply 4-lobed, the lobes nearly truncate, somewhat connivent, rendering the upper and broader part of the organ slightly urceolate.—On moist well shaded northward slopes of the Oakland and Berkeley Hills; also in the counties of Marin and San Mateo.—Feb., March.

Order LIII. LORANTHEÆ.

Half-shrubby parasites on trees and shrubs; color yellowish-green or yellow. Branches dichotomous; the joints swollen. Leaves opposite, either coriaceous, or reduced to more or less distinctly connate scales. Flowers (diœcious in our genera) of 2—5 sepals coherent at base and valvate in æstivation, no petals; anthers as many as the calyx-segments and (in ours) sessile upon them; ovary inferior, 1-celled, 1-ovuled becoming a 1-seeded berry with glutinous pulp.

1. **PHORADENDRON,** *Nutt.* (MISTLETOE). Flowers globose, imbedded in the rachis of jointed spikes. Calyx 3- (rarely 2- or 4-) lobed. Anthers sessile on the base of the lobes, 2-celled, opening by a pore or slit; pollen-grains smooth. Stigma sessile, obtuse, entire or more or less distinctly 2-lobed. Berry globose, pulpy, translucent, crowned with the persistent calyx lobes. Embryo with foliaceous cotyledons.

1. **P. villosum,** Nutt.. Stems much branched, ½—1 ft. long, forming spherical masses on the branches of deciduous trees: herbage of a deep

or dark green, covered with a short almost velvety pubescence: leaves ¾ —1½ in. long, from round-obovoid to spatulate-oblong, short-petiolate: spikes opposite: berries white.—Frequent in the interior valleys from Sonoma and Solano counties southward. Very distinct from the eastern yellow-green *P. flavescens.*

2. **P. Bolleanum** (Seem.), Engelm. Branches 5—8 in. long: leaves very thick, spatulate to linear, obtusish, nerveless, ½—1 in. long: spikes opposite or in fours, with connate ciliolate bracts: berries white.—On Mt. St. Helena, toward the northwestern base, *Jepson.*

2. RAZOUMOFSKYA, *Hoffm.* Small, yellow or greenish, leafless; leaves represented by connate scales. Flowers axillary and terminal, solitary, or several in each axil. Staminate fl. mostly 3-parted, compressed, or the terminal ones globose: anthers sessile on the lobes, orbicular, 1-celled, dehiscent by a circular aperture at base; pollen-grains spinulose. Pistillate fl. ovate, compressed, 2-toothed, subsessile, the pedicel in fruit elongated and recurved. Fruit elastically dehiscent at the circumscissile base, forcibly ejecting the seed. Cotyledons rudimentary, indicated by a notch in the axis of the embryo.

1. **R. Douglasii,** (Engelm.), O. Ktze. Slender, greenish yellow, ¼—1 in. high, much branched, but not verticillately, the accessory branchlets behind (not beside) the primary ones: spikes short, mostly 5-flowered: staminate fl. less than a line wide, with round-ovate acutish lobes, fr. 2½ lines long. Var. **abietina** (Engelm.). Fertile plant larger (1—3 in. high), the sterile smaller, with spreading or even recurved branchlets: fr. smaller (scarcely 2 lines long).—Mt. St. Helena, on *Pinus attenuata, Jepson,* the variety only.

ORDER LIV.   C A P R I F O L I A C E Æ .

Shrubs often trailing or climbing. Leaves opposite, mostly exstipulate. Flowers terminal and cymose or subspicate, or solitary or in pairs in the leaf-axils, regular or irregular. Calyx-tube coherent with the ovary; limb 5-toothed or obsolete. Corolla 4—5-lobed or -cleft; the lobes imbricate in bud. Stamens distinct. Ovary 2—5-celled, or by abortion 1-celled after flowering. Fruit a berry or drupe.

1. **SAMBUCUS,** *Pliny* (ELDER). Shrubs or small trees, with stout thick and very pithy shoots and branches, and pinnate foliage; leaflets 5—11, serrate; young shoots and foliage heavy-scented. Flowers small, white or cream-color, very many, in compound cymes at the ends of terminal and lateral shoots. Calyx with 5 minute teeth. Corolla rotate, 5-lobed. Stamens 5. Stigmas and ovary-cells 3—5. Fruits of the nature of drupelets, though berry-like, each with 3 (rarely 4 or 5) separate seed-like nutlets; each with a single seed.

* *Berries without bloom; shrubs flowering in early spring.*

1. **S. maritima**, Greene. Arborescent, 10—25 ft. high, clustered, and each of the several trunks often a foot in diameter; *bark light brown, more flaky than fissured;* pith of shoots white: young twigs and foliage pubescent with sparse stiff short somewhat retrorse hairs: young leaves with free ligulate callous-tipped stipules 1—3 lines long: leaflets 2—5 pairs, often with conspicuous false stipellæ, or the later leaves on vigorous shoots completely bipinnate, the ordinary leaflets from oval to oblong-lanceolate, abruptly acuminate, closely and rather deeply serrate, thin: cymes rather small but flat-topped: corolla white: *fr. black, without bloom.*—Rare or local shrub of the bay shore at Shell Mound.

2. **S. callicarpa**, Greene. Near the preceding, but not as large; bark darker and fissured; *cymes* small, not flat-topped, rather *low-pyramidal: fr. scarlet.*—Common along mountain streams.

* * *Berries blue with a dense bloom; shrub flowering in summer.*

3. **S. glauca**, Nutt. Aborescent, often 30 ft. high, the solitary trunk a foot thick, covered with a dark close very distinctly and rather finely fissured bark: twigs long and slender; leaves exstipulate, coriaceous, glabrous; leaflets 3—5 pairs, lanceolate, acuminate, sharply serrulate, seldom or never divided: cymes large, flat: fl. white: fr. blue with a dense bloom but black beneath it.—Common in rather dry and sparsely wooded ravines or in open fields.

2. **SYMPHORICARPOS**, *Dillenius* (SNOWBERRY). Low branching shrubs. Leaves small, membranaceous, mostly entire. Flowers small, axillary and terminal, solitary or in dense spicate clusters, white or pinkish. Calyx with globular or oblong tube and 4—5-toothed persistent limb. Corolla short campanulate, slightly gibbous, 4—5-lobed. Stamens inserted on the throat of the corolla and as many as its lobes. Ovary 4-celled; 2 cells containing a few sterile ovules, the other 2 each with a single suspended ovule. Fruit globose, berry-like, containing two seed-like smooth 1-seeded nutlets.

1. **S. racemosus**, Michx. *Usually 3—4 ft. high, slender, with spreading branches:* leaves round-oval to oblong, 1 in. long, glabrous above, pubescent along the veins beneath: axillary clusters mostly few-flowered, the lowest 1-flowered: corolla reddish or pinkish, 2 lines long, slightly gibbous, moderately villous within, cleft above the middle: fr. ⅛—½ in. thick, subglobose, snow-white.—On banks of streams in shady places almost everywhere in the Coast Range. Fl. May; fr. Oct.

2. **S. ciliatus**, Nutt. (?). *Low and diffuse, seldom 1 ft. high,* with many very slender but rather rigid leafy branches, and few-flowered clusters: leaves oval, obtuse, ½—¾ in. long, glabrous above, pubescent along the veins beneath, the margin rather densely ciliate: corolla

rose-red, 2 lines long, slightly gibbous, cleft to the middle or more deeply, scarcely villous within: fr. small, globose, snow-white.—Common in the Oakland Hills on northward slopes. Fl. May. fr. July.

3. **DISTEGIA,** *Raf.* Stems erect. Leaves membranaceous. Flowers in pairs on an axillary peduncle, each pair closely subtended by a pair of ample foliaceous bracts. Corolla salverform or funnelform, gibbous at base. Berries approximate but distinct, black when ripe, their sub-tending bracts then dark red, more or less reflexed.

1. **D. Ledebourii** (Esch.). Stoutish, 5 – 15 ft. high, often with the very long sarmentose branches reclining on other shrubs or small trees: leaves: corolla strongly gibbous at base, strictly salverform above the gib-bosity, the short rounded lobes spreading abruptly, the whole almost scarlet without, yellow within.—Common along streams throughout western California, ranging northward far beyond our borders.

4. **CAPRIFOLIUM,** *Brunfels* (HONEYSUCKLE). Trailing or climbing shrubs, with subcoriaceous leaves occasionally stipulate, the upper pairs usually connate-perfoliate. Flowers larger and showy, verticillate-spicate at the ends of the branches. Calyx-limb small and 5-toothed, or obsolete. Corolla bilabiate. Stamens 5, on the tube of the corolla. Ovary 2—3-celled, becoming a few-seeded red or yellow berry.

1. **C. hispidulum,** Lindl., var. **Californicum,** Greene. Twining, 10 —25 ft. high, the ultimate branches often a yard or two in length and drooping, hispidulous and somewhat glandular as to the upper portion and about the inflorescence; leaves ovate-oblong or elliptical, acutish, 1 —3 in. long, the lower pair without stipules, the intermediate with broadly ovate stipular appendages often ½ in. long and as broad, the one or two floral pairs connate, all very glaucous beneath, pale and glaucescent above: spikes 1—5, each with 3—6 whorls of pink flowers: corolla hispidulous, ½—¾ in. long; anthers exserted, narrowly linear, 2½ lines long.—Common in moist ravines and on shady banks, climbing over small trees, along the seaboard chiefly. May—July.

2. **C. interruptum** (Benth.), Greene. Stoutish, erect and bushy, 4— 7 ft. high, less disposed to twine or climb; bark of branches white and almost shining, glabrous: leaves of a very pallid hue, white-glaucous beneath, glaucescent above, 1 in. or more in breadth, mostly orbicular or round-ovate, never stipulate, several of the uppermost pairs connate: fl. numerous, in several interrupted spikes, corolla ½ in. long, yellow, glabrous.—Common on bushy hills of the inner Coast Ranges.

Order LV. RUBIACEÆ.

Our species herbs (*Cephalanthus* and some species of *Galium* shrubby) with opposite or verticillate mostly exstipulate entire leaves, and 4-merous perfect (or often diœcious) flowers. Calyx-limb obsolete, or of 4 teeth. Stamens distinct, alternate with the corolla-lobes and inserted on its throat or tube. Ovary 2—4-celled, with a solitary ovule in each cell. Fruit indehiscent, dry or baccate.

1. CEPHALANTHUS, *Linn.* (BUTTON-BUSH). Shrubs with opposite or ternate leaves, and flowers in dense globose terminal and axillary peduncled heads. Calyx inverse-pyramidal, 5-toothed. Corolla with long slender tube and small 4-cleft limb. Stamens 4, short, on the throat of the corolla. Style slender, long-exserted; stigma capitate; ovary 2-celled. Fruit achene-like, 1—2-seeded.

1. **C. occidentalis,** L. Shrub or small tree, with ovate-lanceolate leaves 3—5 in. long, rather glossy above, often more or less pubescent: fl. white, in heads 1 in. thick, these solitary or few or several toward the ends of the branches.—River banks of the interior, especially of the lower Sacramento and San Joaquin. June—August.

2. SHERARDIA, *Dillenius.* Annual, slender, rough, with angular stem, and exstipulate leaves in verticels of 6. Flowers umbellate. Calyx-limb of 4—6 accrescent teeth. Corolla salverform, with a slender tube and 4-cleft limb. Stamens 4. Fruit of 2 dry indehiscent 1-seeded carpels, crowned by the calyx-teeth, separating when ripe.

1. **S. ARVENSIS,** L. About 3—6 in. high, hispidulous-roughened or nearly glabrous: leaves obovate-lanceolate, acute: fl. in small subsessile umbellate cymes: corolla bluish.—Berkeley.

3. GALIUM, *Diosc.* (BEDSTRAW. CLEAVERS). Herbaceous or suffrutescent, with slender angular stems, verticillate leaves without stipules, and small cymose flowers. Calyx-limb obsolete. Corolla rotate, 4-parted. Stamens as many as the corolla-lobes, short. Styles 2, short: stigmas capitate: ovary 2-lobed, 2-celled, 2-ovuled. Fruit didymous (biglobular), dry or fleshy, separating into 2 close 1-seeded carpels which are indehiscent, and glabrous, hispid, or hirsute.

* *Fruit dry when ripe.*

+ *Annuals.*

1. **G. SPURIUM,** L. Branching chiefly from the base; diffuse, 1—2 ft. high, glabrous except the retrorsely scabrous angles of the stem and veins and margins of the leaves: leaves 6—8 in the whorl, linear-oblanceolate, cuspidate: fl. 3—9 in axillary umbellate cymes; corolla pale green, the segments acuminate: *pedicels recurved after flowering:* fruit

large, coarsely tuberculate, more or less uncinate-hispid.—Mostly in the mountains back from the seaboard; less common than the next.

2. G. APARINE, L.   Taller and more slender, 3—5 ft. high (or often only a few inches), climbing by the retrorse prickliness of the angles and leaf-margins: corolla minute, white: *pedicels straight in fruit:* surface of carpel smooth but densely uncinate-hispid.—Very common in shady or open places in woods and along the salt marshes.

3. G. ANGLICUM, Huds.   Slender, erect or diffuse, glabrous, but with small hooked prickles on the angles of the stem: leaves firm, mostly 6 to the whorl, narrowly oblanceolate, rough on the margins with minute prickles: fl. greenish-white in small cymes: *fr. small, glabrous, granulate* with small tubercles.—Plentiful in certain wooded districts in Sonoma Co., *Bioletti.*

+ + *Perennials.*

4. **G. triflorum,** Michx.   Stem flaccid, 1 ft. long or more, reclining or at least decumbent, retrorsely aculeate-scabrous on the angles, or smoothish: *leaves in sixes, thin, elliptic-lanceolate, acute at both ends,* or cuspidate-acuminate, the margins and often the midrib beneath beset with very short usually retrorse and hooked prickles: peduncles few, once or twice 3-forked; pedicels divergent; corolla greenish: fr. hirsute with slender hooked bristles, or when ripe merely roughened.—In woods; not common.

5. **G. trifidum,** L.   Erect or reclining, rather slender, 5—20 in. high, glabrous, except the retrorsely scabrous angles of the stem, and the more hispidulous but sparse roughness of the margins of the leaves and the midrib beneath: *leaves* (in our forms) usually *in fours or fives, linear or oblanceolate,* or lanceolate-oblong, *obtuse,* 4—7 lines long: peduncles slender, scattered, 1—several-flowered: fl. minute, white, often 3-merous: fr. small, smooth, glabrous.—In wet grounds.

* * *Fruit fleshy, berry-like.*

6. **G. Californicum,** Hook. & Arn.   Herbaceous from slender creeping rootstocks, in low tufts, or diffuse with slender stems a foot long, hispid or hirsute, rarely glabrate in age: *leaves thinnish, ovate or oval, apiculate-acuminate,* ¼—½ in. long, margins and midrib hispid-ciliolate: fr. blackish, glabrous, on recurved pedicels.—In shady places.

7. **G. Nuttallii,** Gray.   Suffrutescent, tall and climbing, often 3—4 ft. high, mostly glabrous, except the minutely aculeolate-hispidulous angles of stems and margins of leaves, these also sometimes naked: *leaves small, oval to linear-oblong,* mucronate, mucronulate, or obtuse: fr. smooth and glabrous, purple.—In thickets.

8. **G. Andrewsii,** Gray.   *Small and densely matted;* nearly or quite glabrous, the herbage bright green and shining: leaves crowded, acerose-

subulate, either naked or sparsely spinulose-ciliate, 2—4 lines long: fl. diœcious, the sterile in few-flowered terminal cymes; fertile solitary, subtended by a whorl of leaves which are longer than the deflexed fruiting pedicel: berry smooth, blackish.--Dry summits of Mt. Diablo, Mt. Hamilton, etc.

ORDER LVI. VALERIANEÆ.

Herbs with opposite leaves, no stipules, and mostly complete flowers, in a cymose or thyrsoid inflorescence. Calyx-tube adherent to the ovary; limb either obsolete, or composed of teeth which develop as a pappus or feathery crown upon the fruit. Corolla more or less irregular; the limb bilabiate, the lobes imbricate in bud. Stamens 1—3, epipetalous. Filaments and style filiform; stigma undivided and truncate, or minutely 3-cleft. Fruit an achene; seed pendulous.

1. **VALERIANELLA**, *Vaillant.* Rather small spring annuals, with small pinkish flowers in cymes which form a more or less interrupted thyrsiform terminal inflorescence. Corolla more or less bilabiate, spurred or gibbous at base. Calyx-limb none, therefore no pappus to the variously winged often meniscoid glabrous or pubescent fruit.

1. **V. macrocera** (T. & G.), Gray. Corolla only a line long, with spur sometimes as long as the body, sometimes shorter; limb somewhat equally spreading, hardly bilabiate, or equally 4-lobed and the posterior lobe emarginate-bifid: fr. glabrous or puberulent, *obtuse or lightly lineate-sulcate on the dorsal angle*, the broad wing, circumscribing the ventral face of the achene, spreading or incurved.—On hillsides. April—June.

2. **V. congesta**, Lindl. Corolla 3—4 lines long, with obviously bilabiate 5-cleft limb, the lobes oblong, obtuse; tube very gibbous, spurred at base, the spur short, arcuate, obtuse: fr. pubescent, the *keel prominent, obtuse*, circumscribing ventral-face *wing broad, involute.*— Common on dry hills, or in shady places.—April, May.

3. **V. samolifolia** (DC.), Gray. Corolla a line long, obscurely bilabiate, with short obconic-saccate spur: *fr. triquetrous, wholly destitute of wing*, glabrous or a little pubescent.—Near the coast, from Sonoma Co· northward.

ORDER LVII. DIPSACEÆ.

Herbs with opposite leaves, and flowers in dense involucrate peduncled heads; each flower in the head enclosed within a tubular involucel and subtended by a bract. Calyx-tube adherent to the ovary; limb entire, or toothed, or with bristle-like segments that persist upon the fruit. Corolla inserted at summit of calyx-tube, 4- or 5-lobed. Stamens 4, epipetalous, alternate with the corolla lobes. Style filiform; stigma simple, longitudinal or subcapitate. Fruit achene-like, crowned with the calyx-limb, 1-seeded. Seed pendulous; albumen fleshy.

1. DIPSACUS, *Diosc.*,(Teasel). Tall coarse biennials with muricate or prickly stem and foliage; the cauline leaves connate. Involucre of rigid spreading unequal bracts; bracts of receptacle rigid, acuminate. Involucel sessile, 4-angled, 8-ribbed, terminated by 4 short teeth. Calyx-limb cup-shaped, quadrate or 4-lobed. Corolla funnelform, 4-cleft.

1. **D. fullonum,** Mill. Stout, erect, very rough with short prickles, 4—6 ft. high; radical leaves 8—12 in. long, elliptic-lanceolate, arcuate; cauline connate-perfoliate: heads large, ovoid or oblong, on stout naked peduncles: bracts of receptacle rigid, recurved at the tips, as long as the flesh-colored corollas: stamens exserted.—Very common coarse weed in low lands of Alameda and Contra Costa counties.

2. SCABIOSA, *Brunfels.* Soft unarmed plants, with peduncled globose or hemispherical heads, the flowers of the outer circle often larger than the others. Receptacle bearing hairs or soft scales among the flowers. Calyx-limb a cup-shaped border with 4 or more teeth or bristles. Corolla funnelform or salverform, often slightly irregular.

1. **S. atropurpurea,** L. Suffrutescent, freely branching, 2—3 ft. high: radical leaves lyrate; cauline pinnate, the segments oblong, toothed or incised: heads low hemispherical, in fr. ovate: corollas dark maroon to rose-purple, flesh-color, and white, the outer circle of them larger and exceeding the involucre; calyx-limb pedicellate, in fruit bearing 5 pappus-like bristles —An escape from the gardens of old-fashioned flowers, and become a luxuriant street and wayside weed in various sections about the Bay. The almost black flowers of one variety have given rise to the common name *Mourning Bride.*

Order LVIII. COMPOSITÆ.

Herbs or shrubs with watery or resinous (never milky) juice, foliage various, the individual flowers small, in dense closely involucrate heads, the head often resembling a simple flower. Calyx wholly or partially adherent to the 1-celled, 1-ovuled ovary; the limb represented, if at all, by one or more scales, awns or bristles called the *pappus*. Corollas tubular, palmatifid or ligulate; the tubular ones 4—5-toothed or -cleft, often called *disk-corollas;* the ligulate commonly toothed at apex, known as the *ray-corollas.* Stamens mostly 5, syngenesious, their anthers thus forming a tube around the style. Pollen-grains globose, echinate. Style in all fertile flowers 2-cleft at summit (except in one suborder), stigmatose on the margin, the upper portion of the forks usually not stigmatose, often variously hairy or appendaged. Fruit 1-seeded, indehiscent, commonly crowned by its pappus of capillary or plumose bristles, or of scarious scales; at the insertion on the common receptacle often subtended by a bract; this collection called the *chaff:* the receptacle described as *naked* when the chaff is wanting: the surface of the receptacle being diagnosed as *alveolate, foveolate,* or merely *areolate,* according as the insertion of the achenes forms deeper or shallower depressions; or *fimbrillate* when the receptacle around these scars rises in teeth, or awns.—Our largest natural order, so-called, of flowering plants; the genera and species most conveniently considered under subordinal, natural, by not easily definable groups.

Rays none; style branches elongated, usually clavate-thickened upward and obtuse: stigmatic only below the middle.............................**1. EUPATORIACEÆ.**

Rays usually present; anthers not caudate; style-branches of perfect flowers flattened, and with a distinct terminal appendage.........................**2. ASTERACEÆ.**

Rays none; anthers caudate; style-branches of perfect flowers with no appendage, the stigmatic lines reaching almost to the naked truncate or obtuse summit.
....**3. GNAPHALIACEÆ.**

Rays none; fertile fl. apetalous or nearly so; the staminate involucres forming a raceme above the axillary pistillate one; pappus none.......**4. AMBROSIACEÆ.**

Rays seldom wanting; anthers not caudate; involucre not scarious; receptacle chaffy; pappus never of capillary bristles.................**5. HELIANTHACEÆ.**

Rays present, fertile, the achenes of each more or less enfolded by its involucral bract; receptacle chaffy, style-branches subulate, hispid...........**6. MADIACEÆ.**

Rays present; receptacle naked, or merely fimbrillate; pappus paleaceous or aristiform, or when bristly rigid.....................................**7. HELENIOIDEÆ.**

Anthers not caudate; bracts of involucre more or less scarious; style-branches truncate; pappus a scarious crown, or a circle of small scales, or wanting.
....**8. ANTHEMIDEÆ.**

Anthers not caudate; receptacle naked; involucres not imbricated, mostly cylindrical, the bracts not scarious; pappus of many soft-capillary bristles.
....**9. SENECIONIDEÆ.**

Rays none; anthers caudate; style branches united, stigmatic to the obtuse summit, smooth and naked, but often with a pubescent node below; receptacle densely setose.
....**10. CYNAROCEPHALÆ.**

Subordo 1. EUPATORIACEÆ.

Heads rayless. Corollas all tubular and regular, never yellow, though sometimes cream-color. Anthers without tails. *Style branches elongated,* usually clavate, *minutely papillose or puberulent,* the stigmatic lines only near the base.

Achenes 4-angled; pappus partly squamellate......................TRICHOCORONIS 1
Achenes 10-striate; pappus a single series of scabrous bristles.....COLEOSANTHUS 2

1. TRICHOCORONIS, *A. Gray.* Weak and flaccid fibrous-rooted perennial of muddy shores. Leaves opposite or attenuate, sessile. Heads few, peduncled, terminating somewhat corymbose branches. Flowers flesh-color. Style-branches scarcely clavate; rather linear and flattish. Pappus of small awns and intervening paleæ.

1. **T. riparia,** Greene. Stems assurgent, hardly a foot high, sparsely pubescent: leaves linear-lanceolate, remotely serrate, slightly

auricled at base: heads 2½ lines broad: achenes ¾ line long, sharply 4-angled, the sides dark brown, the angles hispid-ciliolate toward the summit; pappus of 4 barbellate bristles and as many intervening minute fimbriate-lacerate scales.—Banks of the lower San Joaquin; perhaps not within our range.

2. COLEOSANTHUS, *Cassini.* Perennial, often suffrutescent. Inflorescence of terminal and subterminal short clusters of narrow heads. Involucre of striate-nerved scales, the outer shorter. Corollas slender, 5-toothed. Style bulbous at base. Achenes 10-striate or -ribbed. Pappus of numerous but uniserial scabrous or barbellate bristles.

1. C. Californicus (T. & G.), O. Ktze. Shrubby at base, 2—3 ft. high, paniculately branching: leaves alternate, broadly ovate or triangular, irregularly crenate-toothed, about 1 in. long, 3-ribbed and roughish, and, with the whole plant, somewhat glandular-puberulent, heads spicate or racemose along the leafy branches, mostly nodding, ½ in. long, 10— 15-flowered: scales of involucre with mostly obtuse straight tips.— Usually along stream banks, in gravelly places, and chiefly in the inner Coast Ranges. Sept.—Dec.

Suborder 2. ASTERACEÆ.

Plants with a watery (never balsamic) juice, destitute of aromatic and bitter qualities, the leaves and heads only, in some, resinous. Leaves mostly alternate. Receptacle seldom chaffy. *Anthers obtuse and entire, or only emarginate at base. Style-branches not clavate,* often with filiform, or shorter and broader, papillose or hispid appendage. Pappus in most of ours of rather firm scabrous bristles. Disk-flowers yellow, in some of the genera with cyanic rays changing to red or purple.

* *Flowers of both ray and disk permanently yellow.*

Pappus of several short scales.................................XANTHOCEPHALUM 3
Pappus of a few stout deciduous awns........................ .GRINDELIA 4
Pappus of many and persistent slender bristles;
    Heads with ligulate ray flowers;
        Heads few, ½ in. high, somewhat spicate...............PYRROCOMA 5
        Heads many, panicled;
            Ray-achenes with no pappus...................HETEROTHECA 6
            All the achenes pappose { pappus double........CHRYSOPSIS 7
                             { pappus simple..........ERICAMERIA 9
        Heads solitary, peduncled; pappus soft, white...........STENQTUS 8
        Heads very small { corymbose: achenes silky...........EUTHAMIA 11
                    { panicled; achenes not silky.........SOLIDAGO 12
    Heads with palmatifid ray-corollas........................LESSINGIA 15
    Heads with no ray-corollas;
        Corollas ventricose..............................ISOCOMA 10
        Corollas not ventricose;
            Shrubs, with corymbose small heads .... ERICAMERIA 9
            Herbs, with larger heads....................CHRYSOPSIS 7

3. XANTHOCEPHALUM, *Willd.* Nearly glabrous, somewhat resiniferous freely branching herbaceous or suffrutescent plants. Leaves
alternate, narrow, entire. Heads small, spherical, hemispherical, or
narrower, usually corymbosely arranged at summit of stem and branches.
Involucral bracts coriaceous., the outer successively shorter, often with
greenish but usually appressed tips. Flowers of both ray and disk
permanently yellow. Style appendages slender. Achenes angled or
striate, mostly silky. Pappus minute, paleaceous or coroniform.

1. **X. Californicum** (DC.). Stems tufted, ascending from a woody
base, 1½ ft. high, loosely paniculate: leaves linear, acute, scabrous:
heads few, solitary or in pairs or threes at the ends of the branchlets,
turbinate or obovate, 3 lines high; fl. of disk and ray each 8—10: achenes
densely silky: pappus of about 12 unequal acutish scales, none longer
than the achene.—Marin Co., and Oakland Hills. June—September.

4. GRINDELIA, *Willd.* Coarse herbs or suffrutescent plants, with
sessile rigid mostly serrate leaves, and rather large hemispherical heads
terminating corymbose branches. Bracts of involucre imbricated in
many series, with usually narrow herbaceous squarrose-recurved tips.
Flowers of both disk and ray very numerous, permanently yellow.
Style-appendages lanceolate or linear. Achenes short, thick, compressed
or turgid, truncate, glabrous. Pappus of 2—8 deciduous stout awns or
bristles.

* *Herbaceous perennials, flowering in early summer.*

1. **G. camporum.** Stems white and shining, tufted from a perennial
root, 2 ft. high, *glabrous, very leafy up to the loosely corymbose heads, even
the branches of the corymb conspicuously leafy-bracted;* radical leaves
almost wanting, cauline oblanceolate-spatulate, sessile and clasping,
2 in. long, saliently serrate-toothed; bracts of flowering branches nearly
entire, spreading, involucres ½—¾ in. wide, their bracts with long
linear recurved tips: ray-achenes obscurely triquetrous, with 3 or more
pappus-awns; disk-achenes compressed, obliquely biauriculate or uni-
dentate at summit, and with pappus of 2 bristles.—Common on rich
plains east of the Mt. Diablo Range. June—September.

2. **G. rubricaulis,** DC. Rather slender, ascending, 2 ft. high, stems
from brownish to dull red, herbage scarcely glutinous, roughish-

pubescent or even somewhat hirsute: radical leaves numerous, tufted, oblanceolate, coarsely serrate; cauline reduced, few and remote: *heads solitary or few, nodding in the bud: inner bracts of involucre* closely imbricated and very glutinous, *without spreading tips:* achenes mostly thin and flat, with obcordate summit and only 2 pappus-awns. Var. **maritima**, Greene. Stouter, often depressed; leaves broader, firmer: pappus-awns 2—5, compressed, barbellate-scabrous on the margins.— Open glades among the wooded hills; the variety on bluffs near the sea.

3. **G. patens**, Greene. Foliage and pubescence of the preceding, nearly, but stem stouter, erect, the flowering branches at no stage nodding at summit; heads larger, ½—1 in. broad; *bracts of involucre mostly linear- or lanceolate-foliaceous, straight and widely spreading,* some of the inner with shorter and recurved tips: disk-achenes with obcordate summit and only 2 awns.—Hillsides and plains about San Francisco Bay.

* * *Late-æstival and autumnal species*

+—*Herbaceous perennial.*

4. **G. procera**. Strictly erect, 5—7 ft. high, simple up to the corymbose-paniculate summit, *the stout white stem scabro-puberulent,* plant otherwise glabrous, slightly glutinous: lower leaves unknown; upper cauline lanceolate, attenuate-acute, entire, 2—3 in. long; involucres small, low-hemispherical; bracts with appressed base and short slender recurved tips: rays short: pappus-awns 2.—Bottom lands of the lower San Joaquin, in places inundated in spring and early summer.

+— +— *Suffrutescent species.*

5. **G. cuneifolia**, Nutt. Bushy, 2—4 ft. high, glabrous: *leaves thickish and rather fleshy,* 3—4 in. long, cuneate-spatulate to linear-oblong, entire or sharply denticulate, clasping though not auricled at the broad base; involucre ½ in. high, glutinous, the bracts all with squarrose green tips: pappus-awns usually several, compressed, barbellulate.— Borders of salt marshes and along tidal sloughs about S. F. Bay and southward along the coast. August—December.

6. **G. paludosa**. About 5 ft. high, sterile leafy shoots a foot high, or more, surviving the winter, the plant otherwise herbaceous: herbage glabrous except the scabrous-ciliolate leaf-margins; only the involucres glutinous: leaves slightly fleshy, oblong-lanceolate to spatulate-oblong, 2—3 in. long, conspicuously serrate, at least *those of sterile shoots with a broad cordate-clasping base, the lobes surrounding the stem:* involucre squarrose: achenes with prominent turgid angles, those of the ray triquetrous, of the disk compressed: awns 2 only, even in the ray, stout, strongly flattened.—Abundant in brackish marshes of Suisun Bay. August—October.

5. PYRROCOMA, *Hook.* Rigid perennial herbs, with coriaceous mostly radical leaves from a fusiform caudex. Stems leafy-bracted, bearing racemose or panicled middle sized heads. Bracts of hemispherical involucre many, rigid, with herbaceous more or less squarrose tips. Flowers yellow; those of the ray rather numerous, short, pistillate; of the disk tubular, slightly dilated upwards. Style-appendages subulate-linear, pubescent. Achenes more or less flattened and striate, glabrous or pubescent. Pappus of copious reddish or brownish slender but rigid unequal bristles.

1. **P. elata.** Stout, erect, 1—3 ft. high, glabrous: radical leaves long-petioled, 6—8 in. long, lanceolate, entire; cauline 1—3 in., sessile, ascending: heads ½ in. high and as broad, disposed in an interrupted spike or narrow panicle: involucral bracts rigid, imbricated in several series, the green tips acute, spreading: achenes flattened, closely costate, pubescent —A somewhat rare plant of subsaline soils at Calistoga and near San Jose. July—October.

6. HETEROTHECA, *Cassini.* Tall hairy herbs, with alternate leaves, and a terminal corymbose panicle of middle-sized heads. Involucres ovate; their bracts closely imbricated in many series, without spreading tips. Flowers yellow; those of the ray pistillate,'of the disk perfect, the later with ovate or lanceolate style-appendages. Achenes compressed, pubescent, those of the ray thin-triquetrous with caducous pappus or none; pappus of disk achenes of an outer series of sparse short bristles, and an inner, more copious series of longer ones.

1. **H. grandiflora,** Nutt. Annual or biennial, 3—6 ft. high, hirsute, the inflorescence viscid and strong-scented by a coat of short gland-tipped hairs: cauline leaves oval or oblong, coarsely toothed, partly vertical by a twist in the petiole, this at base bearing 2 stipuliform lobes: involucre ½ in. high: ray achenes without pappus, those of the disk with but faint traces of the outer and shorter bristles.—Frequent along railways in Contra Costa Co.; an immigrant from S. Calif. July—Dec.

7. CHRYSOPSIS, *Elliott.* Perennials, leafy-stemmed and of rather low growth. Leaves sessile, entire or nearly so. Heads middle-sized, terminating corymbose or fastigiate branches. Involucres ovate or broader, of narrow regularly imbricated bracts in several series. Style-appendages linear-filiform to slender-subulate. Achenes compressed, obovate to linear-fusiform; pappus fuscous, of many capillary scabrous bristles, with or without an outer series of short bristles or paleæ.

＊ *Heads radiate; outer pappus setose-squamellate.*

1. **C. sessiliflora,** Nutt. *Slender, sparsely pilose-hispid, viscid-glandular:* leaves oblanceolate, sharply pointed: corymbose branches ending in about 3 subsessile *heads* ½ in. high, *leafy-bracted at base:* bracts of

involucre not pubescent but very viscid-glandular: achenes slender-fusiform, silky-pubescent; outer pappus slenderly squamellate.—Santa Clara Co., and southward. June.

2. **C. Bolanderi,** Gray. *Stoutish,* ½—1 ft. high; *pubescence long-silky: heads few and subsessile:* bracts of involucre not glandular, silky-villous: outer pappus of narrow paleæ nearly half as long as the achene.—On stony hilltops toward the sea; flowering in summer.

3. **C. echioides,** Benth. *Rigid, brittle,* 2—3 ft. high, often suffrutes-cent, *hoary with a dense hirsute and hispid pubescence:* leaves rigidulous, small: heads less than ¼ in. high, in short fastigiate corymbs; bracts hirsutulous: achenes silky but the hairs not appressed: setulose outer pappus not conspicuous.—Sandy plains, and banks of streams, from Solano Co. southward, east of the mountains. Aug.—Oct.

* * *Rays none; outer pappus obsolete.*

4. **C. rudis.** Erect or decumbent, 1—3 ft. high, rigid, brittle, rough-hairy but not hoary, glandular, heavy-scented: involucres in a narrow leafy panicle: bracts of involucre acute, midrib prominent, margin scarious: achenes oblong, pubescent; pappus copious, slender, scabrous, seldom a trace of the short outer series.—Common along stream banks in Napa Co.; heretofore referred to *C. (Ammodia) Oregana,* from which it is altogether distinct. July—Oct.

8. **STENOTUS,** *Nutt.* Glabrous resiniferous evergreen shrub. Leaves alternate, linear, entire. Heads solitary at the ends of the branches. Involucre hemispherical; bracts in 2 or 3 series, membranaceous, scarious-margined, closely appressed. Flowers yellow; rays few; disk-corollas dilated above, deeply 5-toothed. Style-appendages filiform, flattened, puberulent: Achenes oblong, somewhat compressed, densely villous; pappus very slender, permanently white.

1. **S. linearifolius** (DC.), Torr. & Gray. Very leafy, 1—4 ft. high; leaves 1 in. long, acute, spreading, punctate, 1-nerved: head about 1 in. broad, on a peduncle: rays 12—14: achenes densely white-villous; pappus copious, fragile or deciduous.—Mt. Diablo towards the summit, and southward. May—July.

9. **ERICAMERIA,** *Nutt.* Evergreen shrubs of low stature, with linear entire subterete punctate leaves, and terminal cymose or corymbose clusters of small heads. Involucre turbinate; bracts mostly lanceolate, very regularly imbricated, margins subscarious. Flowers yellow. Disk-corollas slender-tubular with subcampanulate throat and deeply cleft limb. Style-appendages filiform, acuminate, hirsutulous. Achenes prismatic. Pappus of scabrous slender bristles dull-white or yellowish, becoming reddish.

1. **E. microphylla,** Nutt. *Diffusely branching,* ½—1½ ft. high, the branches fastigiate-corymbose, very leafy throughout: *leaves linear,* terete, those of the branches ½—¾ in. long, deflexed, *bearing in their axils very short branchlets hidden by two-ranked closely imbricated shorter ones:* involucres ¼ in. high; bracts tomentose-ciliolate: rays about 5, short: achenes subcylindrical, striate, glabrous.—Sandy hills and beaches from Bolinas Bay southward; plentiful at San Francisco. Aug—Dec.

2. **E. arborescens** (Gray). Erect, fastigiately branching, 3—10 ft. high, densely clothed with *very narrow-linear subterete leaves* 1½—3 in. long, 1 line wide: *heads in a terminal cymose-corymb,* 20—25-flowered: turbinate involucre scarcely 3 lines high; bracts lanceolate, acute: rays seldom present: achenes short, apparently quadrangular, silky-pubescent. —At considerable elevations in the mountains of Sonoma, Marin and Contra Costa counties. Sept.—Dec.

10. ISOCOMA, *Nutt.* Rather rigid tufted erect suffrutescent plants, with thick slightly succulent toothed leaves, and a corymbose terminal cluster of smallish rayless heads. Bracts of several-flowered involucre coriaceous, closely imbricated, the tips herbaceous but appressed, obtuse or acutish. Corollas permanently yellow; tube slender; limb ventricose, the segments being more or less strongly connivent about the style, the pubescent appendages of which are ovate or somewhat narrower. Achenes short, compressed or subterete, silky-pubescent. Pappus-bristles numerous, unequal, the inner longest and often perceptibly flattened and awn-like, hardly scabrous.

1. **I. vernonioides,** Nutt. Glabrous or loosely pubescent, 2—4 ft. high, erect: leaves oblanceolate, more or less serrate, 1—2 in. long, often with many fascicled ones in the axils: heads 4 lines high, campanulate; bracts of involucre obtusish: pappus-bristles stout, none very perceptibly flattened.—Common shrub of S. Calif., found at Black Point, San Francisco, where it may have been introduced accidentally.

2. **I. arguta.** Branches 6—10 in. high, more or less pubescent or hirsute below, glabrous above, leafy throughout; leaves diminishing upwards, the lowest 1 in. long, all broadly oblanceolate, *of coriaceous texture, with saliently spreading coarse and acute or mucronate teeth:* heads ¼ in. high, turbinate, 12—15-flowered: inner pappus-bristles distinctly flattened and tapering very gradually from base to apex.—Subsaline plains east of the Vaca Mts., in Solano Co., *Jepson.*

11. EUTHAMIA, *Cassini.* Erect glabrous perennials, very leafy, the branching more or less distinctly corymbose. Leaves nearly linear, entire, pellucid-punctate. Heads small, clustered at the ends of the branches. Involucral bracts firm, imbricated, glutinous. Flowers permanently yellow; those of the ray about twice as many as those of the disk. Achenes short, turbinate, villous-pubescent.

1. E. occidentalis, Nutt. Somewhat paniculately branching, 3—6 ft. high: leaves lanceolate-linear, obscurely 3-nerved: bracts of involucre linear-lanceolate, acute: rays 16—30; disk-flowers 8—14, their style-tips obtuse.—Common in low grounds along rivers and on the borders of marshes. Aug.—Oct.

12. SOLIDAGO, *Vaillant* (GOLDEN ROD). Strict simple-stemmed perennials, with alternate more or less serrate leaves. Inflorescence a terminal cluster of many small heads, usually disposed in scorpioid racemes and forming a panicle; otherwise forming a thyrsus. Involucre narrow; bracts in 2 or more series, neither herbaceous tipped or glutinous. Flowers all permanently yellow; the outer and ligulate short, the inner narrow-funnelform. Style-appendages flattened, lanceolate. Achenes terete or prismatic, 5—10-nerved, glabrous or pubescent. Pappus a series of unequal scabrous permanently white bristles.

* *Heads numerous, panicled.*

1. S. sempervirens, L *Bright green and glabrous,* leafy throughout, 2—8 ft. high: *leaves rather fleshy,* lanceolate to linear, the upper acute, lower obtuse, *all entire:* panicle narrow, dense, virgate: heads 3—4 lines long: bracts of involucre lanceolate, scabrous-ciliolate: rays 8—10, rather large, golden yellow: achenes minutely pubescent.—Attributed to marshes about San Francisco, at Laguna Honda, etc. Aug.—Nov.

2. S. elongata, Nutt. *Puberulent,* 1 to 2 ft. high equably leafy up to the long panicle: *leaves thinnish, lanceolate, acute, sparingly serrate,* 2—3 in. long: branches of panicle scarcely secund, ascending; heads small; bracts of involucre linear, acutish or obtuse: rays 10—16, narrow: achenes pubescent.—Oakland Hills, in open ravines, etc. July—Oct.

3. S. Californica, Nutt. *Roughish with an almost cinereous short pubescence;* commonly 2—4 ft. high: leaves ampler and more numerous below, passing from obovate to oblong-lanceolate and lanceolate, and from obtuse to acute, the lower and broader more or less serrate: panicle usually virgate but loose, 4—12 in. long, the racemiform clusters secund but seldom recurved: heads 3 lines high; bracts lanceolate-oblong or oblong-linear, obtusish pubescent, rays 7—12, pale yellow: achenes pubescent.—Very common, in dry and even sandy soil. July—Oct.

* * *Heads fewer and larger, somewhat thyrsoidly congested.*

4. S. spathulata, DC. *Glabrous, slightly glutinous,* with the order of *Grindelia,* 1—2 ft. high: stems decumbent and even suffrutescent at base: *lower leaves spatulate,* 2—4 in. long, *rounded at apex, serrate:* heads 4 lines high, almost as broad, about 25-flowered, disposed in short racemes thyrsoidly crowded at and near the summit of the stem: bracts of involucre oblong or broadly linear, all but the inmost series obtuse and green-herbaceous almost throughout, the inner acutish and with a green midvein: rays short: achenes pubescent.—On bluffs near the sea at Point Lobos; also in the Mission Hills. Aug., Sept.

13. PENTACHÆTA, *Nutt.* Slender almost glabrous small vernal annuals. Leaves alternate, linear, entire. Involucres solitary, hemispherical or campanulate, of thin scarious-margined appressed mucronulate bracts in 2 series. Rays white, yellow, or wanting. Disk-corollas yellow, very slender. Style-appendages filiform-subulate, hispid. Achenes pubescent. Pappus of 3—5 slender bristles.

1. **P. bellidiflora,** Greene. Sparingly branching, the peduncles somewhat scapiform: *involucre hemispherical, many-flowered: rays 8—14, white or reddish;* achenes oblong-turbinate, villous: pappus-bristles 5 or none.—Open hills and sterile slopes in Marin and San Mateo counties, not common. April, May.

2. **P. alsinoides,** Greene. Dichotomously branching, only 2—5 in. high: *involucre turbinate,* of 5—7 bracts and *3—7-flowered: rays 0: disk-corollas filiform, not deeply cleft:* achenes obovate-clavate; pappus-bristles 3, very slender.—An obscure hillside plant, but not rare. April—June.

3. **P. exilis** (Gray), Greene. Very slender, only 2—3 in. high, usually simple and monocephalous: *whole plant purplish,* the peduncle white-villous under the small head: *outer series of corollas rose-red, claviform-urceolate, i. e.,* widening upwards, the throat abruptly contracted under the minute teeth: pappus of 3—5 short bristles or cusps, or obsolete.—Frequent on open hills in San Mateo and Contra Costa counties, thence northward. April, May.

14. BELLIS, *Pliny* (DAISY). Low herbs. Involucres broad, many-flowered; bracts of nearly equal length. Rays many, white or reddish. Disk-corollas yellow. Style-appendages short, triangular. Achenes obovate, compressed, nerved on the margins. Pappus none.

1. **B. perennis,** L. Perennial, acaulescent: leaves obovate: scapes several, each with a single head.—Escaped from gardens, and naturalized in the seaboard counties from at least Marin northward.

15. LESSINGIA, *Chamisso.* More or less floccose-woolly annuals with alternate more or less serrate leaves and small cymosely panicled heads of yellow, whitish or purplish flowers, these all perfect. Corollas with slender tube and long narrow lobes; those of the marginal row more deeply cleft on one side and imitating a palmatifid ligule. Involucre campanulate or turbinate; bracts much imbricated and appressed, herbaceous-tipped. Anthers with slender-subulate appendages. Appendages of style-branches obtuse or truncate, densely hispid, often with a setiform cusp amid the hairs. Achenes turbinate or cuneiform, silky-villous. Pappus-bristles rigid, scabrous, red or brownish.

* Yellow-flowered species.

1. **L. Germanorum,** Cham. Low, slender, *branching and spreading from the base;* branchlets at length glabrate, purple: lower leaves sinuate-pinnatifid, those of the branches narrowly oblanceolate: involucre hemispherical, its *bracts more or less green-herbaceous not glandular.*—San Francisco and southward in sandy soil near the sea.

2. **L. glandulifera,** Gray. *Erect, stoutish, diffusely branched above:* leaves more irregularly and deeply toothed or cleft, those of the stem more numerous, ovate or oblanceolate, and of the branchlets minute and almost crowded, rigid, beset along the margin with yellowish large glands: involucre campanulate to turbinate, its *bracts more or less glanduliferous.*—Plains of the lower San Joaquin and southward.

* * Flowers pale- or deep-purplish.

3. **L. ramulosa,** Gray. Erect, 1—2 ft. high, very loosely branching, the glabrate branchlets and upper leaves more or less hirtellous and glandular: involucre campanulate or turbinate, 10—20-flowered: *corollas short, purple:* style-appendages with minute setiform tip.—Dry hills from Sonoma Co. southward.

4. **L. leptoclada,** Gray. Taller and more slender, with almost filiform branchlets bearing few or solitary 5—20-flowered heads: *corollas elongated:* style-appendages with a conspicuous subulate tip.—Same range.

5. **L. virgata,** Gray. More densely woolly: *stem and virgate branches rigid:* upper leaves appressed, concave, carinate-nerved: *heads spicately sessile in the axils* of the leaves: involucre cylindrical, 5—7-flowered: fl. pale or whitish: style-appendages with conspicuous subulate tip.—Plains of the lower San Joaquin and Sacramento.

6. **L. nana,** Gray. Stems *very stout, short, depressed,* the whole plant white-woolly: heads large (½ in. high), oblong, 10—20-flowered; outer bracts linear-lanceolate, mucronate-acute or cuspidate, scarcely herbaceous in any part, *inner scarious-chartaceous, white, tapering into a rigid subulate point:* fl. crimson; style-appendages with no cusp: achenes short and turgid; pappus red.—Foothills of the Mt. Diablo Range and eastward. July—Sept.

16. **CORETHROGYNE,** *DeCandolle.* Genus very nearly allied to *Lessingia;* distinguished chiefly by the numerous and altogether ligulate violet ray-corollas: the habit in our species (the typical), quite different, the roots being perennial, the branches often subscapiform and monocephalous; the heads large; involucres hemispherical. Style-appendages strongly hairy but not cuspidate.—June—Aug.

1. **C. Californica,** DC. *Suffrutescent* and diffusely branched from the base, *densely white-floccose;* the assurgent flowering branches numerous:

lowest leaves oblanceolate-spatulate, few-toothed; upper linear entire.—At Crystal Springs, San Mateo Co., and southward.

2. **C. obovata,** Benth. *Stems fewer, erect or ascending, hoary,* as also the obovate or spatulate obtuse leaves which are serrate above; those of the branches oblong or narrower: heads 1 in. broad or more, including the purple rays.—Sandy hills from Lake Merced to Humboldt Co., plentiful on Point Reyes.

17. ASTER, *Tourn.* Leafy-stemmed autumnal herbs with panicled or somewhat corymbose heads. Involucre hemispherical to campanulate, of several series of unequal imbricate bracts with herbaceous tips. Rays many, not very narrow, white, pinkish, or bluish. Disk-corollas yellow changing to red-purple; tube slender; limb funnelform. Style-appendages from triangular-lanceolate to slender-subulate. Achenes compressed. Pappus copious, dull-white, or rarely more deeply colored, scabrous.

* Perennials; inflorescence corymbose.

1. **A. radulinus,** Gray. Stoutish, roughish-pubescent, 1 ft. high or more, usually bearing an open corymb of middle-sized heads: leaves rigid, obovate-oblong, acute, sharply serrate above, tapering to a narrow entire base, scabrous both sides: involucre obconic, 4—5 lines long: bracts rigid, appressed, acutish or mucronate, the tips green: rays white; disk-corollas becoming red: achenes minutely pubescent; pappus rather rigid.—Borders of woods and thickets; early-flowering. July—Sept.

* * Perennials; paniculate or racemose.

2. **A. Menziesii,** Lindl. Strictly erect, 2 ft. high, usually simple and very leafy up to the mostly *simply racemose or racemose-paniculate inflorescence,* the whole plant cinereously and roughly pubescent: leaves oblong-lanceolate, acute, 1—3 in. long, remotely serrate or entire, sessile by a broad auriculate-clasping base: involucre broadly turbinate, ¼ in. high; bracts somewhat spatulate, well imbricated, the broad green tips obtuse: rays light violet, rather short.—Vaca Mountains, *Jepson,* and southward. Sept.—Dec.

3. **A. invenustus.** Stout stems 2 ft. long or more, ascending from a decumbent base; herbage cinereous with scabrous and short-hirsute pubescence: lower cauline leaves lanceolate-spatulate, 2—3 in. long, with remote and slight serratures: *heads very numerous in an ample cymose panicle;* involucres nearly hemispherical, ¼ in. high, the almost wholly green-herbaceous very obtuse spatulate-linear bracts in rather few ranks; rays dull pale purplish.—Collected only by the author, and in the upper part of Napa Valley, near Calistoga, August, 1888.

4. **A. Chilensis,** Nees. Erect, stoutish, 2—4 ft. high, glabrous or somewhat hirsute, the stem occasionally with strongly hirsute lines:

leaves lanceolate, acute, entire, 2—5 in. long, entire, or obscurely serrate, the whole margin scabrous: heads ½ in. high, in a more or less ample *panicle of short loose leafy racemes; bracts* of campanulate or broadly obconic involucre much imbricated, *linear or linear-spatulate*, with short and rounded green tips: rays 25—30, purple or violet, ½ in. long.— Common and variable; some forms very showy. Aug.—Oct.

5. **A. Sonomensis.** Slender, decumbent at base, 1—1½ ft. high, *glabrous and glaucescent*, only the leaf-margins scabrous-ciliolate: leaves mainly radical, narrowly lanceolate, tapering to a long petiole, this with a dilated and strongly ciliate basal part: *heads rather few in a terminal corymbose panicle;* involucres ¼ in. high, broad-campanulate to broad-obconic, the well imbricated *bracts narrowly oblanceolate, acute;* rays purplish, rather narrow, ½ in. long.—In open plains of the Sonoma Valley, in low subsaline ground. Sept., Oct.

6. **A. lentus.** Erect, slender, 4—6 ft. high, *slightly succulent, glabrous* except a slight pubescence under the heads, and a delicately serrulate-scabrous margin to the leaves: lowest leaves 3—5 in. long, lanceolate-linear, slightly falcate, those of the flowering branches straight and half-clasping at the sessile base: panicle loose and ample, often a yard long, the branches loosely racemose: heads 4—5 lines high; *involucres oblong; bracts linear, acute, appressed, green-herbaceous and somewhat succulent* almost throughout: rays many, ¾ in. long, light purple.— Plentiful along tidal streams in western part of the Suisun marsh; the largest and most beautiful of Californian species. Oct., Nov.

* * * *Biennial or annual, paniculate; heads small.*

7. **A. exilis,** Ell. Glabrous, slender, 2—6 ft. high, with narrow lanceolate or linear entire leaves, and a diffuse panicle of very small heads: bracts of narrow involucre lanceolate-subulate, mainly green herbaceous, the margins scarious: rays white, short and inconspicuous: pappus fine and soft: achenes little compressed.—Borders of Suisun marshes, and elsewhere in low subsaline land. Aug.—Dec.

18. **ERIGERON,** *Linn.* Involucre of narrow, usually almost equal bracts which are never coriaceous or distinctly herbaceous-tipped. Rays very narrow, commonly extremely numerous and in several series, but in several of our species wanting. Style branches with short roundish appendages. Achenes compressed, 2-nerved. Pappus of scanty fragile bristles, a short outer series sometimes manifest.

* *Perennials; stems leafy at base; cauline leaves reduced.*

1. **E. glaucus,** Ker. *Monocephalous branches several from a stoutish leafy caudex,* the whole plant more or less villous or hirsute, especially above; leaves obovate or spatulate-oblong, 2—4 in. long, entire or with

few teeth; those of the flowering branches more bract-like and fewer: heads 1½ in. broad including the not very narrow light-violet rays.—On hills and cliffs along the seaboard.

2. **E. Philadelphicus, L.** Hirsute, 1—3 ft. high from a perennial root: radical leaves obovate or spatulate, the scattered cauline ones oblong or oblong-lanceolate, with broad clasping base, all irregularly toothed: *heads less than 1 in. broad, in an ample loose terminal cymosecorymb:* rays very many and extremely narrow; flesh-color to bright pink.—Along streamlets and the borders of boggy places.

* * *Perennials; stems simple, brittle, equably leafy up to the corymb; leaves narrow, entire: involucral bracts unequal and more or less imbricated; outer pappus of few short bristles.*

+—*Heads with bluish rays.*

3. **E. foliosus,** Nutt. Scabrous, more or less strigose-pubescent, 1—2½ ft. high: leaves narrowly oblanceolate, 1—2 in. long, those of the branches reduced: hemispherical heads ½ in. broad; rays about 30; achenes with a few coarse bristly short hairs.—Dry hills of Sonoma and Contra Costa counties, and southward. June—Sept.

+ + *Rays wanting; involucre much imbricated.*

4. **E. angustatus,** Greene. Stems tufted, 2 ft. high, rigid and brittle; herbage *glabrous except a few short incurved hairs on the margins and midvein of the leaves,* and a somewhat granular minute indument on the much imbricated turbinate involucres: leaves narrowly spatulate-linear, entire: corymbose panicle ample: bracts of the involucre with reddish tips: corollas of a deep golden yellow: achenes setose-hirsute.—Dry hills on either side of Napa Valley. July—Oct.

5. **E. Biolettii.** Size of the preceding: whole plant *scabrous-puberulent: leaves oblanceolate,* obtuse, with sparsely but rigidly *hispid-ciliate margins:* corymb with branches less divergent: achenes appressed-pubescent.—On Hood's Peak, *Bioletti,* and Howell Mountain, *Jepson.*

6. **E. petrophilus,** Greene. Half the size of the last, more leafy, the whole *plant except the somewhat* glandular heads *canescently hirsute;* the corymb less ample; bracts of involucre not as numerous.—Rocky summits of the inner Coast Range, from Mt. St. Helena to Mt. Hamilton; above Wild Cat Creek near Berkeley. July—Oct.

* * * *Annuals, with thyrsoid-paniculate inflorescence.*

7. **E. Canadensis,** L. Sparsely hispid or nearly glabrous; stem stout, erect, 1—6 ft. high, with countless small subcylindric heads in a rather dense panicle: lowest leaves spatulate, upper linear: heads only 2 lines high: rays white, very short.—A common weed in cultivated lands, or by waysides.

19. BACCHARIS, *Linn.* Diœcious shrubs or herbs, with striate or angled branches, alternate simple, often glutinous leaves, and small clustered discoid heads of white unisexual flowers. Involucre of scale-like imbricated bracts. Fl. of staminate heads with tubular-funnelform 5 cleft corolla; of the pistillate slender-tubular, truncate or minutely toothed. Style-appendages ovate to lanceolate, rarely coalescent. Achenes 5—15-costate, glabrous or pubescent. Pappus of fertile flowers very fine and soft, often becoming elongated in fruit.

* Herbaceous perennial.

1. **B. Douglasii,** DC. Erect, 3—4 ft. high, simple up to the terminal corymb; leaves very glutinous, ovate-lanceolate, nearly or quite entire, 3—6 in. long: bracts of involucre erose-ciliate: pappus of pistillate fl short, soft; of staminate clavellate and barbellate at summit.—In moist lowlands. Sept.—Nov.

* * Suffrutescent or shrubby.

2. **B. glutinosa,** Pers. Shrub 6—12 ft. high: *leaves lanceolate, acute,* entire, denticulate or repand-dentate, 2—3 in. long: heads in ample cymose panicles at the ends of long willowy leafy branches.—On banks of streams, from Napa and Solano counties southward. May—Dec.

3. **B. consanguinea,** DC. *Compactly branching evergreen 8—12 ft. high:* branchlets green, angular from the leaf-bases: *leaves subcoriaceous, glutinous, 1 in. long and less, cuneate-obovate,* coarsely toothed: heads sessile singly or in pairs or threes in the leaf-axils: bracts of involucre oblong-linear, obtuse, with subscarious fringed margins.—Hillsides and banks of streams everywhere. Oct.—Dec.

4. **B. pilularis,** DC. *Low, slender, the depressed or prostrate diffusely branching stems 1—2½ ft. long;* branchlets angular: *leaves seldom ¼ in. long,* cuneate-obovate, angular-toothed or subentire, heads mostly solitary in the leaf-axils and at the ends of the broom-like fastigiate branchlets: involucral bracts acutish, fringed toward the tips.—Sandy soils along the seaboard and about San Francisco Bay. Aug.—Oct.

Suborder 3. GNAPHALIACEÆ.

Plants mostly white with floccose wool, the herbage apt to be more or less pleasantly or unpleasantly scented. Heads discoid: *bracts of involucre various, often scarious and white or yellowish. Anthers caudate. Style-branches of perfect flowers blunt, unappendaged,* the stigmatic lines running almost to the summit, which is sometimes papillose or penicillate. Pappus finely capillary or none.

* *Involucral bracts few or none: receptacle chaffy.*

Fructiferous chaff or bract quite enclosing its achene;
  Achenes gibbous...GNAPHALODES   21
  Achenes straight or curved;
    Receptacle columnar...................................ANCISTROCARPHUS   22
    Receptacle globular or ovoid.........................PSILOCARPHUS   23
Fructiferous bract scarcely enclosing its achene;
    Receptacle columnar; pappus none....................EVAX   24
    Receptacle convex; pappus present...................FILAGO   25

* * *Involucral bracts many; receptacle not chaffy.*

Involucre dry, but hardly scarious...........................PLUCHEA   20
Involucre herbaceous; achenes large.........................ADENOCAULON   27
Involucre scarious; achenes small...........................GNAPHALIUM   26

20. PLUCHEA, *Cassini.* Herb with alternate ample leaves, and a terminal cymose cluster of smallish heads; these many-flowered, the flowers largely pistillate only, their corolla reduced to a slender truncate or 2—3-toothed tube, that of the hermaphrodite (but sterile) flowers regularly 5-cleft. Achenes small, 4—5-angled or sulcate. Pappus a series of capillary bristles.

1. **P. camphorata** (L.), DC. Annual, stoutish, leafy, 2 ft. high; minutely and somewhat viscidly pubescent: leaves oblong-ovate to oblong-lanceolate, acute at both ends, toothed or denticulate, the larger (3—5 in.) petioled: heads short-pedicelled, dull-purple, crowded in a corymbiform cluster: involucral bracts ovate to lanceolate, often tinged with the dull pale purple of the corollas.—Borders of brackish marshes about Suisun Bay, etc. Aug.—Oct.

21 GNAPHALODES, *Tourn.* Low floccose-woolly annuals, with alternate entire leaves. Heads scattered, several-flowered. Pistillate flowers on a small receptacle, each enclosed in a conduplicate bract, the tip of which is scarious-appendiculate; the few hermaphrodite-sterile ones mostly naked. Involucre outside of the fruiting bracts scanty and scarious. Achene gibbous, obovate, enclosed in its bract and falling away with it. Pappus none.

1. **G. Californica** (F. & M.). Slender, erect, 6—12 in. high: leaves mostly linear: *fructiferous bracts 5 or 6, firm-coriaceous,* somewhat semi-obcordate or semiobovate, straight anteriorly, the erect beak-like tip largely scarious.—Open ground; very common. May.

2. **G. amphibola** (Gray). *Fructiferous bracts about 10,* somewhat imbricated on an oblong receptacle, *membranaceous or merely chartaceous* at maturity, the beak, an ovate almost hyaline appendage, at maturity porrect.—Hills of Contra Costa Co. May.

22. ANCISTROCARPHUS, *A. Gray.* Low canescently flocculent annual, erect, branched from the base, with alternate entire leaves, and more or less glomerate heads. Fertile flowers 5—9, loosely disposed on a slender receptacle, their enclosing bracts cymbiform, firm except the narrow hyaline tip. Sterile flowers involucrate by 5 larger bracts, these ovate-lanceolate, tapering into a rigid incurved-uncinate cusp, persistent and at length stellate-spreading. Achene ovate-fusiform, obscurely decompressed, the pericarp distinct from the seed and faintly nerved. Pappus none.

1. **A. filagineus,** Gray. Leaves linear to spatulate: heads capitate-glomerate, the hooked empty bracts at maturity ¼ in. long.—In open grounds; not common.

23. PSILOCARPHUS, *Nutt.* Small usually depressed much branched floccose annuals, with opposite leaves and globose heads sessile in the axils or at the forks. Fructiferous bracts numerous, on the globular or oval receptacle, cucullate-saccate, semiobovate or semiobcordate, rounded at top, herbaceo-membranaceous, apex introrse, the ovate or oblong hyaline appendage inflexed or erect. Achene loose within the bract, oblong or narrower, straight, slightly compressed. Pappus none.

1. **P. tenellus,** Nutt. Prostrate, forming a dense mat 3—6 in. wide: *heads very many:* leaves spatulate, ¼—½ in. long: fructiferous bracts scarcely a line long: *achene ovate-oblong.*—In rather low or shaded grounds among the hills. May.

2. **P. brevissimus,** Nutt. Dwarf, *with very few and rather large woolly heads:* leaves oblong or lanceolate, 2—5 lines long, seldom surpassing the heads: *achene cylindrical or slightly clavate,* 1 line long.—Plains of the interior, in low places. May.

24. EVAX, *Gærtner.* Low but rigid, leafy, with heads axillary and terminal. Bracts of the involucre and those of the receptacle subtending the pistillate flowers from oblong to obovate, becoming coriaceous, persistent, concave. Receptacle slender columnar from a broader base, sparsely villous, the pistillate flowers and their bracts crowded at its base; the summit bearing a whorl of 3—7 coriaceous obovate or rounded open bracts subtending a few sterile flowers; these with cleft style but no ovary. Achenes pyriform-obovate, somewhat obcompressed, very smooth. Pappus none.

1. **E. caulescens** (Benth.), Gray. *Branching from the base, erect,* 2—4 in. high; leaves spatulate, the blade ½ in. long, tapering to a slender petiole as long: *heads mostly solitary in the axils,* a number glomerate at summit, which is not specially leafy.—Very common on sterile hills along our northern border. May.

2. **E. acaulis,** Greene. Stout and low, *the very short branches horizontal:* leaves with short blade and greatly elongated petiole: *head glomerate at the ends of all the branches,* none in the axils.—Moist plains near Antioch, and southward. April—June.

3. **E. involucrata.** Stout, *strictly erect, simple,* or rarely with one or more ascending long branches from the base, 8 in. high or more: *heads only* in a *terminal* hemispherical cluster ¾ in. broad, *surrounded by a conspicuous whorl of 15 or 20 leaves,* these with spatulate-obovate cuspidate blade ½ in. long, only a third the length of the slender petiole, this abruptly dilated at base to half the width of the blade; cauline leaves shorter and narrower.—Plains of the lower Sacramento; collected only by the author. May.

25. FILAGO, *Tourn.* Erect rather slender floccose-woolly herbs, with alternate and entire leaves, and small heads in capitate lateral and terminal clusters. Rays 0. Receptacle plane, hemispherical or subconical; its naked summit bearing both sterile and fertile flowers having a pappus of capillary bristles. Base of receptacle bearing pistillate flowers, the achenes from these being destitute of pappus and enfolded by a concave bract. Achenes terete or slightly compressed, sometimes roughish-papillose.

1. **F. Californica,** Nutt. A span high or more: *heads ovate, slightly angular:* convex: pistillate fl. 8—10, their bracts broadly ovate, deeply boat-shaped, incurved; inner bracts oblong, concave: achenes almost terete, obscurely pappillose-granular.—Dry hills. May.

2. F. GALLICA, L. Receptacle nearly plane: *heads pentagonal-conical:* outer achenes completely enclosed in their conduplicate at length indurated bracts.—Introduced from Europe, but not rare with us.

26. GNAPHALIUM, *Diosc.* Floccose-woolly. Leaves sessile, entire. Heads cymosely clustered, white, yellowish or rose-tinted. Receptacle flat, naked. Bracts of involucre scarious, imbricated. At least the outer flowers (usually all of them) fertile. Achenes terete or flattish. Pappus a single series of scabrous capillary bristles.

* *Pappus-bristles not united at base, falling separately.*

+*Plants diœcious.*

1. **G. Americanum,** Clusius (1601). G. *margaritaceum,* L. Erect, 1—2 ft. high, growing in tufts from a perennial root, equably leafy up to the terminal cymose corymb; white-floccose, except the glabrate upper surface of the broadly lanceolate leaves; broadest leaves 3-nerved: bracts of involucre pearly-white, radiating in age.—Common about the Bay, on wooded or bushy northward slopes of hills.

+ + *Heads heterogamous; all the flowers fertile.*

++ *Involucre woolly at base only.*

2. **G. microcephalum,** Nutt. Biennial, slender, with several erect branches 2 ft. high or more, loosely corymbose-paniculate above, the

whole *herbage white with a persistent wool, not at all glandular or heavy-scented: leaves linear,* or the lower spatulate, with slenderly decumbent base: involucres small, ovate, bright white; bracts ovate or oblong, obtuse.—Hillsides. Aug.—Oct.

3. **G. Chilense,** Spreng. Annual and biennial, stoutish, 1—2½ ft. high, cymose-corymbose at summit: *leaves lanceolate, more thinly floccose* than in the last, the short decumbent leaves rather broad: involucre hemispherical, with a greenish-yellowish tinge; bracts thin, oval or oblong, obtuse. Var. **confertifolia.** Very stout and low: leaves linear, densely clothing the stem up to the sessile dense cluster of heads.—Very common and variable; the variety biennial; both flowering at almost all seasons.

4. **G. Californicum,** DC. Stoutish, 2—3 ft. high, biennial, the leaves diminishing in size towards the broad cymose terminal loose cluster of large rather dull white heads: *leaves lanceolate, glabrate above,* glandular and balsamic-scented, *very obviously adnate-decurrent:* outer bracts of the involucre ovate or oblong, the inner acute.—Common on dry hills in places partly shaded.

5. **G. ramosissimum,** Nutt. Biennial, erect, 3—5 ft. high, the fastigiate panicle often 2 ft. long and more, of small reddish heads: *leaves green and glandular on both faces, linear. decurrent, the herbage very sweet-scented:* heads only 2 lines high: involucral bracts rather few, oblong-lanceolate, acutish.—Wooded hills; late-flowering. Sept.—Nov.

++ ++ *Involucres deeply embedded in loose wool.*

6. **G. palustre,** Nutt. Low branching annual, floccose with long wool: leaves spatulate to oblong and lanceolate: heads glomerate, leafy-bracted, a line high: tips of linear involucral bracts white, obtuse.—In low moist lands. May—Aug.

* * *Pappus-bristles united at base, deciduous in a ring.*

7. **G. purpureum,** L. Biennial, simple or branching, erect or decumbent, 6—10 in. high, canescent with a dense coating of close wool: leaves spatulate, obtuse, usually becoming glabrate and green above: heads crowded in an elongated more or less interrupted spiciform inflorescence: involucre brownish: achenes sparsely scabrous.—In open grounds. March—May.

27. ADENOCAULON, *Hook.* Perennial, with alternate dilated leaves on long margined petioles; the slender stem naked and paniculate above, bearing small heads of whitish flowers; the peduncles beset with stalked glands. Involucre of few thin-herbaceous bracts. Receptacle flat, naked. Achenes ovoid-oblong or subclavate. far exceeding the involucre, the upper part beset with stout stipitate glands.

1. **A. bicolor,** Hook. Stem 2 ft. high: leaves ample, deltoid-cordate, coarsely sinuate-dentate or slightly lobed, green above, white-cottony beneath: involucral bracts 4 or 5, in one series, ovate, reflexed in fruit, small by the side of the 4—6 clavate achenes.—Redwood forests.

Suborder 4. AMBROSIACEÆ.

Heads small, greenish, the fertile flowers without corolla, or this reduced to an obscure rudiment. Rays none. Staminate involucres mostly forming a raceme above the axillary and few pistillate ones. Anthers but slightly united or quite distinct. Pappus none.

Heads all alike, and only in the leaf-axils................................IVA   28
Staminate heads racemose above the others..........................AMBROSIA   29
Staminate heads glomerate; fertile head becoming a bur............XANTHIUM   30

28. IVA, *Linn.* Perennial herb with simple mostly alternate leaves, and discoid heads nodding on short pedicels in their axils. Involucre of few scales in 1 series, commonly joined into a cup. Marginal fl. pistillate and with short tubular corolla; the other and more numerous fl. staminate, with funnelform 5-lobed corolla and undivided style: anthers nearly distinct. Receptacle with linear or. spatulate scales subtending the sterile fl. Achenes thick, naked.

1. **I. axillaris,** Pursh. Branching sparingly, 1—1½ ft. high: leaves from obovate and spatulate to broadly linear, sessile, entire, 1 in. long or more: heads hemispherical: scales of involucre about 5, united at base, or beyond the middle. Var. **pubescens,** Gray. Villous with loose spreading hairs; the involucre turbinate, almost entire.—Solano Co. and southward, mostly on subsaline plains, or near the coast.

29. AMBROSIA, *Dodoens.* Weedy aromatic coarse perennials with mostly alternate and pinnately divided leaves. Flowers unisexual, the staminate heads several-flowered and arranged in erect spikes or racemes resembling aments. Pistillate heads mostly in the axils of the upper leaves, 1—4-flowered, their involucres closed and achene-like, in maturity bearing protuberances or prickles. Achene ovoid or obovate, thick.

* *Fruit 1-seeded, more or less roughened, but not spinescent.*

1. **A. psilostachya,** DC. Stems erect, from horizontal rootstocks, 2 ft. high or more, with strigose pubescence and somewhat scabrous: leaves once or twice pinnatifid: fr. mostly solitary in the axils, turgid-obovoid, less than 2 lines long, obtusely short-pointed, rugose-reticulate, either unarmed, or with 4 short blunt or sharp tubercles.—Borders of fields in uncultivated land near the Bay; plentiful on Point Isabel.

* * *Fruit often more than 1-seeded, spinescent.*

2. **A. bipinnatifida** (Nutt.). Stems very stout, procumbent, 2—3 ft long, somewhat hirsute: *leaves ovate,* 1—3 in. long, *twice or thrice pin-*

nately parted into oblong-linear divisions and small oblong lobes, canescent with a silky pubescence: sterile raceme dense, the heads large: fruit ovate-fusiform, armed with short thick flattish spines, their tips often incurving.—Sandy beaches; very common. June—Dec.

3. **A. Chamissonis** (Less.). Size, habit etc., of the last, but *leaves cuneate-obovate*, or oblong-ovate with cuneate base, *obtusely serrate*, only some of *the lower laciniate-incised:* fruiting involucre ovate, the spines broad and channeled. Habitat of the preceding, but less common.

30. **XANTHIUM,** *Tourn.* (COCKLE-BUR). Coarse annuals, with branching stems, alternate lobed or toothed leaves, and clustered heads of greenish flowers; the staminate clusters uppermost, the pistillate in the leaf-axils. Involucre of staminate heads 1 or 2 series of narrow bracts. Stamens monadelphous but anthers merely connivent. Fertile head a closed ovoid bur-like 2-celled and 2-flowered involucre, 1—2-beaked at apex: each flower a single pistil, becoming a thick ovoid achene, the two enclosed in the hardened prickly involucre.

1. X. SPINOSUM, L. Widely branching from the base, 2 ft. high: *leaves ovate-lanceolate, more or less lobed* or pinnatifid, glabrate and green above, *white-tomentose .beneath:* burs ¾ in. long, armed with short weak prickles.—By waysides, common; native of tropical America.

2. **X. Canadense,** Mill. Stout, branching above only: *leaves broad-ovoid, slightly lobed, scabrous:* bur an inch long, densely beset with stoutish prickles, and at apex strongly 2-horned.—In low fields, where it may have been introduced. Aug.—Oct.

Suborder 5. HELIANTHACEÆ.

Plants commonly with balsamic-resinous juice, and coarse roughish or woolly foliage. Rays conspicuous; *receptacle strongly chaffy;* anthers not caudate; *.involucre not scarious;* pappus never of capillary bristles.

Involucre of 1 or 2 series of similar bracts; rays small,
     white..ECLIPTICA   31
Involucral bracts imbricated in several series;
   Pappus none; achenes oblong.............................BALSAMORRHIZA   32
    "   of more or less united awns or paleæ...............WYETHIA   33
    "   of two or more thin caducous paleæ.................HELIANTHUS   34
    "   none; achenes obovoid, compressed................HELIANTHELLA   35
Involucre double; outer series of bracts spreading, inner erect;
   Pappus not retrorsely barbed.............................LEPTOSYNE   36
    "   retrorsely barbed or aculeolate.......................BIDENS   37

31. **ECLIPTICA,** *Rumphius.* Flaccid low riparian herbs with opposite leaves, and scattered small heads of whitish flowers. Involucre broad, of 1 or 2 series of herbaceous bracts. Bracts of flattish receptacle reduced to awn-shaped chaff or bristles. Rays short, fertile. Achenes thick, those of the ray triquetrous, of the disk compressed, margined. Pappus of 2—4 short teeth or awns or none.

1. **E. alba** (L.), O. Ktze. Annual, 1—3 ft. high, decumbent, minutely strigose-pubescent: leaves lanceolate or oblong, sparingly serrate, sessile, or the lower short-petioled: peduncles from the upper axils long or short; rays about equalling the disk: disk-achenes corky-margined, truncate.—Banks of the lower Sacramento, *Jepson.* Sept., Oct.

32. BALSAMORRHIZA, *Nuttall.* Rather coarse but low acaulescent perennials, with thick roots which exude a terebinthine balsam, and bear a tuft of long-petioled leaves and several monocephalous scapes. Involucre broad; bracts large, imbricated. Chaff of receptacle linear-lanceolate. Rays large, fertile. Achenes destitute of pappus, those of the ray oblong, of the disk quadrangular.

1. **B. Hookeri,** Nutt. Canescent with a fine appressed pubescence: leaves a foot long, once or twice pinnately parted, lanceolate in outline: scape often 2-leaved near the base: involucral bracts linear or lanceolate, acuminate.— Hills of Sonoma and Alameda counties; rare.

33. WYETHIA, *Nuttall.* Vegetative characters of *Balsamorrhiza*, but the stout stems in our species leafy. Achenes prismatic-quadrangular, crowned with a short pappus of united or nearly distinct rigid scales or awns.

1. **W. helenioides** (DC.), Nutt. *Hoary-tomentose when young,* glabrate in age: radical leaves oblong, 1—1½ ft. long, 4—8 in. wide: cauline half as large, all contracted into a short petiole: heads leafy at base: outer involucral bracts ovate, or ovate-lanceolate, sometimes toothed: achenes more or less pubescent at summit; pappus more or less united at base. —Common on elevated hillsides. March—May.

2. **W. glabra,** Gray. Size and habit of the preceding; *green and glabrous;* achenes and pappus glabrous, the lobes of the latter ciliolate. —Hills of Marin Co. April, May.

3. **W. angustifolia** (DC.), Nutt. Stems scapiform, with a few reduced leaves toward the base, 1—2 ft. high, more or less *hirsute: radical leaves* 1—1½ ft. long, *elongated-lanceolate,* acuminate at both ends: head naked; bracts of involucre many, broadly linear or lanceolate, foliaceous, loose, ciliate with villous or hirsute hairs: achenes crowned with 1—4 stout hirsute awns, with some short intervening scales.—Very common on dry plains and low hills. May, June.

34. HELIANTHUS, *Linn.* (SUNFLOWER). Annuals and perennials. Leaves simple, the lowest of them opposite. Heads peduncled. Rays conspicuous, yellow. Disk-corollas yellow or dark purple, with short tube and long cylindric throat. Chaffy bracts of receptacle partly embracing the compressed-quadrangular or 2-edged achenes. Pappus a pair of caducous thin scales, with occasionally a few intervening ones.

* *Annuals 3--6 feet high.*

1. **H. annuus, L.** Robust, hispid or scabrous: stem often 1 in. thick at base, mottled or spotted with purple: leaves ovate, acute or acuminate, more or less regularly serrate, 4 –10 in. long, petiolate: involucral bracts broadly ovate to oblong, aristiform-acuminate: *dark-purple disk 1 in. or more in diameter:* rays often 2 in. long.—Plains of the San Joaquin, but probably introduced from the Rocky Mountain region. July—Oct.

2. **H. Bolanderi, Gray.** Not as stout, a yard high, scabrous-hispid: leaves ovate to oblong-lanceolate, entire or coarsely serrate, 2—5 in. long: *disk* 1 in. wide or less, *brownish-yellow;* rays about 1 in. long: *chaff of receptacle subulate-aristiform,* equalling the disk-flowers.—Sonoma Co., and northward and eastward.

3. **H. exilis, Gray.** More slender, seldom a yard in height: leaves lanceolate to ovate-lanceolate, sparingly denticulate, tapering into a slender petiole: *cusp of the chaff a slender awn surpassing the disk-flowers.* —Lower Sacramento plains; thence northward.

* * *Perennial from a tuberiform root, 6—10 feet high.*

4. **H. Californicus, DC.** Stem very leafy throughout: leaves lanceolate, entire or serrate, 6—12 in. long, short-petioled: heads about ⅔ in. high in a terminal corymbose panicle: *involucral bracts linear-subulate,* often somewhat hirsute: rays over an in. long: disk-corollas canescently puberulent toward the base: achenes glabrous; paleæ of the pappus broadly lanceolate.—Plentiful along streams, and borders of marshes. Aug.—Nov.

5. **H. Douglasii, Torr. & Gray.** Stems branching, hispidulous; upper rhomboid-oblong to spatulate-lanceolate, tapering into winged petioles, obtuse, entire 1—2 in. long: heads ½ in. high: *involucral bracts mostly foliaceous,* hispidulous; *outer narrowly oblong, obtuse,* reflexed or spreading, longer than the disk; innermost shorter, erect, acute or acuminate: rays ½ in. long; chaff entire.—Obscure and long lost species, collected by *Douglas* near Santa Clara sixty years since.

35. HELIANTHELLA, *Torr. & Gray.* Low subacaulescent perennials, with habit of some eastern *Helianthi;* differing from that genus in the more compressed and thin-edged achenes, which, in our species, have no pappus.

1. **H. Californica, Gray.** Minutely hirsute-pubescent, slender, 2 ft. high, sometimes branching: all save the radical leaves opposite, all tapering into petioles and of spatulate-lanceolate outline: heads foliaceous-bracted, the disk ¾ in. wide: rays ¾ in. long: *achenes black, obovate-oblong,* smooth and glabrous, obcordate at summit, *narrowly margined.*—Common at considerable elevations among the Coast Range hills of Marin, Napa and Contra Costa counties. May—Aug.

2. **H. castanea,** Greene. Stouter, seldom a foot high, rough-pubescent with short spreading hairs: leaves scabrous, lanceolate, nearly equalling the stem; heads nearly 2 in. wide; rays 1 in. long: *achenes* cuneate-obovate, *neither strongly compressed nor thin-edged,* those of the ray thicker and triquetrous, all dull-black at base, chestnut-brown above the middle; apical notch short and deep.—Summit of Mt. Diablo.

36. LEPTOSYNE, *DeCandolle.* Low glabrous annuals, with an apparently radical tuft of leaves cut into linear lobes, and long scapiform erect peduncles bearing each a showy head of yellow flowers. Involucre double; an outer series of narrow foliaceous spreading bracts, and an inner of broad membranaceous erect ones. Rays broad. Chaff of receptacle linear, thin, scarious, deciduous with the fruit. Achenes flat, margined. Pappus a minute callous cup, or a pair of paleæ.

1. **L. Douglasii,** DC. Peduncles slender; head an inch wide: achenes sparsely beset with capitate rigid bristles, the margin at length corky; *cup-like ring in place of the pappus entire.*—Attributed to the vicinity of San Francisco; perhaps erroneously.

2. **L. Stillmani,** Gray. Less strictly acaulescent: achenes nearly smooth and glabrous, *the corky margin rugose;* terminal cup *sometimes 2-lobed.*—At Alma, Santa Clara Co., according to *Dr. Behr.*

3. **L. calliopsidea** (DC.), Gray. Stoutish and somewhat leafy above the base: bracts of outer involucre thick, broadly ovate, little shorter than the narrowly ovate inner ones: rays often 1 in. long and ¾ in. wide: ray-achenes distinctly thin-winged; *disk-achenes cuneate-oblong, long-villous on the margins* and inner face; pappus of 2 long and conspicuous paleæ.—Santa Clara Co., *Behr.*

37. BIDENS, *Cæsalpinus.* Branching herbs with opposite leaves, the heads with double involucre as in *Leptosyne.* Achene bearing a pappus of 2 or more rigid retrorsely hispid or aculeate awns.

1. **B. frondosa,** L. Somewhat hairy, 2—6 ft. high: leaves pinnately 3—5-divided into lanceolate-serrate petiolulate leaflets: *involucre often very leafy: rays inconspicuous:* achenes obovate or oblong, 2-awned.— Fields of the lower Sacramento. Aug.—Oct.

2. **B. lævis** (L.), BSP. Glabrous, stout, more or less decumbent, 1 — 2 ft. high: *involucre not leafy, surpassed by the oval inch-long yellow rays:* achenes often with more than 2 awns.—In very wet grounds only, near lakes and rivers. Aug.—Nov.

Suborder 6, MADIACEÆ.

Herbs with watery juice, but herbage mostly viscid and glandular. Involucre of a single series of equal bracts. Ray-flowers fertile, *the achene of each embraced by or enfolded within its involucral bract.* Chaff of receptacle mostly in a single row between ray and disk, and often united into a cup. Disk-corollas usually hairy. *Style-branches subulate, hispid.*

    * *Involucral bracts quite enclosing each.its ray-achene.*

Achenes compressed, *i. e.*, flattened laterally...................MADIA	38
" obcompressed;	
Rays very short, erect; pappus showy........................ACHYRACHÆNA	43
" broad, spreading; pappus none................... ....LAGOPHYLLA	44
" ampler;	
Pappus of ray short, cup-like; of disk scanty............HOLOZONIA	45
" " none; of disk copious....................BLEPHARIPAPPUS	46

    * * *Involucral bracts half enclosing their achenes.*

Receptacle chaffy throughout; leaf-lobes pungent..............CENTROMADIA	40
" " " leaves not pungent...............HEMIZONIA	39
Receptacle chaffy only between disk and ray;	
Rays few (often 1 only); plants with large saucer-shaped	
glands...CALYCADENIA	41
Rays 5 or more { pappus plumose..........................BLEPHARIZONIA	42
{ " paleaceous or 0....................HEMIZONIA	39

38. MADIA, *Molina* (TARWEED). Glandular and viscid heavy-scented herbs, with at least the upper leaves alternate, entire, or merely toothed. Heads axillary and terminal, the yellow flowers vespertine, closing in sunshine. Involucre angled by the salient carinate backs of the uniserial involucral bracts; these completely enfolding the laterally compressed smooth achene, and having free herbaceous tips. Receptacle flat or convex, bearing a single series of chaff united and forming a cup which separates between rays and disk-flowers. Ligules 3-lobed. Bracts of involucre deciduous with the mature achenes (except in n. 2); these smooth, beakless, (except in n. 7).

    * *Annuals; rays short, inconspicuous; pappus none.*

1. **M. sativa,** Mol. Stout, 1—4 ft. high, pubescent with slender hairs and beset with pedicellate very viscid glands, ill-scented: leaves lanceolate, nearly entire: heads ½ in. high, short peduncled or sessile in the upper axils and at the ends of some short branches: *cup of receptacle broadly campanulate,* enclosing many disk-achenes, these cuneate oblong and 4-angled; ray-achenes falcate-obovate.—By waysides and in cultivated lands. July—Sept.

2. **M. capitata,** Nutt. Size of the last, but more expressly viscid herbage honey-scented, the loose hairs hispid: leaves linear, sessile by a broad base: heads longer and narrower, capitate congested at the ends of stout ascending short branches: involucre very hispid: *cup of receptacle narrow and nearly closed,* the achenes within it very few: ripe involucral bracts and achenes semipersistent.—Marin Co., and far northward. Very distinct; early-flowering. April—June.

3. **M. dissitiflora** (Nutt.), Torr. & Gray. Slender, loosely branching, 2 ft. high, viscid: heads scattered. broad-ovate, ¼ in. high: cup of receptacle ovoid but not closed: *achenes thin, but none angular.*—Borders of thickets and along mountain roads; not in open plains or cultivated lands. May—July.

4. **M. anomala**, Greene. Lower and stouter than the last, otherwise of the same aspect: *chaff of receptacle not joined into a cup:* achenes of ray 3—5, of disk 3 only, none either compressed or angled, somewhat gibbously obovate.—Mountain districts from Marin Co. northward; perhaps rare.

5. **M. exigua** (Sm.), Greene. Slender, 1 ft. high more or less, hirsute, glandular above, paniculately branched, the *small heads on long filiform naked peduncles:* leaves linear: involucral bracts 5—8, lunate, almost destitute of free tips, hispid-glandular: *cup of receptacle prismatic* and very narrow, enclosing a single straight obliquely obovate achene; ray-achenes obovate-lunate, pointed by a small disk.—Open woods and glades, in the higher hills. June—Aug.

* * *Rays ample and showy; annuals (except n. 6).*

6. **M. madioides** (Nutt). *Perennial,* slender, 2 ft. high: *leaves opposite,* linear-lanceolate, remotely serrate: heads loosely panicled, 4 lines high, slender-peduncled: bracts of involucre 8—12, with short tips: rays ½ in. long: cup of receptacle deeply cleft, enclosing *many sterile flowers, these with a pappus* of small paleæ.—Borders of redwood forests, and elsewhere in damp shades of the coast mountains. July—Oct.

7. **M. radiata**, Kell. Stout, 2—3 ft. high: leaves broadly lanceolate, denticulate, hirsute and viscid: bracts of involucre 10—20, with short tips: rays light yellow, ¾ in. long, obtusely 3-toothed: *disk-flowers very numerous,* all but the central ones fertile, *their achenes somewhat clavate and 4-angled, without pappus; ray-achenes* narrowly obovate-falcate, flat, *tipped with a minute reflexed beak.*—Plains of the lower San Joaquin.

8. **M. elegans** (DC.), Don. Clothed with long hirsute and shorter gland-tipped spreading hairs: stem stout, 3—6 ft. high, simple up to the ample corymbose panicle: *leaves at base of stem alternate, crowded, narrowly linear,* 6—10 in. long, *acute,* entire, the midvein beneath very prominent: bracts of involucre with slender linear tips: rays 12—15, 1 in. long, 3-lobed at apex, yellow throughout or dark red at base: disk-flowers sterile, without rudiments of pappus: achenes obliquely obovate-cuneate, beakless, black.—Valleys of the Coast Range. July—Oct.

9. **M. hispida**, Greene. Clothed with long almost hispid hairs, and a short little more than scabrous indument, a few gland-tipped hairs of intermediate length interspersed: stem 1—2 ft. high: lowest leaves

scattered in pairs (opposite), *oblanceolate, obtusish,* not even the scattered reduced and linear upper ones acute: corymbose panicle ample: rays narrow, ½—¾ in. long, yellow: achenes lunate-clavate, brown dotted with black.—Mountain sides and summits; Tamalpais, Hamilton, Diablo, etc. May, June.

39. HEMIZONIA, *DeCandolle.* Very heterogeneous assemblage of annuals, of which the typical species are only empirically separated from *Madia* on account of the broader ray-achenes, which are but partly enclosed by their bracts. But the ray-achenes in the second group are rugose, and the flowers diurnal. Receptacle flat; chaff mostly as in *Madia.*

* *Habit, pubescence, and vespertine rays of* MADIA; *ray-achenes obovate-triangular, smooth, the terminal areola nearly central; disk-achenes abortive, without pappus.*

+*Corymbosely paniculate-branching.*

1. **H. congesta,** DC. Soft-hirsute but not lanate, 2 ft. high, the inflorescence glandular: bracts of the involucre with lanceolate foliaceous tips little surpassed by the white rays: *margined bracts of the receptacle lightly connate or distinct:* achenes with conspicuous inflexed stipe.— Marin Co., near Olema.

2. **H. citrina.** Lowest *leaves* opposite, *oblong-lanceolate, 3-nerved, obtusish, glandular-pubescent, not lanate:* heads at first few, in a simple corymbose-panicle: flowers lemon-yellow: tips of involucral bracts short and broad; bracts of receptacle joined into a cup: achenes with inconspicuous stipe.—Northern part of Marin Co., a vernal species, flowering in April and May.

3. **H. luzulæfolia,** DC. Lowest leaves opposite, narrowly *linear-lanceolate, acute or acuminate, silvery-canescent with a fine appressed silky wool:* inflorescence at length diffuse, very glandular and ill-scented: rays white or pinkish: involucral bracts, achenes, etc., as in the last. Var. **lutescens,** Greene. Flowers from rich cream-color to lemon-yellow, the branches more slender; leaves narrower.—The type abundant in fields and waste lands generally. The variety in Contra Costa Co., near San Pablo, and in Marin, about San Rafael. June—Dec.

+ +*Heads racemosely arranged along simple branches.*

4. **H. Clevelandi,** Greene. Lower leaves narrowly linear, 1-nerved, silky-lanate as in the last; *racemose or spicate flowering branches villous* with long spreading hairs: fl. white: achenes nearly as in the last.— Common from Marin and northern Napa Co. to Oregon.

＊ ＊ *Rays more numerous, very narrow, diurnal; achenes turgid,
indistinctly rugose, with a beak-like apiculation at
summit of the inner angle.*

5. **H. corymbosa** (DC.), Torr. & Gray. Pubescent, viscid and glandular, 1 ft. high, corymbosely and widely branching: radical leaves pinnately divided into linear segments, the uppermost linear entire: *pappus* of sterile disk ovaries of paleæ *cut into chaffy bristles, or nearly obsolete.*—Plains and hills near the Bay. July—Nov.

6. **H. angustifolia,** DC. Hirsute and viscid-glandular, widely branching from the base: leaves mostly entire, linear, less than 1 in. long: rays 12—15; *pappus* of sterile disk-achenes *none, or a row of minute bristles* rather than scales.—Less frequent than the preceding.

＊ ＊ ＊ *Rays 5 only, broader; disk-flowers also few; tips of involucral bracts short, rigid, erect: achenes as in the preceding group, but more rugose.*

←*Receptacle chaffy only next to the rays; disk-ovaries with a pappus.*

7. **H. fasciculata** (DC.), Torr. & Gray. Hirsute or hispid below, glabrous and viscid-glandular above, 6—18 in. high: *heads small subsessile, usually faciculate-clustered:* involucral bracts glabrous or glandular-hispidulous, those of the receptacle slightly united: *pappus of disk-ovaries of 6—10 linear paleæ.*—Hills of the Mt. Diablo Range, near Walnut Creek, Livermore, etc. June—Aug.

8. **H. Wrightii,** Gray. Slender, diffusely and widely branching; the filiform *branchlets terminating in a single head:* lower leaves laciniate-pinnatifid: *disk-ovaries with pappus of 8 or 9 oblong firm paleæ.*—Native of San Bernardino Co., but found on the Oakland mole in 1881; at that time a species still undescribed. It has not reappeared in this district.

9. **H. Kelloggii,** Greene. Hirsute below, loosely paniculate above, 1—3 ft. high, the heads on slender pedicels: lower leaves pinnately parted: involucre ¼ in. high; bracts glandular on the back: *ray-achenes with a slender curved beak; pappus* of the sterile ones of the disk *long*, almost equalling the corolla, *lacerately truncate and united into a tube* from base almost to summit.—Abundant in fields of grain on the lower San Joaquin from Antioch southward.

←←*Virgately racemose species; all the flowers subtended by bracts; disk-achenes with no pappus.*

10. **H. Heermanni,** Greene. *Viscid and pubescent,* heavy-scented, 1—3 ft. high: minute *leaves of the flowering branchlets scattered:* bracts of hemispherical involucre viscid-pubescent and beset with stalked glands; terminal gland inconspicuous: disk-flowers 10—15.—Mt. Diablo and southeastward. Aug.—Oct.

11. **H. virgata,** Gray. Nearly or quite glabrous, 2—4 ft. high: flowering *branchlets very leafy; their leaves short-linear*, a line long, *glandular-truncate:* bracts of oblong involucre also ending in a truncate gland, and stipitate-glandular on the back: disk-flowers 7—10.—Plains of the lower Sacramento and San Joaquin. July—Sept.

* * * * *Rays very many, narrow; receptacle chaffy throughout.*

12. **H. macradenia,** DC. . Stout, hirsute, viscid-glandular, 1—2 ft. high, leafy below, parted abruptly above the middle into few and widely diverging spicate branches: leaves linear, sharply laciniate-toothed or entire, the chaff of receptacle, floral bracts and uppermost leaves linear-subulate, abruptly gland-tipped and more or less beset with smaller gland-tipped hairs: heads often sessile and glomerate, ½ in. thick: ray-flowers very many, with short yellow ligules: achenes dull-black, scarcely rugose or granular, with an angle on the ventral face and 5 dorsal nerves; the apiculation very short.—Rich open ground in the Bay region and southward. Aug., Sept.

40. CENTROMADIA. Rigid corymbosely or diffusely branching annuals, with alternate pinnatifid or entire spinescent foliage and involucral bracts; the whole plant more or less resiniferous or glandular and scented. Receptacle convex, chaffy throughout and the bracts distinct, persistent (!). Bracts of involucre subulate, pungent, embracing the ray-achenes, persistent (!). Ray-flowers 30—40, small, ligulate, bifid, neither vespertine nor matutinal but open all day; their achenes destitute of pappus, triangular, the inner angle terminated by a short erect apiculation, the whole surface nearly smooth, or faintly rugose-tuberculate. Disk-achenes mostly sterile and with or without a paleaceous pappus.—In point of habit this is the most distinct genus of the suborder, after *Madia* and *Blepharipappus;* and the duration of the corollas, as well as the persistence of the involucral and receptacular bracts are characters of the best kind.

* *Herbage yellowish-green, scentless, or with aromatic or sweet odor.*

+—*No pappus to disk-achenes.*

1. **C. pungens** (H. & A.). Erect, 2—4 ft. high, stout and with rigid ascending branches; hirsute or hispid, scarcely viscid and nearly or quite scentless: lower leaves doubly, the upper simply pinnatifid, all the lobes pungent-tipped; chaff of receptacle rigid and pungent: *ray-achenes nearly black, rather glossy, about a line long,* not strongly compressed, the ventral angle carinate, and with a short apiculation, the plane sides and rounded back *faintly tuberculate-rugose.*—Plains of the lower San Joaquin. July—Oct.

2. **C. maritima.** Stout as the last, but only 1—2 ft. high, less rigid, darker green, more villous or hirsute, and with widely spreading and

divaricate branches at summit of the erect stem; leaves nearly all pinnatifid and softer, but setose-pungent; chaff of receptacle sharply mucronate, scarcely pungent: *ray-achenes dull greenish-brown, scarcely ¾ line long,* not compressed, though with angled face, more prominently and acutely apiculate, *the summit and back quite prominently and sharply rugulose, the whole surface obscurely roughened.*—Borders of salt marshes about San Francisco Bay. July—Nov.

＋ ＋ *Disk-achenes with 3 or more slender linear paleæ.*

3. **C. rudis.** With the aspect of *C. pungens*, but only 1—2 ft. high, commonly branched from near the base, and the branches ascending, sparsely hispid-hairy and scabrous-pubescent, slightly resinous and distinctly honey-scented: earliest cauline leaves pinnatifid, all the others linear-subulate, entire hispid-ciliate, the margins in age revolute: *achenes of ray black, about a line long, strongly compressed,* semi-obcordate in outline, the surface nearly smooth, the apiculation infra-terminal and rather prominent, though short.—Sacramento Valley, near Vacaville, *Jepson.* Long supposed to be mere *C. pungens,* to which it bears a very close general likeness. But the disk-achenes have the pappus of the next species, and those of the ray are altogether peculiar. May—Aug.

4. **C. Parryi** (Greene). Widely branching, 1—2 ft. high, sparsely hirsute, minutely resinous-glandular, aromatic: lowest leaves pinnatifid, the cauline linear, entire, sharply pungent, spreading, the uppermost pilose-ciliate toward the base: heads scattered rather than glomerate: *ray-achenes dull black,* ¾ lines long, somewhat compressed, smooth on the sides, but with a few coarse tuberculations on the back: *those of the disk with 3 or more paleæ exceeding the corollas:* chaff of receptacle not pungent.—Plentiful about the warm springs at Calistoga; herbage with the fragrance of *Wintergreen.* June—Aug.

＊ ＊ *Herbage dull and dark, ill-scented: disk-achenes with pappus.*

5. **C. Fitchii** (Gray). Stout, widely branching, 1—2 ft. high, villoushirsute, somewhat viscid, more or less beset with stalked glands: leaves mostly entire, linear-acerose, the very lowest pinnately divided into about 3 pairs of linear-acerose segments: bracts of the involucre conspicuous, subulate; those of the receptacle soft, pointless, long-villous: ray-achenes obovate-triquetrous, light-brownish, obscurely if at all tuberculate, indistinctly angled; those of the disk with 8—12 linear pappus-paleæ.—Very common on plains and among the foothills of the interior. Aug.—Oct.

41. CALYCADENIA *DeCandolle.* Rigid strict virgate more or less hispid annuals. Lowest leaves opposite, the others alternate, all narrowly linear, entire, revolute; those of the axillary fascicles and about the heads subulate, but obtuse, commonly ending in a large saucer-shaped gland. Receptacle small, flat, the chaff herbaceous and only encircling the disk flowers. Rays 1–5, white or yellow, vespertine, palmately 3-lobed or -parted; the head as a whole narrow and small. Ray-achenes obovoid-triangular, the terminal areola nearly central. Disk-achenes turbinate-quadrangular, the outer fertile, all bearing a conspicuous chaffy pappus.

1. **C. truncata,** DC. Slender, 1–2 ft. high, glabrous, or some of the lower leaves sparsely hispid; herbage keenly benzine-scented, though rigid and dry: *heads sessile and scattered along the virgate branches:* short uppermost leaves and bracts truncate by a large sessile flattish gland: *fl. yellow; rays 5* (rarely more); disk-fl. 10—24; receptacle-chaff distinct, or nearly so, truncate: pappus of disk-achenes of 7—10 oblong fimbriate-toothed pointless paleæ.—Dry hills of Marin, Sonoma and Napa counties. July—Oct.

2. **C. cephalotes,** DC. Stem simple, or with few ascending branches, ½—1½ ft. high: herbage sweet-scented: few-flowered *heads densely glomerate and sessile at the summit and in all the upper axils:* small floral leaves densely hispid-ciliate: involucral bracts 1—3, only a third the length of the cup of the receptacle: *fl. white,* with a reddish tinge, ray-achenes with prominent turgid angles and faintly rugose sides: disk-achenes appressed-villous, their pappus as long as the achene and of about 8 chaffy scales which are alternately acuminate-pointed and pointless.— Marin to Napa Co., and northward. June—Aug.

42. BLEPHARIZONIA, *Greene.* Stout and rather coarse glandular-viscid and hirsute heavy-scented annuals, with linear entire lower, and oblong upper leaves. Ray-flowers 7—10, with 3-lobed white ligules. Disk-flowers 10–30, the outer ones subtended by linear chaff. Achenes silky-hirsute, 10-striate, those of the ray partly embraced by the involucral bracts and without pappus; those of the disk surmounted by many densely plumose awns.

1. **B. plumosa** (Kell.), Greene. Somewhat paniculately branching from the base, the branches bearing, racemosely, many heads, these 15—20-flowered: ray achenes with a minute crown of short scales, those of the disk 20 or more erect plumose bristles half as long as the achene. —Plains near Antioch. Aug.—Oct.

2. **B. laxa,** Greene. Larger, 3—6 ft. high, loosely paniculate above, the large heads borne singly at the ends of the branches, 20—25-flowered: pappus of ray- and disk-achenes alike, short and spreading, less plumose than in the preceding, only a fifth as long as the achene.—Habitat of the other species.

43. ACHYRACHÆNA, *Schauer.* Soft-pubescent sparingly branching annual, with narrow leaves, and rather large oblong-campanulate seemingly rayless heads terminating pedunculiform branches. Rays 6—8, very short, erect; their achenes slightly obcompressed, enclosed, without pappus. Disk-achenes chiefly fertile, clavate, 10-costate, bearing a showy pappus of 10 elongated-oblong obtuse silvery-scarious paleæ. Chaff of receptacle in a single row, not united, between ray and disk.

1. **A. mollis,** Schauer. Erect, 1 ft. high or less; branches fastigiate: heads 1 in. long: rays very short and involute, light yellow, soon changing to scarlet: globose heads of mature achenes with expanded pappus very showy.—May, June.

44. LAGOPHYLLA, *Nutt.* Slender annuals, rigid and brittle, paniculately branching, with many small heads of pale salmon-colored or yellow vespertine flowers. Ray-achenes 5 only, obcompressed, enclosed completely by their involucral bracts, their terminal areola not protuberant. Disk-achenes slender, abortive; no pappus to any, whether of ovary or disk. Receptacle bearing a circle of chaff between ray and disk, this and all the achenes and bracts deciduous at maturity.

1. **L. ramosissima,** Nutt. Canescent with a loose silky pubescence, 1—4 ft. high, diffusely paniculate: lowest leaves spatulate-obovate, cauline lanceolate and linear, all entire: heads $\frac{1}{4}$ in. high, $\frac{1}{2}$ in. broad in expansion of rays: achenes $1\frac{1}{2}$ lines long.—Mountain districts north and south of the Bay. June—Oct.

2. **L. congesta,** Greene. Tall as the preceding and robust, not paniculate, but the *heads twice as large,* densely glomerate on short branches: lowest leaves oblanceolate, remotely serrate: achenes 2 lines long.—Habitat of the last.

45. HOLOZONIA, *Greene.* Perennial, spreading by creeping rootstocks. Leaves opposite. Heads small, on slender pedicels, in an ample panicle. Flowers diurnal, white. Ray-achenes 6—8, obcompressed, completely enclosed, smooth, surmounted by a short saucer-shaped hyaline entire persistent pappus. Disk-achenes with a pappus of 2 slender, deciduous paleæ. Receptacle flat, with a circle of united chaff between disk and ray.

1. **H. filipes** (H & A.), Greene. Stems decumbent, 2 ft. high; slender branchlets and filiform peduncles glabrous or glandular: cauline leaves linear, minutely villous; those of the branches with some short-stipitate dark glands: involucre loosely villous: rays white or rose-tinted, deeply cleft into 3 linear lobes.—By streamlets in the hills east of Napa Valley; also in low fields along Napa River. July—Oct.

46. BLEPHARIPAPPUS, *Hook.* Vernal annuals, with alternate leaves, and mostly showy broad heads of white or yellow diurnal flowers. Bracts of involucre flattened on the back, with dilated and thin margins, the whole completely enfolding its obcompressed achene. Rays 8—20; their achenes obovate-oblong or narrower, without pappus. Disk-flowers with cylindraceous-funnelform 5-lobed corollas; their achenes linear-cuneiform, usually with a pappus of bristles or awns. Receptacle flat, bearing a series of chaffy bracts between ray- and disk-flowers, these, with the involucral bracts, mostly deciduous when mature, leaving a naked receptacle.

     * *Pappus of 10—20 bristles which are stout, and, below the middle, long-plumose.*

     +*Hairs of pappus-bristles not interlaced.*

     1. **B. heterotrichus** (DC.), Greene. Erect, 1 ft. high or more branching from the base, rough-hirsute or hispid and glandular: *lower leaves lanceolate: laciniate-pinnatifid or incised,* the upper entire: rays large, white: long-villous hairs of the pappus-bristles all erect and straight.—Eastern base of the Mt. Diablo Range, on sandy plains. April, May.

     2. **B. graveolens,** Greene. Stout, erect, 2 ft. high or more, sparingly branching, hirsute, and with numerous rigid gland-tipped hairs interspersed: *leaves all entire:* heads very large, rays of a creamy white: achenes slenderly clavate; *pappus* when mature *deciduous in a ring,* the villous wool of the bristles all straight and erect and two-thirds their length.—Said to occur on Mt. Tamalpais; but this may be doubted. It may be looked for on the plains of the lower San Joaquin. Apr.—June.

     3. **B. carnosus** (Nutt.), Greene. Dwarf, depressed, branched from the base, pubescent; *leaves succulent, 1 in. long, linear-oblong or spatulate,* entire, or the lowest sinuate-pinnatifid: heads small: *rays* white, reduced and *inconspicuous:* pappus-bristles sparsely plumose with straight villous hairs.—Sands of the sea beaches in Marin Co., etc. April—June.

     4. **B. hieracioides** (DC.), Greene. Erect, *rather strict,* 2—3 ft. high, *stoutish,* hispid: leaves linear to oblong, laciniate-dentate: *rays* yellow, *short,* little exceeding the disk: hairs of the pappus all straight and erect. Var. **anomala,** Bioletti. Involucral bracts open-boat-shaped, not enfolding the achenes, and persistent on the receptacle after the falling of the fruit.—A coarse weedy species of wooded or bushy hills, in half shady places. May, June.

     5. **B. gaillardioides** (H. & A.), Greene. Freely branching below, 1 ft. high or more, hispid: leaves commonly laciniate-pinnatifid: *rays orange-yellow,* ½—¾ in. long: pappus dull-white or sordid, the bristles about twice as long as their copious straight villous basal hairs.—Said to be common near San Francisco; which we doubt.

     6. **B. nemorosus.** Rather slender, sparingly branched above, 1—2 ft. high, hispidulous: foliage and heads much as in the preceding, but

rays pale yellow below the middle, white above it: pappus short, the bristles often scarcely surpassing their copious brownish villous hairs.—A beautiful species of shaded slopes on Tamalpais, Mt. Diablo, and the Berkeley Hills. May, June.

++ *Villous hairs of pappus-bristles more or less interlaced.*

7. **B. hispidus**, Greene. Scarcely 1 ft. high, branching from the base, rather densely hispidulous; a few dark stipitate glands on the involucre: leaves all narrow, entire: *rays white, rather short* and not conspicuous: pappus white, of 10 aristiform bristles, with copious short-villous hairs, the innermost of which are interlaced.—Mt. Diablo and southward, near the higher summits. May, June.

8. **B. elegans** (Nutt.), Greene. Habit of the last but much larger, more or less stipitate-glandular throughout: lower leaves pinnately toothed; upper entire: *rays yellow*, ½ in. long: *pappus white, its copious villous hairs much shorter than the aristiform bristles.*—More widely diffused than the last, and at lower elevations. May, June.

* * *Pappus of naked aristiform bristles.*

9. **B. platyglossus** (F. & M.), Greene. Sparingly branching, 1 ft. high more or less, hirsute and stipitate-glandular: lower leaves pinnatifid into linear lobes: rays ½ in. long, yellow, with white tips: pappus of 15—20 upwardly scabrous stout awn-like bristles.—Common in open grounds. April—June.

. * *Pappus paleaceous or none.*

10. **B. Fremonti** (T. & G.), Greene. Strictly erect; branching above the base, 1 ft. high, minutely pubescent, not glandular: leaves pinnately cut into short lobes: rays ½—¾ in. long, yellow at base, white above it: *pappus-paleæ ovate to oblong-lanceolate, tapering into a subulate awn*, entire at the margins and with a few long-villous hairs.—Plains of the lower Sacramento, etc. April, May.

11. **B. Douglasii** (H. & A.), Greene. Habit and flowers of the last, but plant nearly or quite glabrous: pappus of 10—18 very unequal rigid subulate awns, these slightly hirsute near the dilated base, Var. **oligochœta** (Gray). *Pappus reduced to 2 marginal awns and some rudiments* of intervening ones.—About the Bay and northward. April—June.

12. **B. chrysanthemoides** (DC.), Greene. Aspect of the preceding; flowers the same: *achenes destitute of pappus and all wholly glabrous.*—Habitat of the last. April—June.

13. **B. nutans**, Greene. Low, slender, with divergent branches above the base, 3—6 in. high: leaves all linear, entire, the lower pairs opposite,

all hirsute-ciliolate: branches, peduncles and involucres glandular-pubescent: rays 5—7, yellow, ¼ in. long: achenes hispidulous: pappus of some 10 unequal linear-lanceolate acuminate white paleæ, their margins barbellate: *heads small, nodding* both in bud and in fruit.—Mountains of Sonoma Co. April—June.

Suborder 7, HELENIOIDEÆ.

Herbs seldom viscid or balsamic. *Receptacle naked.* Bracts of involucre *herbaceous mostly uniserial and equal,* sometimes concave behind the ray-achenes, but never enfolding them. Style branches of perfect flowers with either truncate or appendiculate tips. *Pappus* mostly *paleaceous* or none.

48. HELENIASTRUM, *Vaillant.* Herbs erect, with sessile mostly decurrent leaves, and long-peduncled heads; the herbage more or less resinous-dotted. Rays numerous, cuneate, disk-flowers very many. Involucre of 1 or 2 series of small herbaceous bracts. Receptacle globose, naked. Style branches with capitate truncate tips. Achenes turbinate. Pappus in ours of awn-pointed paleæ.

1. **H. puberulum** (DC), O. Ktze. Minutely cinereous-pubescent, 2—4 ft. high, with slender widely spreading monocephalous branches: leaves lanceolate, entire, all but the radical strongly decurrent: involucre and reflexed *rays very short and inconspicuous: globose disk of red-brown flowers* ½ in. *thick:* paleæ of pappus ovate, short-awned.—Banks of streams, and in other moist places. July—Dec.

2. **H. occidentale.** Stout, 2—4 ft. high, erect, parted above into several stoutish very erect pedunculiform monocephalous branches: leaves lanceolate, 6 in. long or more, thickish and somewhat fleshy, almost gummy to the touch, and with some tomentose pubescence: *rays showy,* ¾ in. *long:* disk brownish-yellow: pappus-paleæ ovate-lanceolate, tapering into a long awn.—Suisun marshes. July—Oct.

49. BLENNOSPERMA, *Lessing.* Low annual, with pinnately parted leaves, and pedunculiform branches bearing solitary radiate yellow heads. Involucral bracts uniserial, equal, oblong, herbaceous but purplish or yellowish. Receptacle flattish, naked. Rays 5—12, linear. Disk-flowers about 20; their narrow tube abruptly expanded into a campanulate limb; their style undivided, with capitate apex; their ovaries abortive. Achenes (of the ray) pyriform, obscurely 8—10-ribbed, with small areola and no pappus; the surface bearing minute papillæ which develop mucilate when wet.

1. **B. Californicum** (DC.), Torr. & Gray. A span high, diffusely branching, flaccid, glabrous: leaves alternate, pinnately parted into narrowly linear entire lobes: expanded heads ½ in. broad: ligules pale yellow within, brownish without: disk-flowers shorter than the involucre: style-branches of fertile flowers broad.—Early vernal plant of the valleys of the Mt. Diablo Range, and plains beyond. March, April.

50. CHÆNACTIS, *De Candolle.* Compound-leaved herbs, often more or less woolly, with discoid heads mostly solitary and pedunculate. Involucre campanulate, the linear bracts equal, uniserial, herbaceous. Receptacle flat, naked. Corollas (yellow in ours) with short tube, long narrow throat, and short teeth; but those of the outer circle more ample, approaching the nature of rays. Achenes slender, smooth. Pappus of hyaline nerveless paleæ.

1. **C. glabriuscula,** *DC.* Annual, seldom a foot high, thinly floccose, at length glabrate; peduncles long, stout: heads ¾ in. high: bracts of involucre thickish, glabrate, obtuse: marginal corollas ample, much exceeding the others: pappus of 4 equal narrowly oblong acutish paleæ. —Plains of the lower San Joaquin. April—June.

51. RIGIOPAPPUS, *A. Gray.* Heads with short narrow ligulate rays; all the flowers yellow. Involucre turbinate-campanulate, its bracts nearly linear, equal, rather rigid, involute in age. Disk-corolla small, with short tube, long narrow throat, and 3—5 short erect teeth. Style branches with slender-subulate hispidulous appendage and short linear stigmatic part. Pappus alike in disk and ray, of 3—5 opaque paleaceous awns; the linear pubescent achene transversely rugose.

1. **R. leptocladus,** Gray. Slender erect annual, 6—12 in. high, simple below, above with few slender corymbose branches: leaves alternate, narrowly linear, sessile, erect, entire, hirsutulous or glabrate, those of the filiform branchlets subulate: paleæ of pappus ½—⅔ as long as the long slender rugulose achene.—Open glades among sparsely wooded hills. May.

52. LASTHENIA, *Cassini.* Mostly annuals and low, from glabrous to slightly flocculent. Leaves opposite. Heads middle-sized, on slender peduncles. Receptacle conical to subulate, muricate with projecting points on which the achenes are inserted. Involucres hemispherical, campanulate, or oblong Rays oval or oblong. Disk-corollas with slender tube and campanulate 5-toothed limb. Achenes linear, subclavate, or linear-cuneate, more or less flattened or angled, naked at summit, with or without a paleaceous or more or less awn-like pappus.

* Bracts of involucre joined into a toothed cup.

1. **L. glaberrima,** DC. Stems weak, decumbent, 1 ft. long or less,
very glabrous: leaves linear, entire: *heads nodding in bud:* involucre
about 15-toothed: *rays very short;* all the corollas shorter than their
broadly linear pubescent achenes: *pappus of 5—10 firm chaffy scales,*
2 or 3 of them subulate-pointed or short-awned, the others not so.—
Subaquatic herb of shallow winter pools on low plains or in depressions
among the hills; not very common. May, June.

2. **L. glabrata,** DC. Stout, sparingly branching, 1—2 ft. high,
peduncles few, elongated, erect: *leaves,* at least the upper pairs, *ovate-
lanceolate,* coarsely but irregularly toothed, *conspicuously connate at the
dilated base* and forming an open cup rather than sheath: heads very
large, 1 in. wide: achenes dark, smooth, without pappus.—Borders of
salt marshes only; not common. June.

3. **L. Californica,** Lindl. More slender, almost diffusely branching,
the many and not elongated peduncles forming a corymbose top to the
herb as a whole: leaves narrow, entire, divaricately spreading, *not dilated
or manifestly connate below:* heads ¾ in. wide.—Abundant on low plains,
and even moist slopes of hills. Much more common than the last, and
inexcusably confounded with it for many years. May, June.

4. **L. chrysantha** (Greene). Habit, foliage and flowers of the last,
but plant smaller: *achenes* obovate-oval, *much compressed, surrounded by
a border of very short clavate* closely packed *hairs.*—Plains of the San
Joaquin; perhaps altogether outside of our district. April.

5. **L. conjugens,** Greene. Only a few inches high; leaves narrowly
linear, only the lowest entire, the others cleft into several pairs of long
linear segments, these entire or toothed: *involucral bracts united below
the middle only:* achenes very small (1 line long) olive-green and polished:
pappus none.—Subsaline soil near Antioch; closely connecting this
group and the following. April, May.

* * Involucral bracts distinct.

+—*Herbage often slightly villous- or floccose-tomentose; pappus
of paleæ or awns, or both, or none.*

6. **L. Fremonti** (Torr.). Erect, slender, 8—10 in. high, only hirsute-
pubescent: leaves mostly palmately parted into linear lobes: involucre
broad; bracts 10 or 12; rays as many, the oval ligules not longer than
the width of the disk: *pappus of 4 slender awns and as many* or more
numerous *small paleæ,* or rarely none.—Moist plains from Napa and
Solano counties southward. May.

7. **L. Burkei** (Greene). Much like the last, but taller, 1—2 ft. high:
*pappus of 8—10 minute entire acute paleæ and a single very long and
slender awn.*—Southern Mendocino Co.; to be expected in Sonoma.

8. **L. tenella** (Nutt.). Erect, sparingly branching, 4—6 in. high, somewhat canescent with deciduous woolly hairs: leaves linear, entire, or *some of the lower laciniate:* rays oval or oblong, short: paleæ and awns each usually 2, but pappus not rarely wholly wanting.—Low plains of Contra Costa Co. April, May.

9. **L. uliginosa** (Nutt.). Stouter than the last, freely branching, often decumbent, somewhat flocculent: leaves linear-ligulate, *the lower, if any, entire, the upper laciniate-pinnatifid* into linear entire or cleft segments: involucral bracts and long-exserted rays.10—13: pappus of about 4 slender awns, and as many or twice as many broad truncate laciniate-fimbriate paleæ.—Common in low grounds. April, June.

10. **L. microglossa** (DC.). Slender, seldom 6 in. high, branching from the base but erect: leaves entire, pubescent: *involucre narrow and nearly cylindrical:* rays few, very short and inconspicuous: achenes linear-fusiform, flattish, minutely and sparsely hispid; pappus of 2—4 rather rigid awned or awn-like scales, or in the ray sometimes none.— Valleys among the coast mountains southward. April, May.

    + + *Pappus usually present and uniform, of awned paleæ or paleaceous-dilated awns; leaves mostly entire.*

    ++*Somewhat succulent; pubescence woolly, if any.*

11. **L. platycarpha** (Gray). Purplish-stemmed and very wiry, 5—8 in. high, with erect or ascending branches: leaves linear, or pinnatifid into filiform segments: *involucral bracts* 6 or 7, *3-nerved,* the middle nerve carinately prominent: *pappus-paleæ* bright-white, *ovate, slender-awned,* the awn as long as the achene.—Subsaline plains of Solano and Contra Costa counties. April, May.

12. **L. carnosa** (Greene). Leaves all filiform, entire: bracts of involucre with a single strongly carinate nerve: pappus of 4 or 5 *subulate-awned ovate paleæ.*—Border of salt marsh north of Vallejo: rare or local. April.

  ++ ++ *Not succulent; leaves entire; (except in n. 15) pubescence not woolly.*

    = *Pappus invariably none; achenes subclavate.*

13. **L. chrysostoma** (F. & M.). Hirsutulous, 1 ft. high more or less, *freely branching: leaves narrowly linear:* head 3—4 lines high: bracts of involucre, and the rays, 7—12, the latter 3—4 lines long: achenes glabrous.—Rich fields and sunny slopes. April, May.

14. **L. macrantha** (Gray). *Perennial, stout and nearly simple,* decumbent at base, with peduncles 4—8 in. long: leaves somewhat 3-nerved and obtuse, linear, 4 8 in. long, hispid-ciliate toward the base: heads ½ in. high and as broad: involucre of about 12 hirsute-pubescent thickish herbaceous bracts: ligules ½—34 in. long.—Moist lowlands of western Marin Co., and northward. June.

$== =$ *Pappus seldom if ever wanting; achenes somewhat cuneate.*

15. L. hirsutula. Stout and low, from a strictly annual root, mostly branching very freely: the whole herbage *rather roughly short-hirsute:* leaves broadly linear, often with saliently projecting scattered teeth, the lower conspicuously connate, sheathing the stem: involucral bracts obovoid, obtuse or acutish: rays oblong: achenes mostly very smooth, rounded at summit, manifestly compressed; *pappus of 2 brownish very slender-subulate aristiform bristles.*—Plentiful on open rocky and grassy hills along the seacoast, from Marin Co. southward. May, June.

16. L. gracilis (DC.). Whole habit and aspect of n. 13, but achenes linear-cuneate, with *pappus of white lanceolate or ovate slender-awned paleæ,* or the paleæ sometimes almost obsolete.—Very plentiful and variable; often very small, slender and simple, not rarely as large as *L. chrysostoma.* April—June.

53. MONOLOPIA, *De Candolle.* Annuals white with floccose wool. Leaves alternate, not linear, toothed or entire. Ray-corollas with ample ligule, bearing at base, and opposite the ligule, a rounded denticulate appendage. Achenes black, without pappus. Genus otherwise like *Lasthenia,* and too near it; but also as easily referable to *Eriophyllum.*

1. M. major, DC. Stout, nearly simple, or with several pedunculiform naked monocephalous branches, 2 ft. high; expanded heads more than 1 in. wide: *bracts of involucre joined into a broad-campanulate toothed cup:* achenes 2 lines long.—In rich fields, or on hillsides. May.

2. M. gracllens, Gray. Slender, diffusely paniculate, bearing scattered short-peduncled heads less than 1 in. wide: *bracts of involucre distinct to the base:* achenes 1 line long.—Plentiful on Mt. Diablo near the summit, thence southward to the Santa Cruz seaboard. May—July.

54. ERIOPHYLLUM, *Lagasca.* Ours mostly suffruticose, floccose, and with divided leaves. Involucres oblong or campanulate, the bracts of firm texture, permanently erect. Rays few, short and broad. Disk-corolla with distinct slender proper tube. Style-tips truncate, obtuse, or obscurely conical: Achenes clavate-linear to cuneate-oblong, mostly 4-angled. Pappus of nerveless and mostly pointless hyaline paleæ.

　　　* *Suffruticose; heads smallish, terminally clustered.*

1. E. stæchadifolium, Lag. Stem and lower face of leaves white with a close pannose tomentum; shrub much branched, 2—5 ft. high, very leafy throughout: *leaves subcoriaceous,* cut into linear pinnate divisions: heads compactly corymbose-cymose; *involucres oblong, angular,* $\frac{1}{4}$ in. high or more, of linear bracts: receptacle convex: rays 6—8: pappus-paleæ 8—12, the 4 over the angles of the achene somewhat longer.—Sandy hills, and slopes of bluffs near the sea only. May—Dec.

2. **E. confertiflorum** (DC.), Gray. Smaller, 1—2 ft. high; *leaves* on the flowering branches reduced and scattered, *membranaceous, hoary-tomentose on both faces*, ternately 3—7 parted into linear divisions; heads 2 lines high, short-peduncled or sessile in a dense terminal cluster: *involucre* obovoid-oblong, *of broadly oval bracts:* rays 4 or 5: paleæ of the pappus 8—14. Var. **discoideum.** More condensed and leafy; heads broader, with more numerous flowers, but no rays.—Very common on all hills; the variety in Sonoma Co. June—Dec.

3. **E. Jepsonii,** Greene. Suffruticose, 2 ft. high; stem white with pannose tomentum; leaves hoary on both faces, pinnately divided into 5—7 narrowly linear revolute segments: *inflorescence loosely cymose-corymbose, the peduncled heads 3—4 lines high,* and, with 6—8 oblong rays expanded, 1 in. broad: bracts of involucre 6—8, coriaceous, ovate: achenes with a few short hispidulous hairs, and 2 unequal sets of pappus-paleæ, those of the inner circle exceeding the others.—Mountains of Alameda Co., south of Livermore, *Jepson.* May.

* * *More herbaceous; heads large, solitary or scattered.*

4. **E. arachnoideum** (F. & L.). Loosely branching from a decumbent base, 1—2 ft. high, clothed with long floccose wool: leaves broad, from rhombic or cuneate in outline to oblong-lanceolate, thinnish, 3—5-lobed or -incised, the lobes or coarse teeth triangular or oblong: involucre hemispherical, 3—4 lines high: rays 10—13, large; disk-corollas with very glandular-hirsute tube: achenes short, thickish: pappus-paleæ short.—In the redwood districts of Marin Co., etc. June—Oct.

5. **E. achilleoides** (DC.). Leaves mostly basal, opposite, pinnately parted into 3—5 divisions, these incised or pinnatifid: *heads somewhat corymbosely collected* and short-peduncled; involucres hemispherical, the bracts and rays 9—13.—Hills of Napa, Sonoma and Marin counties and northward. June—Sept.

55. JAUMEA, *Persoon.* Procumbent very succulent perennial herb with opposite subterete leaves and solitary terminal short-peduncled heads. Involucre campanulate, the outer bracts shorter. Corollas glabrous. Style-branches papillose or hairy. Achenes 10-nerved. Pappus 0.

1. **J. carnosa** (Less.), Gray.—Common denizen of sandy salt marshes, associated with *Salicornia;* flowering during summer and autumn.

Suborder 8, ANTHEMIDEÆ.

Mostly *aromatic-scented plants, with a very bitter juice.* Leaves often much dissected. *Bracts of involucre imbricated, more or less scarious.* Receptacle either naked or chaffy. Anthers not caudate. Style-branches of perfect flowers truncate, sometimes penicillate. *Achenes small and short, with no pappus, or a paleaceous crown.*

56. ANTHEMIS, *Diosc.* Herbs with pinnately dissected leaves, and rather large heads on peduncles terminating leafy branches. Involucre hemispherical. Ray-flowers white; disk-flowers yellow. Chaff of receptacle bristly. Achenes not flattened, glabrous; the truncate summit with or without a short coroniform pappus.

1. A. COTULA, L. (MAYWEED). Strong-scented annual weed 1—2 ft. high, somewhat freely branching: receptacle conical, destitute of chaff near the margin: achenes 10-ribbed, rugose or tuberculate.—In waste grounds. July—Oct.

57. ACHILLEA, *Diosc.* Erect perennial herb, with pinnately multifid lanceolate leaves, and small heads in a dense terminal cymose corymb. Chaff of receptacle membranaceous. Rays few, short and broad. Achenes obcompressed, callous-margined, glabrous, destitute of pappus.

1. A. Millefolium, L. (YARROW). Stoutish, 1—2 ft. high, villous-lanate to nearly glabrous: heads crowded in a fastigiate flat-topped cyme: involucre oblong; rays 4 or 5, white.—Very common in sandy soils toward the sea.

58. MATRICARIA, *Tourn.* Our species annual herbs, with finely dissected sweet-scented foliage and rayless heads of green flowers terminating the branches. Receptacle conical or ovoid, naked. Achenes glabrous, 3—5-nerved on the sides, rounded on the back, nearly destitute of pappus.

1. M. discoidea, DC. Low often *diffusely branching, mostly less than 1 ft. high, sweet-scented:* heads short-peduncled: bracts of involucre broadly oval, scarious, with green centre, not half the length of the ovoid disk: achenes oblong, somewhat angled, with an obscure coroniform margin at summit.—By waysides everywhere, mostly in hard sterile soil. April, May.

2. M. occidentalis, Greene. Erect, very stout, 1½—2½ ft. high, *corymbosely branched at summit,* the herbage *nearly scentless:* heads more than twice as large as in the last and 6—8 lines high: receptacle

somewhat fusiform: achenes sharply angled and with a broad coroniform margin below the summit.—Grain fields and by roadsides, chiefly in the interior; only occasional near the Bay. May, June.

59. CHRYSANTHEMUM, *Diosc.* Herbs of various habit and foliage. Heads large, with yellow or white rays. Receptacle flat or hemispherical, naked. Achenes glabrous, 5—10-ribbed all around.

1. C. SEGETUM, Lobel (1570). Annual, erect, 1—2 ft. high, with few ascending leafy monocephalous branches: leaves *somewhat fleshy*, glabrous, glaucous, *oblong, incisely serrate*, the upper sessile by a clasping base: *heads yellow*, 2 in. wide including the rays: disk-achenes compressed, corky-winged.—Rather common in fields and by waysides above West Berkeley, and east of Oakland.

2. C. PARTHENIUM, Pers. (FEVERFEW). Biennial, erect, 1—2 ft. high, stoutish, branching, puberulent: *leaves thin, pinnately parted*, the oval or oblong divisions incised or toothed: *rays white*, obovate, 2—3 lines long: pappus a minute crown.—Only occasionally spontaneous; an escape from the gardens.

60. SOLIVA, *Ruiz & Pavon.* Small depressed herb with rigid short branches, petioled pinnately dissected leaves, and heads of greenish flowers sessile in the forks. Involucre of 5—12 nearly equal bracts. Receptacle flat. Outer and only pistillate flowers apetalous, their achenes pointed with the hardened and persistent style, obcompressed, with rigid callous margins. Pappus none.

1. **S. sessilis,** R. & P. Plant depressed, seldom 6 in. broad, villous or glabrate: primary divisions of the leaves 3—5, petiolate, parted into 3—5 narrow lobes: heads depressed: achenes broadly obovate, thinwinged, the wings entire or panduriform-incised near the base, spinulosepointed at summit: persistent style long and stout.—In moist open ground, or, less frequently, in shady places. May.

61. COTULA, *Linn.* Low herbs with alternate lobed or dissected leaves, and slender peduncled discoid short-hemispherical heads. Outer series of flowers pistillate only, and apetalous, the style deciduous. Disk-flowers 4-toothed. Bracts of involucre greenish, in about 2 ranks. Mature achenes pedicellate, obcompressed, thick-margined or narrowly winged, in ours nearly or quite destitute of pappus. Both species of our flora supposed to have come from Australia, in recent times.

1. C. CORONOPIFOLIA. L. Somewhat succulent glabrous stoutish and decumbent usually subaquatic perennial: leaves ligulate-linear, laciniate-pinnatifid, or the upper entire, the base clasping or sheathing: head much depressed, ⅓—½ in. broad: pistillate fl. in a single series, their achenes with a thick spongy wing: disk-achenes with wing reduced.—

Abundant in shallow pools and on muddy banks of tidal streams; occasionally on higher lands; flowering throughout the year.

2. C. AUSTRALIS (Sieb.), Hook. f. Slender, not fleshy, very diffusely branched: leaves bipinnately dissected into linear lobes and somewhat pubescent: heads small; pistillate fl. in 2 rows, their achenes pedicelled, those of the disk less so.—Plentiful in gardens, and along some streets in San Francisco and elsewhere. Feb.—June.

62. TANACETUM, *Pliny* (TANSY). Robust aromatic perennials, leafy to the corymbose summit. Leaves ample, 2—3-pinnately dissected into very many divisions or lobes. Heads discoid, erect. Receptacle naked. Flowers yellow. Anther-tips broad, obtuse. Achenes truncate, and with a coroniform-dentate pappus.

1. **T. camphoratum,** Less. Camphoric-aromatic, villous-tomentose, very stout but decumbent, the stems 1—3 ft. long: pinnæ and segments of the very large leaves much crowded: heads in a large corymbose cluster, short-peduncled, the low-convex disk ½ in. broad; achenes 4-angled.—Abundant on sand dunes about Point Lobos, etc. Aug.—Dec.

63. ARTEMISIA, *Diosc.* Bitter-aromatic herbs and low shrubs. Leaves alternate. Heads discoid, small, paniculately disposed, usually nodding. Flowers whitish or yellow, often sprinkled with resinous globules. Anther-tips slender and pointed. Achenes obovate or oblong, usually with small summit and no pappus.

* Shrubby species.

1. **A. Californica,** Less. Branches ascending, 2—4 ft. high; leaves cinereous with a minute appressed pubescence, terebinthine-scented, the lowest parted into a few linear-filiform segments, the upper entire: heads many, in leafy panicles; involucre 2 lines broad: achenes broadish and truncate at summit, with a squamellate or coroniform-dentate pappus.—Southward and westward slopes of hills. Sept.—Dec.

* * Herbaceous species; perennial (except n. 4).

2. **A. pycnocephala,** DC. Stout, erect, simple, very leafy up to the *dense close thyrsoid or virgate panicle, densely silky-villous* even to the involucre: leaves 1—3-pinnately parted into few and short linear or spatulate lobes: heads 2 lines broad, only the marginal fl. fertile; style of the disk fl. undivided and tufted at apex: achenes glabrous; pappus none.—Sandy hills and shores about the Bay. Aug.—Dec.

3. **A. dracunculoides,** Pursh. Stems clustered, ascending, 2—4 ft. high, virgately branched: *herbage glabrous,* pungent-scented when bruised, but *neither aromatic nor bitter:* lowest leaves 3-cleft at summit, the others linear, entire: heads little more than a line broad.—Mostly beyond our limits eastward and southward, but said to occur at Black Point, San Francisco, and near Lake Chabot. Aug.—Nov.

4. A. biennis, Willd. *Annual, erect, virgate,* 1—3 ft. high, leafy to the summit: herbage deep green, glabrous and tasteless, nearly: leaves 1—2-pinnately parted into lanceolate or broadly linear laciniate or toothed lobes, or the uppermost only pinnatifid: *heads small, in close glomerules* on the spiciform short branches and main stems: achenes with small epigynous disk and no pappus.—Mostly in or near cultivated ground, at West Berkeley, etc. Aug.—Oct.

5. A. heterophylla, Nutt. Strictly erect, 3—5 ft. high, simple and leafy up to the short naked dense panicle; herbage bitter and aromatic: *leaves white beneath with cottony tomentum, green and glabrate above,* 2—4 in. long, *lanceolate or broader,* acute, often laciniately toothed or cleft, as often entire, of firm texture: heads very numerous, 2 lines high, seldom as broad.—Very common in low rich land, along streams, etc. July—Oct.

Suborder 9, SENECIONIDEÆ.

Plants herbaceous or suffrutescent, mostly with watery juice, but pungent; some genera bitter and aromatic. *Bracts of involucre herbaceous, in 1 or 2 series.* Receptacle naked. Anthers not caudate; sometimes sagittate. Style-branches of perfect flowers obtuse or truncate; in some penicillate. Pappus of capillary bristles.

Heads subdiœcious; flowers whitish.................................PETASITES 64
Heads large; flowers yellow;
    Leaves palmatifid..CACALIOPSIS 65
    "   ovate or rounded, or pinnatifid;
        Pappus brownish, not copious...........................ARNICA   66
        "   white, fine and copious.........................SENECIO   67
Heads small, mostly cylindric; flowers yellow....................SENECIO   67

64. PETASITES, *Tourn.* Perennial herbs, with creeping rootstocks sending up early scapiform leafy-bracted flowering stems, and later ample long-petioled radical leaves which are cottony-tomentose at least when young. Flowers whitish or purplish, in a racemose corymb, subdiœcious. Achenes narrow, 5—10-costate. Pappus elongating in age, very soft and white.

1. P. palmata (Ait.), Gray. Leaves of round-reniform outline, 7—10 in. broad, palmately 7—11-cleft to beyond the middle; these appearing later than the very early flowering stems.—Wet mountain woods, at Taylorville, Marin Co., etc. Fl. March.

65. CACALIOPSIS, *A. Gray.* Floccose-woolly stout perennial, with ample palmatifid leaves appearing before the flowers. Stem stout, with few large rayless heads of yellow flowers. Involucre campanulate, of many lanceolate-linear acuminate bracts. Corolla with cylindraceous throat longer than the slender tube. Style puberulent below the flattish branches. Achenes 10-striate. Pappus copious, soft, white.

1. **C. Nardosmia,** Gray. Robust, 1 ft. high or more: leaves mostly radical or at the base of the stem: heads 1 in. high, peduncled, corymbosely or racemosely arranged near and at the summit of the stem: flowers honey-yellow, sweet-scented.—Mountain summits of northern Napa, or adjacent Sonoma Co., collected by the author, in 1874; otherwise a far northern plant. June.

66. ARNICA, *Ruppius.* Perennial herbs, somewhat glandular or viscid and aromatic. Leaves opposite. Heads one or several and large, at summit of the stem. Involucre broadly campanulate, not bracteolate at base; the herbaceous bracts lanceolate, equal, in about 2 series. Receptacle flat, naked. Disk-corollas yellow (as also the rays when present), with distinct long tube and funnelform or cylindraceous 5-lobed limb. Achenes linear, angled. Pappus a single series of rather rigid brownish scabrous or barbellate bristles.

1. **A. discoidea,** Benth. Stoutish, very hairy, 2 ft. high or less: leaves ovate or oblong, 2—4 in. long, coarsely toothed, cordate, or truncate, or sometimes slightly cuneate at base; the upper smaller, sessile, often alternate: heads ¾ in. high, rayless: involucre villous and glandular: achenes sparsely pubescent, not glandular.—Northward slopes of the higher coast mountains, in shady places chiefly. June, July.

67. SENECIO, *Matthiolus.* Plants extremely diverse in habit, foliage, etc., but leaves always alternate. Heads clustered cymosely, or solitary; either radiate or discoid. Involucre usually cylindrical, of many equal bracts, and with calyx-like bracteoles at the base. Flowers yellow; those of the disk 5-toothed or -lobed. Receptacle flat or convex. Achenes commonly glabrous, terete, or somewhat ribbed. Pappus of very copious fine white capillary merely scabrous bristles.

* Annual species.

1. **S. vulgaris,** Tragus (1552). Nearly glabrous, slightly fleshy, 1 ft. high more or less, branching, leafy throughout: *leaves clasping at base, pinnatifid;* the lobes and sinuses sharply toothed: scales at base of the small cylindric involucre with black tips: *rays none.*—Abundant in rich shady cultivated grounds; flowering at all seasons of the year.

2. **S. aphanactis,** Greene. Slender, 2—6 in. high, slightly arachnoid when young, glabrate in age, scarcely viscid, scentless: *leaves* ½—¾ in. long, slightly fleshy, erect or ascending, *the lowest linear-spatulate, the cauline linear to oblong, coarsely toothed or slightly lobed:* heads very small, 2 or 3 together at the ends of the few branches: bracts of involucre linear-acuminate, not black-tipped: *rays about 5, minute, recurred:* achenes appressed-silky.—Clayey slopes, or open hilltops of the Mt. Diablo Range. A somewhat rare indigenous plant, strangely mistaken, by some, for the Old World weed *S. silvaticus.* March, April.

** *Perennial species.*

3. **S. curycephalus,** Torr. & Gray. Somewhat floccose when young, at flowering time glabrous: stem stout, 2—3 ft high, leafy: leaves 4—6 in. long, lyrately pinnate, the lobes or leaflets 7—15, cuneate, acutely and incisely cleft: *heads many, in an ample corymb:* involucres ½ in. high or more, with few bracteoles: *rays 10—12, light yellow, long and showy.*—In groves along rocky bases of the coast mountains from Sonoma and Contra Costa counties southward. May, June.

4. **S. Greenei,** Gray. Somewhat floccose, less than 1 ft. high, leafy at base: *heads 1—3, large, peduncled, terminal:* radical leaves roundish, with abrupt or slightly cuneate base and long petioles, coarsely crenate-toothed: heads ⅔ in. long, with no bracteoles at base; *rays deep orange, ½ in. long:* style tips of disk-flowers penicillate and with a central cusp. —Under bushes, and on more open and rocky spaces among the higher mountains of Napa and Sonoma counties and northward.˙ May, June.

5. **S. aronicoides,** DC. Growing parts *loosely woolly, afterwards glabrate:* stem stout, 2—3 ft. high, leafy chiefly at base, the many small heads in a compound terminal cyme: leaves ovate to oblong and lanceolate, 3—6 in. long, irregularly and coarsely toothed, much reduced on the stem, the uppermost only bract-like: involucral bracts lanceolate, acuminate, not black-tipped: flowers 10—20; rays none, or rarely 1 or 2. —Chiefly among thickets, on northward slopes of hills. April, May.

6. **S. hydrophilus,** Nutt. var. **Pacificus,** Greene. Very stout, *succulent, glabrous, glaucescent,* the purplish coarse stems 2—4 ft. high, leafy mostly at base: leaves lanceolate, the lower 5—9 in. long, with stout petiole, the upper successively shorter and sessile, all more or less denticulate: heads small, very numerous, cymose-corymbose: rays none. —Brackish marshes; formerly plentiful at West Berkeley, and on the lower Napa River; still abundant in the Suisun marshes. July—Sept.

*** *Suffrutescent species.*

7. **S. Douglasii,** DC. Usually 3 ft. high, branching from the base, stoutish, loosely leafy; growing parts and young leaves whitish-tomentose, later glabrate—at least the upper surface of the leaves; *lower leaves pinnately divided into about 5 narrowly linear revolute lobes,* the upper linear, entire, all with revolute margins: heads few, large, corymbose; rays conspicuous, light-yellow: achenes glabrous.—Frequent on dry hills of the Mt. Diablo Range from Alameda Co. southward. July—Nov.

8. S. CINERARIA, DC. Stout, 2—4 ft. high, white-tomentose: leaves of firm texture, petiolate, pinnately parted, the *segments oblong, obtuse, more or less distinctly 3-lobed:* heads many, in a terminal˙ corymb: rays 10—12, short, oval.—An ornamental species of southern Europe, not infrequent as an escape from the gardens.

9. S. MIKANIOIDES, Otto. Glabrous, twining and trailing to the height of 20 feet: leaves roundish-cordate-hastate, sharply 5—7-angled, 2—5 in. long, nearly as broad, on petioles as long or longer, these with a reniform stipulaceous lobe on either side at base: heads small, in compound corymbs terminating axillary branchlets: rays none: disk-fl. 9—15.—Plentiful along banks of streams at the base of the Oakland and Berkeley Hills. Native of S. Africa; flowering profusely in January.

Suborder 10. CYNAROCEPHALÆ.

Herbs with watery juice, *usually prickly leaves,* and flowers in *dense ovoid or globose heads;* involucre imbricate. *Receptacle densely setose.* Flowers all alike and perfect, or the marginal larger than the others and neutral. *Corollas* regular or slightly irregular, *deeply cleft into 5 long and narrow lobes;* the marginal in some enlarged and palmatifid but *never ligulate.* Stamens syngenesious and also rarely monadelphous below. Anthers caudate. *Style slightly or not at all cleft, commonly with a pubescent node or ring below the stigmatic part.* Achenes thick, hard, smooth, basally or obliquely inserted. Pappus setose, plumose, or rarely paleaceous, sometimes wanting.

Pappus of many setose bristles.................................CENTAUREA 68
" double, *i. e.* of 2 different sets { of bristles...............CNICUS 69
                                      { of paleæ...............CENTROPHYLLUM 70
" many series of barbellate paleæ.........................SILYBUM 71
" united at base, deciduous in a ring;
       in many series of plumose bristles................CYNARA 72
       in one series of plumose bristles.................. CARDUUS 73

68. CENTAUREA, *Linn.* Annual or biennial herbs of various habit. Leaves unarmed and decurrent, or spinescent and merely sessile. Heads ovate; the imbricated bracts various. Achenes compressed or somewhat tetragonal, the insertion somewhat oblique or sublatral. Pappus of many slender scabrous bristles, mostly in two sets, or occasionally wanting.—All our species naturalized from Europe.

* *Bracts of involucre armed with a rigid spine; marginal corollas not enlarged.* ⁻

1. C. CALCITRAPA, L. Stoutish, rigid, 1—2 ft. high, very widely branching, leaves narrow, laciniate-pinnatifid; the uppermost somewhat involucrate-crowded about the sessile head: principal bracts of the involucre armed with a widely spreading long and rigid subulate spine, which bears 2 or 3 spinules on each side at base: *corollas red-purple: pappus none.*—Common along roadsides at and near Vacaville.

2. C. SOLSTITIALIS, L. Erect, 1—2 ft. high, canescent with cottony wool: radical leaves lyrate-pinnatifid; cauline lanceolate and linear, mostly entire, decurrent on the branches in narrow wings: heads pedunculate: middle *bracts of the involucre with a long rigid spreading spine*

several times their own length, and having 1 or 2 spinules at base; outermost bearing a few small palmate prickles; innermost only scarioustipped: *corollas yellow:* pappus double; outer of short and squamellate, inner of large bristles.—Very common in Napa Valley, less frequent in other parts of middle California.

3. C. MELITENSIS, L. Erect, 2—4 ft. high, cinereous-pubescent, or when young slightly woolly: radical leaves lyrate-pinnatifid; cauline lanceolate, mostly entire, narrowly decurrent: principal *bracts of involucre with a slender spine of about their own length* which is pectinate-spinulose at base; innermost with spinescent tips: flowers yellow: pappus of very unequal rather rigid bristles or squamellate.—A very common field and wayside weed.

    * * *Bracts of involucre unarmed, merely fringed; marginal*
       corollas much larger than the others.

4. C. CYANUS, L. Slender, 1—2 ft. high, not prickly, whitened, at least when young, with floccose wool: leaves linear, entire, or the lower toothed or pinnatifid: heads naked, on slender peduncles: involucral bracts narrow, fringed with short scarious teeth: marginal flowers asexual, much enlarged, irregularly and somewhat palmately cleft and ray-like, blue or pinkish or white: pappus of unequal bristles about equalling the achene.—Escaped from gardens to waysides; also occasional in grain fields; everywhere admired for the beauty of its flowers, and called *Cornflower, Bluebottle,* etc.

69. CNICUS, *Vaillant.* Annual with sinuate-pinnatifid leaves thinnish and reticulate-veiny, only weakly prickly. Heads enclosed within large leafy accessory bracts. Proper bracts of the involucre thin-coriaceous, in few ranks, many or all of them abruptly tipped with an aristiform or spinescent and pectinate-prickly appendage. Receptacle densely setose with long and soft bristles. Achenes terete, strongly many-striate, the corneous margin at summit 10-toothed. Pappus double; each set consisting or 10 aristiform bristles; the outer set longer and naked, the inner short and fimbriolate.

1. C. BENEDICTUS, L. A less rigidly thistle-like annual, with pale yellow flowers in large leafy-involucrate heads; occurring rarely as a ballast waif at San Francisco.

70. CENTROPHYLLUM, *Necker.* Annual or biennial. Leaves amplexicaul, reticulate-veiny, rigid, pinnatifid, and spinescent. Outer bracts of involucre foliaceous, with few spinescent lobes; the inner firmer, appressed, not cleft, but with a dilated and spinescent tip. Receptacle densely setose-paleaceous. Flowers yellow. Achenes obpyramidal; those of the marginal row more plump, somewhat convex and gibbous on one side, more angular on the other; those of the disk more

regularly and sharply rhombic-tetragonal; all truncate at the broad summit and surrounded by a crenulate margin. Pappus of the outer row of achenes reduced or wanting; of the others double, namely with an outer set of elongated but unequal ciliate paleæ in several series, and an uniserial and much shorter inner set.

1. C. LANATUM (L.), DC. & Duby. A very rigid thistle-like yellow-flowered annual, naturalized about South San Francisco, *Bioletti*.

71. **SILYBUM,** *Vaillant.* Coarse and stout annual, with very large sinuate-pinnatifid prickly leaves veined and blotched with white. Heads large, at the ends of long pedunculiform branches. Involucre broadly ovoid; bracts foliaceous, spinose along the margins, tapering into a rigid prickle. Receptacle densely soft-bristly. Corollas all alike, red-purple. Stamens monadelphous below. Achenes compressed and smooth. Pappus multiserial, of narrow almost setiform barbellate paleæ which are united slightly at base, and deciduous in a ring.

1. S. MARIANUM (L.), Gærtn. One of the most common and trouble-some of rank thistles, in all waste lands in California.

72. **CYNARA,** *Galen.* Very stout perennials, with pinnatifid or bipinnatifid sessile though not decurrent leaves; the lobes spinescent-tipped. Heads very large; involucral bracts coriaceous. Receptacle fleshy, fimbrillate. Achenes obovate compressed and somewhat 4-angled. Pappus of many series of plumose bristles.

1. C. SCOLYMUS, L. (ARTICHOKE). Stout and low, with very ample hoary-tomentulose bipinnatifid leaves, the segments spine-tipped: involucral bracts ovate, obtuse or emarginate.—Escaped from gardens near Benicia, West Berkeley, Alameda, etc.

73. **CARDUUS,** *Pliny* (THISTLE). Stout herbs, mostly biennial. Leaves mostly sessile or decurrent, and with sharply spinose lobes or teeth. Heads large, ovoid or subglobose; the pluriserial and imbricated involucral bracts usually prickly-tipped. Receptacle densely villous-setose. Flowers all alike, crimson, purple or white, the segments of the corolla long and linear-filiform. Achenes obovate or oblong, compressed, smooth, not striate; pappus a single series of long and barbellate or plumose slender bristles concreted at base and deciduous in a ring, often clavellate-dilated at the naked tip.

　　* *Bracts of involucre with dilated and squarrose-spreading tips;*
　　　　　　　　　heads short-peduncled.

1. C. **fontinalis,** Greene. Stout, 2 ft. high, the widely spreading branches ending in middle-sized nodding heads: stem and upper surface of leaves glandular-pubescent: bracts of the involucre herbaceous, broad, squarrose-spreading or recurved, abruptly acute, with a short

spinose tip and no glandular spot: flowers dull white: anther tips acute.
—Along streamlets and in springy places about Crystal Springs, San
Mateo Co.; a very peculiar species. July.

* * *Only the outer involucral bracts squarrose; terminal spine in all erect.*

+—*Heads not conspicuously long-peduncled.*

2. **C. remotifolius,** Hook. Loosely arachnoid when young, glabrate
in age, 3—5 ft. high; leaves from sinuately to deeply pinnatifid: heads
on long rather slender peduncles, 1½ in. high, ovate; bracts rather loose
and thin, linear-lanceolate, the outer narrowed to a small weak prickle,
the middle ones more or less *cartilaginous and lacerate as well as somewhat dilated toward the tip;* the inner linear-subulate with narrow fimbrillate scarious margin: corollas ochroleucous, the limb only a third as
long as the throat.—Marin Co.

3. **C. crassicaulis,** Greene. Very stout and tall, 4—7 ft. high: stem
strongly striate, simple to near the summit, there becoming rather close-
paniculate, with 3—7 short-peduncled heads 1½—2 in. high: herbage
permanently hoary-lanate: leaves small, pinnately parted, the *segments
spine-tipped and the whole margin spinulose-ciliate:* involucral *bracts*
rather loose, linear-lanceolate to lanceolate-acuminate, all *tipped with a
slender spine, the outer and middle ones with pectinate-spinescent margins:*
segments of the whitish or pinkish corolla about as long as the throat.—
Abundant in moist grassy bottoms of the lower San Joaquin.

4. **C. amplifolius,** Greene. Somewhat fleshy, green and glabrous
except a sparse arachnoid tomentum on the lower face of the leaves:
stem stout, 3—4 ft. high: *leaves very broad and ample, conspicuously
decurrent,* the ample *lobes* crowded and *overlapping each other,* trifid and
spinose-ciliate: heads short-peduncled and clustered, 1½ in. high, leafy-
bracted at base; the outer bracts loosely spreading and arachnoid, the
inner more appressed and glabrous, their spinose tips often reflexed:
corollas of a rich bright purple, their linear obtuse lobes much shorter
than the throat.—On stream banks back of Point San Pedro.

5. **C. edulis** (Nutt.), Greene. Robust, 3—6 ft. high, pubescent,
leafy up to the short panicle: *leaves oblong or narrower, sinuate-pinnatifid,*
weakly prickly: heads 1½ in. high, depressed-globose, leafy-bracted at
base: *involucre arachnoid* when young: corollas deep but dull purple;
segments shorter than the throat.—Along streams in the Oakland Hills,
etc. June, July.

+— +— *Heads solitary on stout peduncles.*

6. **C. Californicus** (Gray), Greene. Rather slender, 2—4 ft. high,
canescently woolly: leaves sinuate-pinnatifid, moderately prickly, heads
middle-sized, the *lower bracts coriaceous-acerose, spreading and incurved,*

the others straight, all subulate-spinescent at the tip: *corollas lilac-purplish* or reddish, lobes shorter than the throat.—Mt. Diablo and southward. June, July.

7. **C. venustus**, Greene. Stoutish, 3 ft. high, the foliage *permanently arachnoid-tomentose:* heads large (2 in. high), terminating long and almost naked pedunculiform branches: involucre glabrate, the many subcoriaceous bracts with closely appressed base, and long *lanceolate-subulate abruptly short-spinescent tips:* corollas bright crimson, the segments longer than the throat: pappus-bristles barbellate above the plumose part, the tips scarcely dilated.—Foothills of the Mt. Diablo Range. May, June.

8. **C. occidentalis**, Nutt. Stout, 2—3 ft. high; peduncles stout, rather short: leaves deeply pinnatifid, glabrate above, canescently tomentose beneath: involucre subglobose; *bracts straight, subulate-lanceolate*, with short spines, the whole mass *densely festooned with remarkably distinct cobwebby hairs:* corollas red-purple: anthers distinctly bisetose and lacerate at base: pappus somewhat scanty.—Sandy hills along the seaboard only. May—Aug.

　　* * * *Involucral bracts appressed, the slender spine at their tips more or less abruptly spreading.*

9. **C. hydrophilus**, Greene. Rather slender, freely branching above, 3—5 ft. high; when young pale with a fine thin arachnoid pubescence, *in maturity green and glabrate:* leaves not large, deeply cut into uniform 3-lobed segments: heads 1 in. high, somewhat clustered at the ends of the branches; involucre ovate, the appressed-imbricate *bracts with a green and glutinous ridge toward the summit*, and ending in a short slender somewhat spreading spine.—Brackish marshes about Suisun Bay. July—Sept.

10. **C. quercetorum** (Gray), Greene.˙ Sparingly villous-arachnoid when young, soon glabrate: stem stout, 1 ft. high or less, with few rather large heads: *leaves mostly petiolate*, the larger 1 ft. long, pinnately parted, the oblong divisions often 3—5-cleft, prickly: involucral bracts coriaceous, closely imbricated in numerous ranks, the outer with short prickles, the inner obscurely scarious at tip: corollas either dull-purple or white: anther-tips narrow, very acute.—Open grassy summits and higher slopes of hills on both sides of the Bay. June, July.

11. C. LANCEOLATUS, L. More or less villous or hirsute, seldom cottony; 2—4 ft. high, *stem and branches interruptedly winged by the decurrent leaves*, both leaves and wings prickly: heads nearly 2 in. high, arachnoid-woolly at first, the bracts lanceolate, attenuate into slender and rigid prickle-pointed spreading tips: fl. rose-purple.—A most troublesome Old World weed, already abundant on this coast far northward, only lately beginning to be seen occasionally in our district.

Order LVIX. CICHORIACEÆ.

Plants with alternate leaves and a milky narcotic juice; as to inflorescence closely analogous to Compositæ, though not naturally allied to them very closely; much nearer Lobeliaceæ. Flowers in the head all ligulate, the ligules 5-toothed at apex. Anthers appendaged at summit, at base sagittate, or abruptly acuminate-setaceous. Style-branches slender, obtusish or acutish, minutely papillose. Pollen-grains perfectly smooth and distinctly 12-sided.

* *Pappus of plumose bristles.*

Receptacle not chaffy;
Achenes truncate..PTILORIA   2
              { flowers white............................NEMOSERIS   3
"   beaked { "   purple......................TRAGOPOGON   4
              { "   yellow........................PICRIS   9
Receptacle chaffy...HYPOCHÆRIS   5

* * *Pappus paleaceous, awned or awnless.*

Pappus-paleæ awnless...CICHORIUM   1
  " " awned;
Awn from a notch in the palea................................UROPAPPUS   6
Palea tapering into the awn;
    Annuals; achenes oblong or turbinate...................MICROSERIS   7
    Perennials; achenes cylindric.........................SCORZONELLA   8

* * * *Pappus of capillary bristles only.*

Achenes not compressed;
  Achenes truncate;
    Pappus soft, deciduous...............................MALACOTHRIX   10
    Pappus firmer, persistent;
      Pappus dull-white or darker........... .HIERACIUM   11
      " bright-white.................................CREPIS   12
  Achenes slender-beaked;
      10-ribbed or -angled...............................AGOSERIS   13
  " 4–5-angled...TARAXACUM   14
Achenes compressed;
  Beaked or at least attenuate above.... ....LACTUCA   15
  Not beaked or narrowed above.............................SONCHUS   16

1. CICHORIUM, *Theophr.* Perennials, leafy at base; the tall stem and branches with reduced foliage, bearing several heads of blue flowers in the axils. Bracts of the involucre in 2 series; the inner erect, partly enfolding the subtended achenes, the 4 or 5 outer more spreading. Achenes short, truncate, somewhat angled; the broad summit bordered with 2 or more series of short blunt paleæ.

1. C. INTYBUS, L. (CHICORY). More or less hirsute below, 2—5 ft. high: radical leaves runcinate; cauline oblong or lanceolate, dentate; those of the flowering branches scarcely more than bract-like: heads 1 in. broad or more, expanded in the morning, closing by midday.— Common in many places, as an escape from the market gardens.

2. PTILORIA, *Raf.* Ours stoutish and rather rigid tall annuals. Leaves runcinate. Heads small; fl. pinkish or purplish, few in the head, the ligules all equal. Involucre of several longer erect inner bracts, and as many short appressed calyculate outer ones. Achenes truncate, 5-angled. Pappus a series of plumose bristles curving outward.

1. **P. virgata** (Benth.), Greene. Rigid, *virgate*, 1—3 ft. high, *glabrous throughout* and the herbage deep green: leaves runcinate: heads 3—4 lines long, subsessile along the naked upper part of stem and branches, 4—8 flowered: achenes subclavate or oblong, rugose-tuberculate between the ribs: pappus clear white, plumose almost throughout.—Sandy banks and hills. Aug.—Oct.

2. **P. canescens,** Greene. More slender, *paniculate*, 2—4 ft. high; stem and foliage *hoary-tomentose when young,* somewhat glabrate in age: leaves lanceolate, more or less sinuate- or runcinate-pinnatifid: achenes larger than in the last, and less tuberculate; pappus as white, slightly longer and of fewer bristles.—Mountain sides, and clayey banks of streams, in exposed places. June—Sept.

3. NEMOSERIS, *Greene.* Stout annual, near *Ptiloria*, but decidedly inclining to the corymbose in branching: flowers much more numerous in the head: ligules white, unequal. Achenes tapering to a long beak supporting the pappus; the rays of the latter not at all curved.

1. **N. Californica** (Nutt.), Greene. Glabrous, the stem white, 2—3 ft. high, herbage glabrous and with a strong narcotic smell: leaves oblong, pinnatifid, sessile and clasping: heads 1 in. wide when expanded: outer achenes pubescent; plumose pappus sordid.—On clayey banks and slopes of wooded hills; common. June—Sept.

4. TRAGOPOGON, *Theophr.* Stoutish biennials with fusiform edible root, leafy erect stems and large long-peduncled slender conic involucres. Receptacle naked. Achenes muricate, long-beaked; the beak supporting an ample pappus of setaceous bristles which are long-plumose at base and naked above.

1. **T.** PORRIFOLIUS, L. (SALSIFY). Leaves entire, long and grassy: stem 2—4 ft. high: rays deep purple.—Naturalized in waste lands; an escape from the gardens.

5. HYPOCHÆRIS, *Vaillant.* Plants leafy mostly at base of the branching naked or leafy-bracted somewhat corymbose stems. Involucres oblong-conic, erect in the bud; bracts imbricated. Receptacle scarious-chaffy, the chaff deciduous. Flowers yellow. Achenes oblong or fusiform, 10-ribbed, glabrous or scabrous, at least the inner ones tapering to a beak. Pappus a series of fine plumose bristles, often accompanied by some outer naked ones.—Weeds introduced from Europe.

1. **H. GLABRA, L.** *Annual, glabrous,* 1 ft. high more or less: leaves in a depressed radical tuft, oblanceolate, obtuse, sinuate-toothed: scape branching: ligules and expanded head small: *outer achenes truncate* at summit, the inner tapering to a long slender beak: pappus of capillary bristles intricately plumose below the nearly naked apex, and of some fine short naked outer ones.—Very common in all open grounds.

2. **H. RADICATA, L.** Twice as large as the preceding, *perennial, hirsute:* ligules long, and expanded heads 1 in. broad: *achenes all rostrate.*—Common in shaded grassy ground at Berkeley, where it began to appear only a few years since.

6. UROPAPPUS, Nutt. Subacaulescent annuals, nearly or quite glabrous, with laciniately cleft or pinnatifid leaves, and stout scapiform monocephalous peduncles enlarged under the oblong-conic heads, these always erect. Ligules short, in expansion surpassed by some of the bracts of the oblong conic imbricated involucre. Achenes slender-fusiform, 8—12-ribbed, truncate at summit. Pappus of 5 scarious ample bifid paleæ, with an awn or bristle arising from the notch.

* *Achenes brownish; pappus brownish, persistent.*

1. **U. Lindleyi** (DC.), Nutt. Stoutish, 1½ ft. high or smaller: achenes 5—6 lines long, slightly narrowed toward the summit: pappus-paleæ linear-lanceolate, 4 lines long, the awn very little shorter.—The most common species. May, June.

2. **U. leucocarpus,** Greene. Like the preceding in size, etc., but achenes almost white, slenderly attenuate at summit, the narrow part vacant, *i. e.,* not filled by the seed; very light-brown paleæ and slender awn each about 2½ lines long.—With the preceding, but far less common.

3. **U. Clevelandi,** Greene. Smaller than the preceding, furfuraceous-puberulent: achenes 3 lines long, not at all attenuate, the body, and also the pappus, of a deeper brown; paleæ with a very short awn.—Plains at the eastern base of Mt. Diablo, thence southward. April.

4. **U. Kelloggii,** Greene. More slender than any of the preceding, the scapes little enlarged under the heads: achenes slightly attenuate at each end: paleæ of pappus only about 2 lines long, the awn somewhat longer.—Toward the seaboard, and not common: near Tomales, etc. April, May.

* * *Achenes black; pappus clear white, deciduous.*

5. **U. linearifolius** (DC.), Nutt. Stouter than any of the foregoing; the numerous scapiform peduncles much dilated under the involucres: blackish achenes almost rostrate-attenuate. Var. **elatus.** Slender and tall, the stem (not rarely a foot high) and slender peduncles together

sometimes making a height of 2 ft. or more: heads smaller, with fewer flowers.—The type common, chiefly along the seaboard; the variety, of the interior, extending far southward. April, May.

7. **MICROSERIS,** *Don.* Stemless annuals, with an ample radical tuft of mostly pinnatifid leaves, and many slender erect or decumbent monocephalous scapes; the ovoid or subglobose heads nodding in bud, and even in flower; usually erect in fruit. Involucre with short-imbricated bracts at base; the main bracts longer and equal. Ligules short; expanded heads small. Achenes oblong-claviform to turbinate. Pappus of usually short paleæ tapering into a long scabrous awn.

1. **M. Douglasii** (DC.), Gray. Scapes many, decumbent at base, 8—18 in. high: head broad-ovoid: *achenes 2 lines long, thickish, oblong-turbinate,* contracted near the summit, those of the outer circle usually white-villous: *paleæ of the pappus 2 lines long, round-ovate to orbicular,* half or a third the length of the awn, glabrous or villous on the outside. Very common, and in great variety of forms. April, May.

2. **M. attenuata,** Greene. More slender, with fewer scapes, these not as long, more erect: head oblong: *achenes 4 lines long, attenuate-fusiform,* the upper and narrower half not filled by the seed but vacant: *pappus-paleæ* 3 lines long or more, *oblong-lanceolate,* about half the length of the awn, more or less villous externally.—Hills of Contra Costa Co., but first collected at Berkeley, where it is long since extinct.

3. **M. indivisa,** Greene. Stoutish, leaves not pinnatifid, many not even toothed, oblanceolate; scapes quite erect, 1—1½ ft. high: heads subglobose, the fl. and achenes more than 100; outer row of achenes silvery-silky, the others glabrous, all chestnut-brown, 2 lines long, the *pappus* about 5 lines, *of 5 whitish barbellulate not fragile bristles the bases of which are dilated into small triangular lanceolate paleæ.*—Plains of Solano Co., east of the mountains. May.

4. **M. tenella** (Gray). Very slender, and the leaves subentire, or larger and the leaves pinnatifid: heads broad-ovate to subglobose: *achenes dark-brown, oblong-clavate: pappus mostly of only 2 or 3 very slender fragile bristles* which are merely *deltoid-dilated* at base.—Very common and variable; occasionally destitute of pappus and the small heads hemispherical. April, May.

5. **M. elegans,** Greene. Seldom 1 ft. high, slender, the leaves pinnatifid: fruiting head small (less than ½ in. high): *achenes little more than a line long, turbinate: pappus-paleæ ovate-deltoid, a fourth the length of the very slender awn,* these and often the whole *summit of the achene minutely villous.*—Low plains of Solano and Contra Costa counties.

6. **M. Bigelovii,** Gray. Often 1 ft. high and more: broad-ovate head ½ in. high: involucre more imbricated than in the foregoing: achenes

oblong-turbinate, 2 lines long: pappus-paleæ ovate-lanceolate, brownish, only half or a third the length of the similarly brownish awn.—Along the seaboard chiefly, in sandy soil; extremely variable, if not indeed an aggregate of two or more species. April—June.

7. **M. acuminata,** Greene. Only a span high, but heads elongated; leaves deeply pinnatifid into slender lobes: *achenes slenderly fusiform-turbinate,* nearly 3 lines long, the pappus nearly ¾ in. long; *palex narrowly lanceolate, the awn shorter* than the palea.—Plains and mountain valleys from Sonoma Co. to Solano. April, May.

8. **SCORZONELLA,** *Nutt.* Stems erect, from a fusiform perennial root, parted into several pedunculiform monocephalous branches, and leafy mostly at base. Leaves lanceolate, coarsely toothed, or pinnatifid into narrow segments, or some entire. Heads large, nodding in bud; ligules elongated, yellow, the expanded head therefore showy. Involucre cylindraceous-ovoid, more or less imbricated. Achenes cylindric, with 10 or more obscure angles. Pappus of 10 or more scabrous- or barbellulate-awned long or short firm paleæ.

1. **S. silvatica,** Benth. Mostly simple and monocephalous, 1–2 ft. high: head 1 in. high, 30—40-flowered: bracts of involucre in 3 or 4 series, the outer ovate or ovate-lanceolate, abruptly acuminate, the inner lanceolate, gradually acuminate: ligules glandular-puberulent: *palex of the pappus lanceolate, as long as the achene,* tapering into *the subplumose awn which is somewhat longer.*—Hills and plains along the eastern base of the Mt. Diablo Range.

2. **S. paludosa,** Greene. Stems several, 2 ft. high or more, strictly erect: leaves subentire to laciniate-parted into long and very narrow segments: head 50—75-flowered; bracts all tapering from a lanceolate base into a long slender acumination: achene 2 lines long: pappus brownish, the lanceolate *palex little more than a line long, the barbellulate awn 4—5 lines.*—Low moist plains, from Marin Co. to Solano.

3. **S. Bolanderi** (Gray), Greene. About 1 ft. high, the *scapiform branches leafy at base only and decumbent:* leaves linear-lanceolate, entire, or with linear lobes above the middle: bracts all gradually attenuate from a broad base, rather regularly imbricated: *pappus brown, 5 lines long, the ovate palea ½ line.*—Marin Co., in wet places. April.

4. **S. maxima,** Bioletti. Very stout, 2—4 ft. high, leafy-stemmed, the branches long and erect: leaves 1 ft. long, 2 in. wide, entire or barely somewhat toothed: heads very broad, 400-flowered; bracts of involucre from ovate, acute, to lanceolate-acuminate: achenes ¼ in. long, the *pappus almost white; lanceolate palea 1½ lines long, the merely scabrous awn 5 or 6 lines.*—At Los Guilucos. June, July.

9. PICRIS, *Theophr.* Coarse branching leafy herbs, rough-bristly, yellow-flowered; the heads scattered. Involucre double, an outer series of foliaceous spreading bracts and an inner erect series. Achenes subterete, transversely rugose, beaked and bearing a pappus of plumose bristles.

1. **P. ECHIOIDES, L.** Biennial, 2—3 ft. high: stem hispid with hooked hairs; oblanceolate leaves rough with bristles from a pustulate base: outer involucral bracts ovate-acuminate, bristly-spiny on the margins: achenes reddish, slender-beaked; pappus very plumose.—Common wayside weed about Vallejo, West Berkeley, Santa Clara, etc.

10. MALACOTHRIX, *DeCandolle.* Annuals, heterogeneous as to habit, involucre, etc., but known by short glabrous terete and 5—15 costate, or 4—5-angled truncate achenes with denticulate border to the summit, and a soft-capillary deciduous pappus, accompanied by one or more stouter bristles which are more persistent; the inner, more copious deciduous set more or less joined into a ring at base.

* Acaulescent; heads large, pale-yellow.

1. **M. Californica,** DC. Woolly when young, with very long soft hairs: leaves once or twice pinnately pinnatifid into narrow-linear or almost filiform lobes: scapes stoutish: involucre long-conical, ⅔ in. high; outer bracts slender-subulate; receptacle slender-bristly: achenes narrow, striate-costate: outer pappus of 2 persistent bristles, and between them some minute pointed teeth.—Plains of eastern Contra Costa Co.

* * Caulescent, branching freely; heads smaller.

2. **M. parviflora,** Benth. Leaves scattered, only some of the lower pinnatifid; plant 1 ft. high; panicle narrow; heads ¼ in. high, few-flowered, *fl. yellow;* achenes oblong linear, minutely striate-costate, 4 or 5 of the ribs slightly more prominent; *outer pappus of 1 conspicuous bristle and many white setulose teeth.*—Mt. Diablo Range, and plains to the eastward. April—June.

3. **M. obtusa,** Benth. Leaves in a rosulate radical tuft, runcinate-pinnatifid, the lobes sinuate-toothed, the teeth or sinuses bearing tufts of cottony white wool: stem 8—10 in. high, diffusely corymbose-paniculate: heads ¼ in. high; *fl. pure white, or with a pinkish tinge:* achenes obovate-oblong, obtusely 5-angled, the apex slightly narrowed and *border entire; none of the pappus-bristles persistent.*—Coast and Mt. Diablo ranges at middle and higher elevations. May, June.

4. **M. Coulteri,** Gray. Stoutish, erect, 1—2 ft. high, with rather few branches; herbage glabrous, glaucescent: upper cauline leaves ovate or cordate, sessile, clasping, all sparsely toothed: involucres ½ in. high, ovoid, of *silvery-scarious bracts very distinctly imbricated* in several series: fl. white, fading purplish: achenes acutely about 15-ribbed and 4- or 5-angled, the summit obscurely denticulate: 1 or 2 stouter bristles more persistent.—Plains of the lower San Joaquin. April—June.

11. HIERACIUM, *Diosc.* Perennial, often rough-hairy herbs. Involucre subcylindric, of uniserial linear bracts. Receptacle naked. Flowers mostly yellow (in ours white). Achenes linear, 10-ribbed or -striate. Pappus of rigid fragile dull-white or brownish bristles.

1. **H. albiflorum,** Hook. Leaves mostly radical, oblong-spatulate, entire or denticulate, 2—5 in. long, thickly beset with long bristly hairs: stem nearly naked above, ending in a more or less ample panicle of white-flowered heads: involucre 3—5 lines long, achenes 1½ lines long, not tapering, evenly striate-ribbed.—Woods of the Coast Range.

12. CREPIS, *Dalechamps.* Perennial, tomentulose or glabrous herbs. Involucre of linear equal bracts, usually with calyculate short ones at base. Flowers yellow. Achenes columnar to fusiform, 10—20-costate. Pappus of copious soft white bristles.

1. **C. acuminata,** Nutt. *Hoary-tomentulose,* slender, 1—2 ft. high, with open cyme of many slender peduncled narrow heads: leaves radical, runcinately pinnatifid into lanceolate or linear lobes below the middle, the apex prolonged into a narrow entire acumination: involucre 5—7 lines high, narrow-cylindric: *achenes* 10-striate, *tapering at summit.*—Mt. Hamilton; thence southward and eastward.

2. C. VIRENS, L. *Green and glabrous,* slender, 1—2 ft. high: leaves mostly radical, lanceolate or broader, toothed or pinnatifid; cauline sessile, with subsagittate base: heads very small; ·fl. yellow: *achenes oblong,* 10-striate, smooth, narrowed slightly but about equally at both ends.—In shady grassy places about Berkeley; naturalized from Europe.

13. AGOSERIS, *Raf.* Herbs usually quite acaulescent, with tufted radical leaves lanceolate, pinnately toothed or cleft, and simple scapes bearing solitary large heads of yellow or orange-colored flowers. Involucre at first subcylindric, later approaching the conic: bracts imbricated in 2 series, the outer often short, more foliaceous and spreading. Achenes oblong or linear, or slender-fusiform, terete, 10-ribbed; the apex produced in ours into a very slender beak with a dilated terminal areola on which are inserted the copious fine white pappus-bristles.

* *Perennials; achenes tapering into the beak.*

+— *Heads very large, but ligules very short.*

1. **A. plebeia,** Greene. Sparsely lanate-hirsute on the leaves beneath, the involucres woolly at base, otherwise glabrous: scapes often 2 ft. high: leaves not rarely 1 ft. long, narrowly oblanceolate, usually slenderly acuminate, the sides with several pairs of abrupt triangular teeth or subfalcate lobes: *ligules short, suberect, deep yellow:* achenes slender-fusiform, 2—2½ lines long; beak 5 or 6 lines; pappus very soft and white.—Western side of the Coast Range, and about the Bay; the most rank and the least showy species. May, June.

+ + *Heads large, and ligules elongated.*

2. **A. grandiflora** (Nutt.), Greene. Rather more lanate than the last, often as large: leaves more deeply and constantly pinnatifid, the terminal undivided part oblanceolate, obtuse: *ligules light yellow, elongated,* spreading, the *expanded head 2—3 in.* broad: achenes about 3 lines long, the beak 10 lines.—Plains of the eastern part of Solano Co. and far northward; not in the Bay region proper. May.

3. **A. intermedia**, Greene. Size of the last, nearly, pale green and glaucescent, but with some lanate pubescence when young: leaves with a linear rachis, many remote narrowly linear pinnate segments and a long linear-acuminate terminal lobe: expanded ligules forming a head 2 in. broad, fl. pale yellow: *achenes 2 –2½ lines long, very sharply carinate-ribbed, the ribs along their bases closely beset with short stiff setulose hairs;* beak 8—10 lines long.—Mt. Diablo, near the summit, and elsewhere, at considerable elevations of the inner ranges.

4. **A. hirsuta** (Hook.), Greene. Hirsute-pubescent, not rarely caulescent and the depressed or ascending stem 6 in. high: leaves from narrowly spatulate and merely toothed or lyrate-pinnatifid, to pinnately parted into linear lobes: scapes or peduncles slender, 1—1½ ft. high, reddish; the elongated bright yellow ligules also fading reddish: *achenes slender-fusiform,* 1½ *lines long or more, the beak only about twice as long; pappus usually dull* or yellowish white.—Only on open grassy slopes near the Bay and seacoast. May—Nov.

5. **A. apargioides** (Less.), Greene. Very near the last, but every way much smaller, the leaves more remotely and slenderly pinnatifid: heads only ½ in. high: *beak not longer than the body of the achene.*—Sand hills of San Francisco; flowering almost throughout the year.

* * *Perennials; achenes abruptly beaked from a truncate summit.*

6. **A. retrorsa** (Benth.), Greene. Hoary with a woolly pubescence: leaves pinnately parted into linear-lanceolate usually long retrorse lobes, the terminal one long and narrow, all callous-tipped: ligules long, pale salmon-color: *achenes truncate, 3 lines long, the filiform beak nearly an inch.*—Higher elevations of all the coast mountains. June, July.

* * * *Annuals; manifestly caulescent.*

7. **A. heterophylla** (Nutt.), Greene. Slender, seldom 1 ft. high, more or less villous or hirsute: leaves spatulate to oblanceolate, toothed irregularly, or entire: heads ½—¾ in. high; *ligules short and inconspicuous:* achenes about 2 lines long, with beak of about 3 lines; *inner achenes mostly with obsolete ribs* and not filled to the summit by the seed; outer ones extremely variable, normally with ribs developed into broad undulate wings, otherwise merely ribbed and hirsute, or again, inflated to the subcylindric and the ribs not visible.—Very common, and dispersed widely beyond our limits. April –May.

8. **A. major,** Jepson. Twice larger every way; leaves often pinnatifid: *ligules elongated and head conspicuous* (1½ in. broad), when in flower: achenes inclined to vary, as in the last, but not reaching such extremes, *tapering more abruptly into the stipe,* and more or less distinctly *toothed under it:* pappus sordid or almost fuscous.—Plains of the interior, from Solano Co. southward. May.

14. TARAXACUM, *Haller.* Flaccid nearly glabrous acaulescent herb, with a tuft of depressed runcinate-pinnatifid leaves, and hollow scapes bearing solitary heads. Involucre double; outer bracts short, spreading, partly connate at base; inner erect, narrow, equal. Achenes oblong-obovate, 4—5-angled or -costate, muricate or prickly near the summit, abruptly narrowed into a long filiform stipe which bears the soft capillary whitish pappus.

1. **T. OFFICINALE,** Weber (DANDELION). Common weed in parks and lawns, where it has been introduced accidentally, with lawn-grass seed; scarcely naturalized in our district.

15. LACTUCA, *Pliny.* Leafy-stemmed biennials and annuals, with ample though narrowish panicles of small conoidal heads. Flowers yellow. Achenes compressed, oblong, abruptly narrowed to a short slender beak which bears at its dilated summit the soft white capillary pappus.

1. **L. VIROSA,** L. Strict, 3—6 ft. high, glaucescent, glabrous except the lower part of the stem, which is somewhat hispid: leaves horizontal, oblanceolate to oblong, with spinulose-dentate margins, the midrib beneath beset with a row of soft prickles: beak about the length of the striate-nerved achene.—Shady places along Strawberry Creek, Berkeley.

2. **L. SATIVA,** Bauhin (LETTUCE). The common salad plant, with obovate or obovoid thin unarmed foliage, and a short somewhat corymbose panicle of heads, is a common field and orchard weed in Napa Valley, and elsewhere in the Coast Range regions.

16. SONCHUS, *Diosc.* Coarse annuals with pinnatifid leaves and indistinctly panicled or more scattered and irregular inflorescence. Involucre conic, in age broad and thickened at base. Achenes obcompressed, without beak or dilated pappiferous disk. Pappus of very soft fine flaccid bristles, which fall more or less unitedly, and commonly one or more stronger ones which fall separately.—Old World weeds; but common in all parts of California.

1. **S. OLERACEUS,** L. Stoutish, 2—4 ft. high, sparingly leafy, glabrous, or with a few glandular hairs on pedicels and calyx, glaucescent: leaves from obovoid to narrower and runcinate-pinnatifid, toothed but not prickly margined, amplexicaul, the *auricles straight, acute, holding the same plane with the blade:* achenes striate-nerved, *transversely rugulose-scabrous.*—Common everywhere; flowering at all seasons.

2. S. ASPER, Fuchs. (1542). Stouter, the distinctly and acutely angled very leafy stem often an in. thick, the heads irregularly umbellate at its summit: leaves pinnatifid, prickly-margined, the *auricles helicoid and appressed* to the stem: *achenes smooth*, 3-nerved on each side.—Nearly as common as the preceding, but more a plant of spring and early summer.

ORDER LX. LOBELIACEÆ.

Herbs with milky juice, alternate simple leaves, and racemose or scattered irregular flowers. Calyx-tube adnate to the ovary, the limb divided. Corolla bilabiately lobed or cleft. Stamens 5, alternate with the corolla-lobes; filaments joined into a tube; anthers also united; pollen 12-sided, as in the *Cichoriaceæ*. Ovary 1- or 2-celled; style entire. Fruit capsular, many-seeded.

1. **HOWELLIA,** *A. Gray.* Delicate herbs of muddy shores. Calyx with linear-clavate tube wholly adnate to the ovary; limb of 5 segments. Segments of corolla only slightly unequal; tube obsolete. Stamen-tube nearly free, and with the included style, slightly incurved. Capsule membranaceous, 1-celled, few-seeded, bursting irregularly on one side.

1. **H. limosa,** Greene. Weak procumbent branches 6—12 in. long, leafy and floriferous throughout: leaves lanceolate, entire, sessile, ½ in. long: corolla white, the segments 1 line long, cuneiform, the two upper narrower and more widely separated: capsule ½ in. long, surmounted by the 5 triangular erect calyx-teeth.—Margins of pools on the plains of Solano Co. near Suisun. May.

2. **BOLELIA,** *Raf.* Dwarf herbs of low plains or along lake shores. Calyx-tube very long, stalk-like, wholly adherent to the long slender ovary. Corolla with very short tube and ample bilabiate limb; lips spreading, the larger broad and only toothed or lobed, the divisions of the upper (and smaller) usually distinct. Capsule elongated and linear, many-seeded, dehiscent longitudinally into 3 long valves.

* *Larger lip of corolla merely 3-toothed or -lobed at apex.*

1. **B. insignis,** Greene. Stoutish, erect, mostly simple and few-flowered: lower lip of the very large corolla ½ in. broad, obovoid, 3-lobed, the lobes and lateral parts of the body sky-blue marked with darker veinlets, the main portion white, bearing in the middle 2 oblong parallel green spots; upper lip merely bifid, the lobes ascending and parallel; throat of corolla with a pair of bright yellow folds in a field of dark violet.—Low plains of the lower Sacramento and San Joaquin. Very different from its far northern ally, *B. elegans.* May.

* * *Larger lip of corolla trefoil-shaped.*

2. **B. pulchella** (Lindl.), Greene. Freely branching, often 6—10 in. high: lower lip of corolla parted into 3 broad *obtusish and mucronulate*

violet lobes, the undivided part yellow, with white border; divisions of the upper lip spreading and divergent, of broad-lanceolate outline.—With the preceding, and a more profusely flowering species. May.

3. **B. tricolor**, Greene. Branches few, weak, tortuous and reclining: lower lip of corolla parted into 3 equal *broadly obovate truncate* and slightly *cuspidate lobes,* these deep blue at tip, white below, the undivided part with a *transverse somewhat quadrate spot of dark maroon;* upper lip of 2 small segments somewhat recurved, but parallel.—Plains near Suisun. May.

4. **B. concolor**, Greene. Rather firm branches 3—5 in. high, very numerous and *forming a dense tuft;* the herbage very minutely puberulent: *corolla violet almost throughout,* but base of lower lip very dark and this spot circumscribed by lighter blue, the lobes slightly unequal, very obtuse and somewhat cuspidate; upper lip cleft to the middle only, the lobes lanceolate, more or less deflexed.—Low grain fields near Suisun.

5. **B. ornatissima**, Greene. Slender, 6—10 in. high, mostly simple, erect and rather strict: corolla pale throughout; lobes of lower lip somewhat obcordate and cuspidate, the *undivided part with 4 prominent fold-like protuberances* partly filling the throat; *segments of upper lip each deflexed and coiled backward into a ring, the corolla-tube* at base of these segments abruptly *raised into a sharp protuberance.*—Plains of the lower Sacramento, especially about Elmira. May.

*   *   * *Segments of corolla not very dissimilar.*

6. **B. humilis**, Greene. Very dwarf, the stem branching, only about 1 in. high: segments of the calyx-limb linear, unequal, exceeding the *minute white corolla;* this bilabiate, the upper segments being smaller, but all of them ovate-oblong, acute.—Moist plains of Sonoma Co.

Order LXI. CAMPANULACEÆ.

Closely allied to *Lobeliaceæ;* but corolla regular (in our genera); stamens quite distinct. Style 1, with 2—5 introrse stigmas. Flowers in all ours blue, or bluish.

Capsule dehiscent by small valvular openings on the sides;
    Flowers all having corollas...CAMPANULA 1
      "   all but the uppermost apetalous........................TRIODANIS 2
Capsule bursting on the thin sides indefinitely......................HETEROCODON 3
Capsule opening at top, within the calyx.............................GITHOPSIS 4

1. **CAMPANULA**, *Dodoens.* Flowers all complete. Calyx-lobes 5, narrow; tube short and broad. Corolla 5-lobed or -parted. Filaments dilated at base. Capsule short, opening on the sides by 3—5 small uplifted valves.

1. **C. prenanthoides,** Durand. Perennial, 1- 2 ft. high, erect, glabrous or scabro-puberulent: leaves ovate to oblong, 1 in. long or less, sharply serrate, the cauline mostly sessile: *fl. mostly racemose, short-pedicelled:* corolla slender-cylindric in bud, ½ in. long, almost 5-parted, the narrowly lanceolate lobes thrice the length of the tube.—Wooded hills of the Coast Range.

2. **C. Scouleri,** Hook. Perennial, but delicate, decumbent, few-leaved, the stems seldom a foot long; leaves mostly tapering into a marginal petiole: *fl.* somewhat panicled, nodding, *on long filiform pedicels:* corolla oblong in bud, exceeding the slender calyx-lobes, deeply 5-cleft; lobes ovate-oblong.—Redwoods of Marin Co., in moist shades.

3. **C. exigua,** Rattan. Annual, 2—5 in. high: leaves only 1—3 lines long, sessile, the lowest obovate, entire or with few teeth, the upper subulate: *fl. erect at the ends of the numerous divergent branchlets:* calyx-lobes subulate-linear, twice the length of the turbinate tube: corolla light blue, oblong-campanulate; tube longer than the oblong acute lobes: filaments dilated into a broad ciliolate base: capsule urceolate, opening by 3 valves above the middle.—Only at the very summits of our highest mountains, Diablo, Tamalpais and Hamilton. June.

2. **TRIODANIS,** *Raf.* Annuals with very broad closely sessile leaves, and flowers of two kinds sessile in their axils. Only the uppermost flowers complete, these with rotate 5-lobed corolla; the others apetalous, fertilized in the bud. Calyx prismatic, of the apetalous fl. 3-lobed, of the complete 5-lobed. Prismatic capsule opening by 2 or 3 small lateral valves. Seeds ∞, ovoid, flattish, smooth.

1. **T. biflora** (R. & P.). Slender, 6—10 in. high, often with many branches from the base: leaves ovate or oblong, somewhat crenate-toothed; the upper reduced to lanceolate bracts shorter than the flowers: calyx-lobes of complete fl. lanceolate-subulate, of apetalous shorter and broader: valvular openings of capsule just below the summit: seeds lenticular.—Sandy plains and dry hillsides; not common. April, May.

3. **HETEROCODON,** *Nutt.* Habit, and dimorphous flowers of the last genus; but calyx with obpyramidal tube, and broad foliaceous lobes which are veiny and sharply toothed. Capsule 3-angled, thin and membranaceous, bursting on the sides irregularly. Seeds oblong, obscurely triquetrous.

1. **H. rariflorum,** Nutt. Very slender, 2—10 in. high: leaves rounded, with cordate partly clasping base, and coarse sharp teeth.—Mountains of Napa Co. and northward.

4. **GITHOPSIS,** *Nutt.* Low rigid branching annuals, with flowers all alike complete. Calyx with clavate or linear 10-ribbed tube and 5 long narrow rather rigid foliaceous lobes. Corolla narrowly campan-

ulate. Capsule firm, strongly ribbed, crowned with rigid calyx-lobes of its own length, opening by a round aperture formed by the falling away of the base of the style. Seeds oblong to somewhat fusiform.

1. **G. specularioides,** Nutt. Roughish with short stiff hairs, 2—5 in. high, mostly quite *erect, and with few short divergent branches,* each 1-flowered: leaves lanceolate-oblong, sessile, coarsely toothed: corolla blue, usually shorter than the narrow-linear 1-nerved calyx-lobes: *capsule pedicellate:* seeds rather slender-oblong. Dry open grounds. May.

2. **G. diffusa,** Gray. Slender, *diffusely branching,* glabrous: calyx-lobes subulate-lanceolate from a broad base, half the length of the *linear sessile capsule:* seeds short-oblong.—Vaca Mountains, and southward. May, June.

ORDER LXII. **ERICACEÆ**.

Woody plants (*Pyrola* herbaceous), with alternate simple leaves, and symmetrical mostly 5-merous flowers. Calyx either adherent to or free from the ovary, and stamens epigynous or hypogynous, opening by terminal pores. Corolla sympetalous (in *Pyrola* choripetalous), often urceolate. Style 1; stigma sometimes girt with a ring. Ovary with cells equal in number to the petals. Fruit baccate, drupaceous, or capsular.

Corolla sympetalous;
  Calyx-tube adherent to the ovary..VACCINIUM   1
  "   "   free from the ovary;
      Fruit granular, baccate.............................ARBUTUS   2
      "   smooth, drupaceous..........................ARCTOSTAPHYLOS 3
      "   capsular, but berry-like by enclosure in a
          fleshy calyx.......................BROSSÆA   4
      "   woody-capsular.............................RHODODENDRON 5
Corolla choripetalous; stamens hypogynous;
  Flowers umbellate, on a leafy stem......................CHIMAPHILA   6
  "   racemose, on a naked scape...........................PYROLA   7

1. **VACCINIUM,** *Linn.* Shrubs. Calyx-tube adherent to the ovary, and corolla epigynous. Anther-cells separate, tapering upwards into a tube, this opening by a round orifice at the apex. Stigma simple, without a ring. Fruit a many-seeded berry crowned with the 5 small teeth of the calyx.

1. **V. ovatum,** Pursh. Evergreen, erect, 3 —6 ft. high, with spreading branches and hirsute branchlets: leaves coriaceous, smooth and shining above, ovate to oblong-lanceolate, acute, serrate: fl. crowded in very short axillary and terminal racemes: corolla campanulate, pink: calyx-teeth as long as the ovary: berries black, without bloom —Plentiful on wooded hills in the vicinity of the Redwoods. May, June.

2. **ARBUTUS,** *Pliny.* Trees. Calyx 5-lobed, free from the ovary. Corolla urceolate, with 5 small recurved teeth. Stamens 10, included;

anthers with a pair of reflexed awns on the back. Ovary raised on a hypogynous disk. Fruit a berry with granular surface, and many seeds.

1. **A. Menziesii,** Pursh. Small symmetrical large-leaved evergreen tree, with hard wood, and a red-brown exfoliating bark: leaves oval or oblong, deep green and shining above, entire or serrulate: fl. in an ample terminal panicle of dense racemes; corolla white, nearly globular: berries dark reddish.—Common in the Coast Range.

3. **ARCTOSTAPHYLOS,** *Galen.* Shrubs or small trees, with inflorescence and flowers nearly as in *Arbutus.* Foliage coriaceous and evergreen. Cells of ovary only 1-ovuled. Fruit with hard smooth surface, and a mealy or almost powdery pulp between it and the 5 or more hard-woody or almost bony stone-like 1-seeded nutlets; these often more or less firmly consolidated.

* *Ovary glabrous; branches and petioles never hispid.*

1. **A. pumila,** Nutt. *Stems depressed, several feet long,* tomentose-pubescent when young: *leaves obovate, or oblong-obovate, obtuse,* entire, short-petioled: fl. in short racemose clusters, with veiny bracts: corolla pinkish: fr. orbicular, yellowish-brown; the nutlets broadly carinate, occasionally partly coalescing into irregular 2-celled stones.—A rare undershrub, found chiefly about Monterey, but occurring at the base of Lone Mountain, San Francisco. Fr. July.

2. **A. Manzanita,** Parry. Trunks erect, usually clustered, the bush or small tree 6—25 ft. high: bark mahogany-red, exfoliating: *leaves* commonly *vertical by a twist in the short petiole, rigid, ovate,* with broad rounded base and obtusish though often mucronate apex: peduncles and pedicels of the somewhat paniculate inflorescence pubescent; bractlets broad, acuminate: corolla pinkish, broadly urceolate: fr. cinnamon-color, but bright and shining, depressed-globose, 4—6 lines broad: nutlets more or less firmly coalescent, the whole including 5—7 fertile cells.—Very common in the Coast Range: in Napa and Sonoma counties often very large and tree-like. Fl. Nov.—Feb.

3. **A. Stanfordiana,** Parry. Stems 3—5 ft. high, slender, the smooth bark not exfoliating: *leaves narrowly ovate to oblanceolate,* tapering to a short petiole, entire, mucronate, deep green on both faces: *calyx red;* corolla pink: fruit in pendulous racemes, orbicular, much flattened, the nutlets broader than high, carinate, usually 2 or more coalescent, rarely all united into an irregular stone.—Abundant on hills above Calistoga, Napa Co. Fl. March; fr. July.

4. **A. glauca,** Lindl. Erect, 8—20 ft. high, *glabrous, glaucous:* leaves rigid, often vertical, round-ovate to oblong: racemes panicled; pedicels glandular-hirsutulous; *fruit large,* not depressed, but rather *longer than broad; the nutlets completely consolidated* into a 5-celled stone a half-inch thick.—Mt. Diablo to Los Gatos, and southward.

* * *Ovary pubescent; branchlets often hispid.*

5. **A. tomentosa** (Pursh), Dougl. Erect, 4—8 ft. high, tomentose when young, the branchlets and often the petioles hispid with long white rather flaccid-bristly hairs: leaves glaucescent, oblong-lanceolate to ovate and somewhat cordate, entire, or rarely serrulate: fl. in short conspicuously bracted clustered racemes, the pedicels shorter than the bracts: fr. red, puberulent.—Cool slopes of the Coast Range.

6. **A. Andersonii**, Gray. Size of the last, every way similar in aspect, save that the leaves are thinner, longer, and *sessile*, or nearly so, *by a strongly sagittate-clasping base:* drupes reddish, much depressed, clothed with short viscid-glandular bristles.—Oakland Hills and south-ward to the Santa Cruz Mts.

7. **A. nummularia**, Gray. Low, more or less spreading, glabrous excepting the bristly hairs of the branchlets: *leaves of a vivid light green,* not in the least glaucescent, *oval,* rounded at both ends, or at apex acutish, ½—⅔ in. *long, short-petioled:* racemes short and clustered.—Marin Co. and northward.

4. **BROSSÆA,** *Plumier.* Low shrub, with large evergreen leaves, and flowers in axillary racemes from scaly buds. Calyx 5-cleft. Corolla urceolate, 5-toothed. Stamens 10. Fruit closely imitating a berry, but consisting of a depressed 5-celled many-seeded capsule enclosed within the enlarged and fleshy or pulpy calyx.

1. **B. Shallon** (Pursh), O. Ktze. Shrubby stems ascending, 1—2 ft. high: leaves ovate-cordate, acute or acuminate, 2—4 in. long, finely serrate: fl. in terminal and axillary compound viscid glandular racemes: bracts scaly: pedicels 1—2-bracteolate below the middle: corolla ovate: fr. blackish and sweet.—Only in the redwood districts, and not very common with us.

5. **RHODODENDRON,** *Linn.* Shrubs with alternate entire leaves, and showy flowers in terminal umbels. Corolla large, with ample limb and, in ours, 5 exserted stamens. Capsule woody, elongated, septicidally 5-valved from the summit. Seeds many, small, with a loose testa.

1. **R. occidentale** (T. & G.), Gray. Shrub 5—8 ft. high: leaves obovate-oblong or oblanceolate, glabrate, the margins minutely hispid-ciliate: flowers faintly mephitic, appearing after the leaves; *corolla 2 in. long or more,* minutely viscid-pubescent on the outside, *white,* with a long yellow spot on the upper lobe, the narrowly funnelform tube about as long as the limb.—Banks of mountain streams, and on swampy north-ward slopes: plentiful in some parts of Marin Co.

2. **R. Sonomense,** Greene. Shrub 2—5 ft. high; leaves somewhat elliptical, 1 in. long or less, the margin serrulate and ciliolate: *fl. rose-*

color, sweet-scented; corolla 1 in. long or more, one or two of the segments with a narrowly elliptic deep salmon-colored spot.—Mostly on dry slopes of the mountains of Sonoma Co., from Mt. St. Helena,to near Petaluma.

6. CHIMAPHILA, *Pursh.* Low evergreen undershrubs, with rather large serrulate leaves in irregular whorls, and a terminal naked umbel of a few fragrant flowers. Corolla rotate; petals 5, orbicular, concave. Stamens 10; filaments enlarged and hairy in the middle. Style inverseconic; stigma broad, orbicular. Fruit a depressed 5-lobed 5-celled capsule opening loculicidally from the apex.

1. **C. Menziesii** (Don.), Spreng. Leaves ½—1½ in. long, ovate to oblong-lanceolate, acute at both ends, purplish beneath, more or less mottled or veined with white above: peduncle 1—3-flowered: filaments villous.—Mt. Hamilton; and to be expected on Mt. Tamalpais; otherwise a northern plant.

7. PYROLA, *Brunfels.* Low perennial herbs. Leaves when present ample, petiolate and near the ground. Scape scaly-bracted, bearing a raceme at summit. Petals orbicular, concave, but more or less convergent. Stigma 5-lobed. Capsule loculicidal, the valves separating from below.

1. **P. picta,** Smith. *Leaves ovate*, or ovate-oblong, on short or margined petioles, coriaceous, *pale, veined or blotched with white:* petals greenish-white.—Reported from Mendocino Co., and likely to be found within our limits.

2. **P. aphylla,** Smith. *Leafless;* the stems red, 7—10 in. high, from a scaly-bracted rootstock: lobes of the calyx ovate, acute, much shorter than the obovate white petals. --Lower slopes of Mt. Tamalpais.

Division IV. SYMPETALÆ HYPOGYNÆ.

Corolla sympetalous, the stamens attached to its tube, the whole inserted around the base of the ovary.

Order LXIII. PLUMBAGINACEÆ.

Maritime herbs, with radical leaves clasping the stem at their insertion. Flowers regular, 5-merous, perfect. Calyx 5-plaited, 5-toothed, persistent. Petals with long claws united into a ring at base. Stamens opposite the petals and joined to their base. Fruit utricular or achene-like, in the bottom of the calyx; the one seed with straight embryo in mealy albumen.

1. **STATICE,** *Dalechamps.* Leaves narrowly linear, in a close radical tuft. Flowers in a single globose head terminating a simple scape.

1. **S. Armeria,** L. Leaves flat, 1-nerved: scape solitary, stoutish, 1 ft. high or more: fl. dull pink or flesh-color.—Along sandy beaches in wet ground, or occasionally in elevated stations among the hills of the seaboard. May, June.

2. **LIMONIUM,** *Diosc.* Leaves ample and rather few. Flowers in short spikes terminating the many branchlets of a branching scape.

1. **L. commune,** S. F. Gray, var. **Californicum** (Boiss.). Leaves 8—10 in. long, obovate-oblong, entire, fleshy-coriaceous: scape 1—2 ft. high, the spikes corymbose-panicled: calyx-tube more or less hairy on the angles.—In salt marshes, plentiful. Sept.—Nov.

Order LXIV. PLANTAGINACEÆ.

Comprising scarcely more than the genus

1. **PLANTAGO,** *Pliny* (PLANTAIN). Acaulescent herbs with elongated leaves, and spikes of colorless small flowers on naked scapes, each flower subtended by a bract. Calyx of 4 persistent imbricated sepals free from the ovary. Corolla short-salverform, scarious, persistent; its limb 4-parted, imbricate in bud. Stamens 2—4, inserted on the corolla alternate with its lobes. Style filiform, all the upper part pubescent and stigmatic. Fruit a circumscissile capsule, few- or many-seeded. Embryo straight; albumen fleshy.

* *Corolla remaining expanded, i. e. not closed over the fruit.*

1. P. **major,** L. Perennial, stout, glabrous or sparsely pubescent: leaves ovate or oval, 3—10 in. long, very distinctly 5—7-ribbed, entire or somewhat toothed: spikes linear, elongated, the naked part of the scape not as long as the leaves: *capsule ovoid, obtuse, circumscissile near the*

level of the sepals: seeds 8—18, brownish, minutely reticulated.—Mostly in shaded or moist places; not very common; supposed to have been introduced from Europe.

2. **P. Asiatica, L.** Rather more slender, the leaves more rounded; spikes slender, little surpassing the leaves, the flowers of the lower part scattered: *capsule globose-ovoid, circumscissile* not a little *below the level of the sepals.*—Less common, and less confined to the neighborhood of buildings, or to cultivated lands, presumably native.

3. P. LANCEOLATA, L. Perennial: *leaves oblong-lanceolate,* villous or glabrate, tapering into a short petiole, very *strongly ribbed:* scapes slender, deeply sulcate and angled: *spike short, dense:* bracts and sepals broadly ovate, scarious, brownish.—Very common in moist meadow lands and waysides; native of Europe.

4. **P. maritima, L.** Very stout maritime perennial, with many *linear obtuse very fleshy leaves:* spikes cylindric, long and dense: bracts mostly roundish, shorter than the calyx: sepals oval, carinate: corolla with pubescent tube: capsule 2—4-seeded.—Plentiful on rocks and cliffs of the seaboard; also in sandy salt marshes.

5. **P. Bigelovii, Gray.** Stoutish, fleshy and glabrate like the last, but annual and small, the *leaves and scapes erect,* the latter only 2—4 in. high, the *leaves* shorter, *linear, entire:* spike in fruit 1 in. long or more: stamens 2 only: capsule ovoid-oblong, well exserted from the calyx, 4-seeded.—Borders of saline or brackish marshes; quite common about the Bay, and on the lower San Joaquin. April, May.

6. **P. Californica, Greene.** Annual, 3—6 in. high, the *rosulate leaves commonly depressed,* and scapes more or less decumbent at base: leaves linear, entire or *with few remote salient teeth,* glabrous or very sparingly hirsutulous: scapes twice the length of the leaves: spikes linear, 2—3 in. long, rather dense: bracts broadly ovate and with broad scarious margins below the middle, shorter than the sepals: stamens 2: seeds 8—12, irregularly pitted, blackish.—Plains of Alameda and Contra Costa counties and southward, in alkaline soils. March—May.

7. **P. Patagonica, Jacq.,** var. **Californica.** *Annual, slender,* 3—10 in. high, more or less *villous* but not lanate, the *leaves and scapes slender, strictly erect* from the very base: leaves narrowly oblanceolate linear, nearly equalling the scapes, these gradually dilated up to the base of the short cylindric spike: bracts less than half the length of the obtuse scarious-margined sepals: lobes of the corolla roundish, reflexed: seeds oblong-oval.—Abundant on grassy plains and hillsides. March—May.

* * *Corolla in age closed and forming a beak over the fruit.*

8. **P. hirtella, HBK.** Perennial, somewhat fleshy: leaves oblong-ovate or -spatulate, 6—10 in. long, glabrate, sparsely denticulate, 5 7-

nerved, narrowed to a short and broad petiole-like base: *scapes, with their long and dense spike, 1—2 ft. long, stout, hirsute:* corolla-lobes ovate, acute.—In wet places among the hills at San Francisco.

ORDER LXV. PRIMULACEÆ.

Herbs with simple exstipulate leaves, and flowers either axillary and solitary or, more commonly, terminal and umbellate, sometimes racemose. Corolla regular, the stamens as many as its lobes and opposite them, inserted on its tube. Ovary 1-celled, the many ovules borne on a free central placenta. Fruit capsular. Embryo small, in fleshy or corneous albumen.

Flowers umbellate, terminating a scape;
  Segments of corolla long, reflexed..............................MEADIA   1
      "   " short, rotate...............................ANDROSACE   2
Flowers few, at leafy summit of stem............................ALSINANTHEMUM 3
  " solitary in the leaf-axils;
      Pedicellate, red or blue................................ANAGALLIS   4
      Sessile { corolla small....................................CENTUNCULUS   5
           { " none; calyx white..................GLAUX   6
Flowers racemose; corolla small, white.........................SAMOLUS   7

1. MEADIA, *Catesby*. Herbs with tufted simple radical leaves, and a naked scape, from a short perennial often bulbilliferous crown; the roots fleshy-fibrous. Flowers 5-merous (rarely 4-merous or 6-merous), umbellate, nodding. Calyx 5-parted, reflexed. Corolla very deeply cleft, the purplish segments reflexed. Stamens with very short filaments, often connate at base, inserted on the very short tube of the corolla; anthers long, erect, mostly connivent around the style. Capsule oblong, many-sided, in our species circumscissile near the summit.

1. **M. Hendersonii** (Gray), O. Ktze. Leaves ovoid or obovoid, very obtuse to almost truncate and retuse, entire, 2 in. long or more exclusive of the equally long or longer broad petiole; herbage glabrous: scape 8—12 in. high: segments of corolla of a rich rose-purple, the short tube dark maroon encircled by a band of light yellow: anthers about 2 lines long, neither connivent nor divergent but erect: capsule oblong, twice the length of the calyx, circumscissile well below the summit. Var. **cruciata** (Greene). Leaves narrower; scapes taller and more slender: flowers of a darker purple, always 4-merous: the anthers longer and narrower.—The type is common among the higher hills of both ranges of coast mountains; the very marked variety on northward slopes of hills about the Bay. Feb.—April.

2. **M. patula** (Greene), O. Ktze. Low and stout, pale green and very glandular throughout, 3—7 in. high: leaves 1—2 in. long, elliptic, entire, narrowed to a short broad petiole: segments of corolla pale cream-color with sometimes a purplish tinge; tube of a dark velvety maroon-purple

encircled by a band of yellow: anthers deep blue-purple, barely a line long, linear-oblong, with broad retuse apex, at length divergent from the style: capsule short-oblong or subglobose, circumscissile near the summit.—Subsaline plains of the interior; abundant in the eastern part of the Livermore Valley; also on the Oakland Hills, and Bernal Heights, in a modified form. March, April.

2. ANDROSACE, *Matthiolus.* Diminutive herbs, with small rosulate radical leaves, and slender scapes bearing an involucrate umbel of very small white flowers. Corolla salverform, with constricted throat. Stamens 5, included. Capsule subglobose, valvate-dehiscent; seeds few.

1. **A. acuta.** Very slender, 1—4 in. high, rather densely hirsutulous or almost hispidulous throughout: leaves linear-lanceolate, attenuate-acute, entire, hispid-ciliolate, ½—¾ in. long: bracts of involucre ovate-lanceolate, acuminate, 3–4 lines long: umbel few-flowered, the filiform pedicels ¾—1½ in. long: calyx 2—3 lines high including the triangular acuminate almost pungently acute and slightly recurving segments: capsule not known.—Not rare on northward slopes of the hills of Contra Costa and Alameda counties; but an obscure plant, known to us for years, and hitherto mistaken for *A. occidentalis.* March.

3. ALSINANTHEMUM, *Thalius.* Low glabrous perennials, the simple stem, from a tuberous-thickened root or rootstock, bearing an apparent whorl of ample leaves at summit, and among these a few filiform 1-flowered pedicels.· Flowers 7-merous or 6-merous. Corolla rotate, deeply parted, the divisions convolute in bud. Capsule dehiscent by about 5 revolute valves. Seeds few, large, covered with a white pellicle.

1. **A. Europæum** (L.), var. **latifolium** (Torr.). Stem naked below: leaves oblong-obovate, mostly acute, 1½—4 in. long, in a cluster of from 4 to 7: corolla 5—6 lines broad, from pinkish to almost rose-red.—Northward slopes of hills, in shade of trees or thickets. May.

4. ANAGALLIS, *Diosc.* Low annuals with opposite leaves, and flowers in their axils on slender pedicels. Corolla rotate, 5-parted; the rounded lobes convolute in bud. Stamens on base of corolla; filaments pubescent. Capsule globose, circumscissile.

1. **A. ARVENSIS,** L. Glabrous, depressed; stem 6—10 in. long, leafy and flowering throughout: leaves ½ in. long or more, ovate, sessile: calyx-lobes narrow, almost equalling the scarlet, or sometimes bluish corolla, this expanded only under a clear sky.—Very common; flowering at all seasons except mid-winter.

5. CENTUNCULUS, *Dillenius.* Diminutive glabrous annual, with alternate leaves, and small inconspicuous flowers solitary in their axils.

Corolla with globular tube, and 4—5-lobed limb shorter than the calyx. Stamens on the tube of the corolla; filaments short, subulate. Capsule circumscissile.

1. **C. minimus, L.** Stems ascending, 2—6 in. long: leaves usually obovate or spatulate-oblong, 2—3 lines long; fl. nearly or quite sessile, usually 4-merous: calyx-lobes lanceolate-subulate, as long as the capsule. —Obscure little herbs of low plains. May.

6. GLAUX, *Dodoens.* Small erect succulent seaside herb, with leaves mostly opposite, and small flowers sessile in the axils. Calyx campanulate, corolla-like and white or pinkish, 5-cleft. Corolla 0. Stamens 5, hypogynous. Capsule globose, 5-valved from the summit.

1. **G. maritima, L.** Perennial, glaucescent, 3—6 in. high, leaves oblong, ½ in. long more or less, minutely dotted.—Frequent both along the seaboard and in subsaline soils in the interior.

7. SAMOLUS, *Pliny.* Glabrous perennial herbs, with alternate leaves, and small white flowers in a terminal raceme, or racemose panicle. Ovary at base connate with the base of the calyx. Corolla subcampanulate. Stamens 5, on the tube of the corolla; filaments short; anthers cordate. Staminodia 5, in the sinuses of the corolla.

1. **S. Valerandi, J.** Bauh., var. **Americanus,** Gray. Slender, 3—10 in. high, branching and somewhat racemose-paniculate: leaves obovate, thin, narrowed to a petiole: pedicels of the small flowers not bracted at base, but bracteolate near the middle: calyx-lobes ovate, shorter than the corolla.—Not rare among rushes and sedges, in the Suisun Bay marshes. Aug.—Oct.

Order LXVI. OLEACEÆ.

Important family of trees; represented in our district by two very dissimilar species of *Ash.*

1. **FRAXINUS,** *Vergil.* Trees, with unequally pinnate leaves, and very many small flowers in crowded panicles, developed before the leaves in spring, and from separate buds: dioecious or polygamous. Calyx 4-cleft or -toothed, or obsolete. Corolla of 2—4 petals distinct or united at base. Stamens 2—4, hypogynous. Ovary 2-celled; a pair of ovules pendulous in each cell. Fruit a samara, winged from the summit.

1. **F. dipetala,** Hook. & Arn. Tree small: leaves glabrous; leaflets 5--9, oval or oblong, serrate, mostly petiolulate, 1—2 in. long, in age coriaceous: calyx usually 4-toothed, sometimes almost entire: *petals 2, white,* obovate-oblong, with short claw: fr. narrowly spatulate-oblong, mostly retuse, 1 in. long, the base merely sharp-edged, or else the wings continuing to the base.—Near Niles, Livermore, etc.

2. **F. Oregana,** Nutt. Tree of fair dimensions: leaves tomentose, or glabrate in age; leaflets 5—9, oval to oblong-lanceolate, entire, sessile, 2—4 in. long: fl. all with *minute calyx and no petals;* fr. marginless at base, margined upwards into an oblanceolate or spatulate retuse wing, the whole 1—1½ in. long.—Marin and San Mateo counties.

ORDER LXVII. **A P O C Y N A C E Æ**.

Herbs with milky juice, opposite entire exstipulate leaves, and regular 5-merous flowers. Corolla-lobes convolute in bud. Stamens borne on the corolla alternate with its lobes, anthers more or less coherent with the stigma. Ovaries 2, both developing into long follicles filled with flattish seeds usually bearing a tuft of silky down at the end.

1. **APOCYNUM,** *Matthiolus.* Flowers small, in terminal cymes. Calyx small, deeply 5-cleft. Corolla cylindric or campanulate, 5-lobed, bearing within and toward the base 5 triangular-subulate appendages, alternate with the stamens. Stamens on the base of the corolla; filaments short, broad; anthers sagittate, acute. Stigma ovoid, obscurely 2-lobed. Follicles slender, terete. Seeds many, with long coma at apex.

1. **A. pumilum.** Commonly pubescent, stoutish, seldom 1 ft. high, with spreading branches from the base: lowest leaves subreniform to round-ovate, the others cordate-ovate and oval, commonly ¾ in., seldom more than 1 in. long: fl. solitary in the upper axils, and in short cymose clusters at the ends of the branches: *corolla subcylindric, 3—4 lines long; oblong-lanceolate segments little spreading.*—Common in the Mt. Diablo Range and northward. Referred as a variety to *A. androsæmifolium* hitherto, but wholly distinct in size and form of corolla, as well as in other particulars.

2. **A. cannabinum,** L. Herbage of a light almost *yellowish green,* flaccid, glabrous: stem 2—4 ft. high, erect, rather strict: leaves oblong, 2—3 in. long: *fl. small, greenish,* in small dense terminal cymes; *corolla little exceeding the lanceolate calyx-lobes.*—Frequent along streams, and in low lands. June—Aug.

3. **A. vestitum.** Allied to the last, but dwarfish, rather widely branching, and densely soft-pubescent throughout: *leaves all ovate-lanceolate, 1—2 in. long:* cymes ample and compound: *corolla* cylindric, about *twice the length of the lanceolate-subulate calyx-lobes,* its segments a third the length of the tube, little spreading.—Hills west of Napa Valley, in dry soil. July.

2. **PERVINCA,** *Tragus.* Flowers large, solitary in the axils of the leaves. Calyx 5-parted, with acuminate segments. Corolla with narrow-funnelform tube, hairy and angular throat, and rotate limb. Stamens 5, the inflexed anthers longer than the short filaments; these on the tube

of the corolla. Style enlarged above, ending in a reflexed membrane. Stigma obscurely 2-lobed, hispid. Follicles 2, terete, striate. Seeds oblong-cylindric, with no coma.

1. P. MAJOR (L.), Scop. Stems slender, when young and flowering erect, 1 ft. high, at length procumbent or trailing and several feet long: leaves ovate, 2—3 in. long, short-petioled, finely ciliate: calyx-segments subulate, ciliate, toothed below: corolla blue: the limb 1 in. or more in breadth.—Common in moist shady places, as an escape from the gardens; multiplying by rooting at the joints or the ends of the long shoots, but not fruiting with us. Jan.—March.

ORDER LXVIII. ASCLEPIADACEÆ.

Ours perennial herbs, with milky juice, opposite or whorled entire leaves, and smallish flowers in axillary mostly subglobose umbels. Flowers 5-merous, but carpels 2 only. Stamens joined to the stigma. Pollen in waxy masses. Fruit a follicle. Seeds thin, flat, crowned with a coma of silky down.

1. **ASCLEPIAS,** *Diosc.* Calyx very small; corolla larger, both deeply 5-parted, the divisions reflexed. Filaments monadelphous, inserted on the very base of the corolla; a circle of hood-like organs between corolla and stamens being the most conspicuous part of the flower, these with (or sometimes lacking) a horn-like appendage within. Anther cells, with waxy pollen-masses, connected with the stigmatic disk; the body of the anther with a triangular corneous wing widening down' to the base of the organ, *i. e.* to the column. Follicle usually one only from each flower, the other ovary being abortive.

      * *Species hoary-tomentose, or almost white.*

1. **A. speciosa,** Torr. Stout, 2—4 ft. high: leaves oblong-ovate, acutish, 4—6 in. long, very short-petioled: peduncle longer than the white-woolly pedicels: fl. ¾ in. long, of a dull red-purple; *hoods nearly ¾ in. long, ascending* or almost spreading, *lanceolate,* the short inflexed horn not surpassing the anthers: follicle soft-spinous, densely tomentose.—Marin and Contra Costa Co. hills; also at Alameda, and in the hills to the eastward.

2. **A. eriocarpa,** Benth. Stoutish, 2—4 ft. high, the stem often sharply angled: leaves sometimes 3 or 4 in a whorl, oblong-lanceolate or narrowly oblong, acute, 4—7 in. long, short-petioled: umbels on stout peduncles mostly longer than the pedicels: fl. 3½ lines long, creamy-white with slight purplish tinge; *hoods* shorter than the anthers, *oblately semiorbicular,* open to near the middle of the back, the summits produced inwardly into an acute angle or tooth, nearly enclosing the acute falciform horn.—Napa and Santa Clara counties, and southward.

3. **A. vestita**, Hook. & Arn. Densely floccose-woolly, even to the outside of the corolla: leaves ovate to oblong-lanceolate, very acute or acuminate, often subcordate, short-petioled or the upper sessile: only the terminal umbel at all peduncled, the others sessile: corolla purplish; *hoods nearly erect, ventricose*, slightly surpassing the anthers, *somewhat truncate, the inner angles involute*, the vomer-shaped crest-like *horn* attached up to the summit of the hood and *not exserted.*—Alameda and San Francisco counties, and southward.

4. **A. Californica**, Greene. General aspect of the last, but stouter and low, seldom 2 ft. high: leaves ovate or oblong, 2–4 in. long, sharply acuminate: umbels sessile: corolla purplish; *hoods dark-maroon, nearly orbicular*, attached centrally and lying partly below the anthers, 2-valved half way down the back, *destitute of horn.*—Mt. Diablo Range.

* * *Glabrous species.*

5. **A. ecornuta**, Kell. Herbage dark almost purplish green: stem 2–3 ft. high: leaves 3–5 in. long, from deltoid-ovate to ovate-lanceolate, sessile by a cordate base: umbels mostly at naked summit of stem and axillary to mere bracts; pedicels slender: *fl. dark red-purple; hoods oblong, obliquely truncate* and acutely angled, attached near the base, *destitute of horn:* follicles smooth and glabrous.—Napa Co. and northward, in the higher mountains. June.

6. **A. fascicularis**, Desne. Slender, 2–3 ft. high: leaves in whorls of 3–5, or the lower and uppermost opposite, linear to linear-lanceolate, 2–5 in. long, short-petioled: slender-peduncled umbels many, often in whorls; pedicels and calyx often puberulent: *fl. small, flesh-color or purplish;* column of filaments half as long as the anthers; the *slender horn subulate, exserted* from the hood and *incurved* over summit of the stigma.—In low plains of Alameda and Santa Clara counties; perhaps of more general distribution in our district.

2. **SOLANOA**, *Greene.* Low and rather slender perennial, differing from the preceding genus in having lunate rather than triangular anther wings, and the hoods split down both sides into 2 valves, and destitute of horn or appendage.

1. **S. purpurascens** (Gray), Greene. A few inches to 1 ft. high, decumbent; herbage canescently puberulent: leaves cordate-ovate, obtuse, 1 in. long or more: umbels globose and dense, on peduncles exceeding the pedicels: fl. purplish-red, about 2 lines long.—Mountain summits of Sonoma Co., near the Geysers. June.

Order LXIX. GENTIANACEÆ.

Herbs with watery and bitter juice, mostly opposite simple entire sessile leaves, no stipules, and complete regular 5-merous flowers. Calyx persistent. Corolla convolute in bud. Stamens on corolla-tube, as many as its lobes and alternate with them; anthers 2-celled, opening lengthwise. Ovary 1-celled, with parietal placentæ. Style 1; stigma 2-lobed. Seeds very many and small.

Small and slender yellow-flowered annual..........................MICROCALA 1
Pink-flowered annuals..ERYTHRÆA 2
Blue-flowered perennials...GENTIANA 3
Trifoliolate white-flowered perennial...............................MENYANTHES 4

1. **MICROCALA,** *Link.* Plants almost filiform, the branches pedunculiform, terminating in a minute 4-merous yellow flower. Corolla funnelform; tube ventricose. Stamens 4, inserted in the throat of the corolla; anthers cordate-ovate. Capsule 2-valved, septicidal, the placentæ along the sutures.

1. **M. quadrangularis** (Lam.), Griseb. Simple, or with a few erect branches from the base, 1—2 in. high: leaves minute, oval or oblong: peduncles strict, quadrangular: calyx short, quadrangular, appearing truncate at both ends: corolla deep yellow, twice the length of the calyx, open under a sunny sky only.—Frequent along the bases of low hills, and in moist fields; doubtless native. April.

2. **ERYTHRÆA,** *Renealm.* Stems leafy and mostly freely branching. Calyx 5-parted; the divisions slender. Corolla salverform; stamens inserted on its throat; anthers oblong or linear, twisting spirally in age. Capsule elongated. Seeds oblong or spherical, reticulate-pitted.

* Seeds spherical.

1. **E. venusta,** Gray. Simple and cymosely several-flowered at summit, or corymbosely branched, 3—8 in. high: leaves ovate to oblong-lanceolate, ½—¾ in. long, obtusish: calyx-lobes very narrow down to the base: *corolla bright pink with a yellow center, the limb* ¾ *in. broad; lobes oval or obovate:* anthers oblong-linear.—A southern species, but occurring near San Leandro, according to *Dr. Behr.*

2. **E. Douglasii,** Gray. Slender, 2—10 in. high, loosely paniculate: leaves lanceolate to linear-lanceolate, acute, ½—¾ in long: peduncles filiform, erect, the central ones 1 in. long or more: *lobes of the pink corolla oblong, 2 lines long.*—Attributed to San Francisco by *Dr. Behr.*

* * Seeds slightly elongated.

3. **E. Muhlenbergii,** Griseb. Low, fastigiately branched from the base, cymosely flowered at summit: leaves oblong, obtuse; the floral lanceolate: pedicels short or almost none in the forks, but the lateral often as long as the flower, and bibracteolate at the summit: corolla rose-red, its *lobes 2—3 lines long, obtuse or retuse: seeds short-oval.*—Common in the Bay region.

4. **E. floribunda,** Benth. Rather strict and closely flowered, 6—10 in. high, corymbose-cymose at summit: lower leaves oblong; upper lanceolate: fl. short-pedicelled, or those in the forks nearly sessile; *lobes of pinkish corolla* oblong, only *2 lines long, the tube twice or thrice longer:* anthers short-oblong: *seeds round-ovoid.*—San Bruno Mountains, *Behr.*

5. **E. trichantha,** Griseb. Fastigiate-cymose and profusely flowering, 4—8 in. high: leaves oblong-oval to lanceolate: *corolla-lobes oblong-lanceolate, ¼ in. long or more:* filaments slender; anthers linear: *seeds oval-oblong.*—Marin and Napa counties.

3. **GENTIANA,** *Diosc.* Herbs with showy usually blue corollas commonly persistent on the calyx and enclosing the mature capsule. Corolla campanulate or funnelform, the 5 lobes with plaited folds in their sinuses. Stamens included. Style short or 0. Capsule septicidal, sometimes with its whole inner wall placentiferous.

1. **G. Oregana,** Engelm. Perennial, erect, stoutish, 1—2 ft. high: leaves ovate to ovate-oblong, 1—1½ in. long: fl. few at summit of stem: corolla 1 in. long or more, deep blue, funnelform, the ovate lobes not narrowed at base, the plaits extended into a conspicuous laciniate-toothed or -cleft appendage.—Mt. Tamalpais, and hills northward.

4. **MENYANTHES,** *Dalechamps.* Bog perennial, with thick horizontal rootstocks bearing long petioled leaves (in ours trifoliate) and a naked scape with a terminal raceme of white flowers. Corolla campanulate, 5-cleft; the lobes fimbriate-bearded on the face. Anthers sagittate, versatile. Ovary surmounted by a long style. Capsule globular. Seeds few and rather large; the close testa smooth, shining crustaceous.

1. **M. trifoliata,** L. Petioles and scape stout, 1 ft. high or less: leaf divided into 3 oblong-obovate entire or repand leaflets: corolla less than 1 in. long, densely bearded within, white or rose-tinted.—Formerly at San Francisco, but extinct there since 1859, according to *Dr. Behr;* not rare in the Sierra Nevada, and elsewhere northward beyond our limits.

ORDER LXX. POLEMONIACEÆ.

Mostly herbs, with watery juice of no marked properties. Leaves exstipulate, simple or divided. Flowers complete, 5-merous (rarely 4-merous), except the pistil, which is 3-merous; the style 3-cleft, the ovary 3-celled. Fruit capsular, loculicidally dehiscent, with axial placentæ and few or many seeds; these commonly mucilaginous when wet.

* *Calyx herbaceous throughout.*

Calyx slightly accrescent, otherwise unchanged in age.................POLEMONIUM 1

* * *Calyx scarious or coriaceous at base.*

Calyx-segments equal, the sinuses in age distended into a short lobe..COLLOMIA   2
  "   "   unequal; sinuses unchanged in age...................NAVARRETIA 3

* * * *Calyx scarious between the angles.*

Leaves alternate; stamens inserted equally.......................................GILIA 4
Leaves mostly opposite and corolla salverform;
 Leaves mostly digitate-divided; stamens inserted equally.........LINANTHUS 5
 " undivided; stamens unequally inserted....................PHLOX 6

1. **POLEMONIUM,** *Diosc.* Herbs with alternate pinnate flaccid leaves, the leaflets or segments sessile, entire. Calyx herbaceous throughout, neither angled nor costate, slightly accrescent and loosely investing the capsule, campanulate or narrower, cleft to the middle, the segments lanceolate or broader, equal, erect, or connivent over the capsule, or campanulate-spreading, entire, never recurved nor aristate-pointed. Corolla regular, in ours campanulate, blue or purplish. Stamens more or less declined. Seeds angular or winged.

1. **P. carneum,** Gray. Stoutish, 1—2 ft. high, sparingly leafy, loosely cymose-paniculate, the branchlets 3-flowered: leaflets ovate to oblong-lanceolate, 1—1½ in. long: calyx deeply 5-cleft; lobes ovate-oblong: corolla campanulate-funnelform, nearly 1 in. long and as broad in expansion, salmon-color to violet-purple; lobes round-obovate: ovules and seeds 3 or 4 in each cell.—Most beautiful species, of mountain woods in San Mateo and Marin counties, and far northward. June.

2. **COLLOMIA,** *Nutt.* Herbs with alternate leaves, and cymose clustered flowers. Calyx scarious below between the angles, in flower turbinate, in age obpyramidal or nearly cyathiform, not distended by the capsule; segments herbaceous, equal, entire, triangular or lanceolate, erect, never recurved or even spreading; the sinuses at length distended below into a revolute lobe. Corolla with narrow tube, open throat, and a spreading limb. Stamens unequal and unequally inserted on the tube of the corolla. Capsule narrowed at base. Seeds usually 1 in each cell.

1. **C. grandiflora,** Dougl. Erect, 6 in. to 2 ft. high, leafy throughout, flowering only at and near the summit: leaves linear, and oblong-lanceolate, to almost ovate around the flower-clusters, 2—3 in. long, mostly entire: fl. capitate-crowded at summit, and at the ends of a few subterminal short branchlets: calyx and bracts viscid-pubescent: *corolla pale salmon-color, 1 in. long or more, narrowly funnelform,* with limb of oblong lobes spreading to the width of 8—10 lines.—Slopes and summits of the inner coast mountains, at considerable elevations. June, July.

2. **C. heterophylla,** Hook. Low, diffuse, pubescent and viscid: leaves thin, more or less bipinnatifid, or some of the upper merely toothed, or entire: fl. in small clusters in the axils, and at ends of branches; *corolla small, slender-salverform, deep purple:* capsule ellipsoid; seeds 2—3 in each cell.—Common in mountain shades.

3. **NAVARRETIA,** *Ruiz & Pavon.* Annuals, glabrous and scentless, or viscid-pubescent and heavy-scented; the leaves always alternate, even

the lowest, and setaceously or spinosely pinnatifid, or the lowest sub-entire; flowers in crowded bracted clusters at the ends of all the branches. Calyx-tube scarious between the 5 prominent green angles or costæ, not accrescent, prismatical or obpyramidal; segments unequal, erect or spreading, not recurved, pungent-tipped, all entire, or the two larger spinulose-toothed or -cleft. Corolla tubular-funnelform or salverform. Stamens and style usually straight, rarely declined. Pericarp 1—3-celled, 1—many-seeded, partially dehiscent from above, or from below, or occasionally indehiscent.

* *Pericarp hyaline, indehiscent, the walls closely adherent to and trans-parently exhibiting the agglutinated mass of dark-colored seeds.*

1. **N. prostrata** (Gray), Greene. *Primary flower-cluster sessile near the ground,* the few branches radiating from beneath it and prostrate: leaves pinnatifid, the rachis broad and strap-shaped, the segments short and spreading, some of the uppermost occasionally 3-cleft: calyx-tube minutely white-hirsute, thin-hyaline between the stout costæ, constricted over the capsule, the segments spreading, 3 spinulosely trifid, 2 subulate and entire: pericarp a transparent utricle close-fitted to the glutinous seeds, breaking transversely or irregularly when soaked.—Plains of the lower Sacramento. May.

2. **N. leucocephala**, Benth. *Erect, a span high or more,* the stem whitish-puberulent: leaves once or twice pinnatifid, the rachis not broader than the divaricate segments: calyx-tube nearly glabrous with-out, white-hairy in the sinuses, little constricted over the capsule, the segments erect, all entire, or one or two of them cleft: corollas larger than in the preceding, white.—Very common on the plains of Solano Co., beyond Suisun. April, May.

3. **N. intertexta** (Benth.), Hook. Nearly simple, or branching from the base, 3—10 in. high, neither viscid nor glandular: stem retrorsely pubescent: leaves mainly glabrous, their divaricate acerose divisions simple or sparingly divided: *fl. densely glomerate: base of their bracts and tube of calyx densely villous* with long spreading white hairs: corolla pale blue or white, little exceeding the calyx: capsule subglobose, 3-celled, several-seeded, but walls hyaline, adherent to the seeds, breaking irregu-larly.—Marin and Sonoma counties, and northward. June.

* * *Capsule of firm texture, opaque, more or less perfectly dehiscent; seeds not agglutinate in a mass.*

4. **N. cotulæfolia** (Benth.), Hook. & Arn. Stem stout and rigid, 4—8 in. high, puberulent, or villous and somewhat glandular: *leaves twice pinnatifid into slender herbaceous soft and innocuous segments;* the upper-most ones, and the bracts decidedly spinescent: *fl. white, 4-merous:*

staméns exserted: capsule few-seeded.—Plains of the lower Sacramento; plentiful near Fairfield and Suisun. A peculiar soft-leaved and scentless species. April, May.

5. **N. viscidula,** Benth. Only 2—4 in. high, simple or branching from the base: viscid-pubescent: stem-leaves extremely narrow, remotely pinnatifid; the floral rigid, pinnatifid-toothed, and, with the bracts and calyx-teeth spinescent: *corolla large, rather narrow, perfectly regular,* deep blue, *stamens straight:* cells of capsule 1-seeded.—In dry rather sandy soils, from Napa Valley southward. June.

6. **N. heterodoxa,** Greene. Slight pubescence of the preceding, but extremely viscid, mephitic-scented; stem 6—12 in. high, with many slender wide-spread branches: lowest cauline leaves with linear-filiform rachis and pinnate segments, the latter in 5—7 pairs; the upper and floral more lanceolate, and with pinnate teeth toward the base only; bracts broadly ovate, subulate-acuminate: segments of calyx subequal, entire: *corolla blue, with short tube, open-campanulate limb and exserted declined stamens:* seeds small, 4 or 5 in each cell.—Hills of Napa and Sonoma counties, and southward to beyond Santa Clara. June—Aug.

7. **N. parvula,** Greene. Low but stoutish with numerous short branches, 2—4 in. high; glandular-puberulent, very viscid and aromatic: lowest leaves linear, entire, the upper rather broader and with subulate teeth or segments: corolla about 4 lines long, broadly tubular-funnelform, light blue: *stamens very unequal, the 2 posterior included, the 3 anterior long-exserted and declined.*—Near Crystal Springs. June.

8. **N. mellita,** Greene. Slender and low, 2—5 in. high and with ascending or spreading branches, the branches glandular-villous: herbage very viscid and honey-scented: lowest leaves divided pinnately into subulate-acerose spine-like segments, those of the upper leafy-dilated and spine-tipped: *corolla* narrowly tubular-funnelform, *minute,* not exceeding the calyx, *very pale blue: stamens included.*—San Mateo Co., near Belmont, and northward. June—Aug.

9. **N. pubescens** (Benth.), Hook. & Arn. Tall, stoutish, flexuous and branching, soft-pubescent, the herbage with a strong hircine odor: leaf-segments 5—11, the terminal or odd one spatulate-dilated, the others linear, all with numerous sharp and stiff teeth or lobes: calyx-teeth all pungent-tipped, 3 small and entire, 2 twice as large and toothed: *corolla deep blue or purple, ¾ in. long;* throat funnelform; *stamens exserted:* capsule 1-celled and 1-seeded.—Common on low foothills of the inner ranges, and plains adjacent. April—June.

10. **N. squarrosa** (Esch.), Hook. & Arn. Stout, rigid, ½—1 ft. high, branching, very viscid-pubescent, *strongly mephitic-scented:* leaves and bracts pinnately parted, the segments spinescent: *corolla salverform,*

small, deep blue: stamens not exserted: seeds numerous, small, black.-
A noxious wayside weed; *Skunkweed*, of the country people   June—Oct.

11. **N. atractyloldes** (Benth.), Hook. & Arn. Stout and low, with
short branches paniculately arranged: *leaves ovate-lanceolate, rigidly
coriaceous*, in age reticulate, the margins beset with straight spinose-
subulate teeth: segments of the calyx subulate, entire, erect, only
moderately unequal: *corolla narrowly funnelform*, ¾ inch long, deep
purple.—From the lower Sacramento southward. Ill-scented like the
last, but less common. June—Aug.

4. **GILIA,** *Ruiz & Pavon*. All our species (except n. 6) annuals, with
leaves alternate, flaccid, not pungent, 1—3-pinnately dissected into
narrow segments. Flowers more or less clustered at the ends of the
few branches, the clusters not bracteate. Calyx-teeth equal, connivent
or recurved in fruit; tube in some (scarcely in our species) ruptured to
the base (through scarious sinuses) by the splitting of the capsule.
Corolla funnelform. Stamens inserted equally on its throat. Seeds
several in each cell of the capsule.

1. **G. capitata,** Dougl. Slender, 1—2 ft. high: leaves dissected into
filiform lobes: fl. many, in dense capitate long-peduncled clusters: *calyx
nearly glabrous, in fr. globose: corolla very small, light blue;* tube scarcely
dilated at the throat; *segments nearly linear:* stamens inserted in the
sinuses of the corolla.—Species of the higher parts of the Coast Range.

2. **G. achilleæfolia,** Benth. Stoutish, pubescent, 1—3 ft. high, the
head-like flower-clusters larger and less compact: *calyx more or less
woolly;* the somewhat triangular teeth connivent over the growing ovary:
corolla broad-funnelform, deep or pale blue; *lobes obovate or oblong;*
throat abruptly and amply dilated.—Common in sandy soil. May—Aug.

3. **G. multicaulis,** Benth. Slender, in age commonly diffuse, ½—1½
ft. high: leaves twice cleft into narrowly linear lobes: fl. few and in less
dense clusters, on short peduncles: *calyx-teeth recurved in fruit: corolla
nearly salverform,* deep or pale blue, *its lobes obovate:* capsule ovoid.—
Rocky hills and sandy plains; extremely variable. April—June.

4. **G. tricolor,** Benth. Slender, seldom a foot high, cymose-panicled:
leaves and calyx almost as in the last, but fl. much larger and ampler,
distinctly 3-colored: *corolla* ½ *in. long; tube yellow;* ample campanulate-
funnelform *throat marked with deep violet-purple* or darker; broad
rounded *lobes white or lilac.*--Common from Napa and Solano counties
northward and southward, but chiefly in the interior. April, May.

5. **G. gilioides** (Benth.), Greene. Loosely branching, ½—2 ft. long,
viscid-pubescent: *leaves simply cut into few narrowly oblong or lanceolate
divisions:* fl. few in the cluster: *corolla slender-salverform,* deep blue-
purple: stamens hardly equally inserted: capsule globular: ovules and
seeds 1 or 2 in each cell.—Higher parts of the Coast Range. May—July.

6. **G. densifolia,** Benth. *Perennial,* 1—2 ft. high, *more or less floccose-woolly;* stems erect, virgate, from a somewhat woody base: *leaves rigid,* linear, laciniate-pinnatifid or incised, the short lobes subulate: fl. many, in a dense head: corolla ½--¾ in. long, salverform, violet-blue; limb ½ in. broad in expansion: anthers linear.—Mountains of Santa Clara Co., and southward.

5. **LINANTHUS,** *Benth.* Annuals, with opposite leaves mostly divided digitately into 3 or more narrowly linear rigid segments. Inflorescence dichotomous and loose, or cymose-glomerate and terminal only. Calyx prismatic; its segments equal; its tube scarious between the angles. Corolla usually salverform, often with very long and slender tube. Stamens equal and equally inserted on the corolla. Capsule few- or many-seeded.

* *Stems dichotomously branching; flowers scattered.*

1. **L. dichotomus,** Benth. Erect, 4—10 in. high, the nodes few and exceeding the leaves, these or their 3—5 segments linear-filiform, 1 in. long: flowers subsessile: calyx prismatic, scarious except the 5 prominent angles which are prolonged into acerose-linear recurved segments: *corolla salverform; tube scarcely exserted;* limb 1—1¾ inches broad, white, shaded on the outside with dark chocolate-color: stamens inserted below the middle of the throat, the base of each filament set within a nectariferous groove; anthers linear: cells of capsule very many-seeded: seeds roundish, not mucilaginous when wet.—Common. May.

2. **L. liniflorus** (Benth.), Greene. A foot high, slender, glabrous: leaf-segments about 3, filiform: flowers on long slender pedicels in a loose cymose panicle: *corolla with nearly obsolete tube; limb rotate,* ½—1 in. broad, the obovate entire lobes white, marked with 7 deep blue veinlets: stamens nearly as long as the corolla-lobes; filaments with a dense pilose ring just above the base, the very short corolla-tube below them pubescent: ovules 6—8 in each cell: seeds mucilaginous.—Dry hills and sandy plains. April, May.

3. **L. filipes** (Benth.), Greene. Slender, often diffuse, 4—10 in. high, *scabrous-puberulent:* pedicels elongated, filiform: calyx a line long, narrow-campanulate, hispidulous throughout, the subulate-acerose teeth little shorter than the tube: *corolla pale purplish,* 3½ *lines long, the limb campanulate from a short cylindrical tube:* seeds about 8 in each cell of the capsule.—Open hillsides. May.

4. **L. Bolanderi** (Gray), Greene. Slender as the last, differing in a narrowly cylindraceous calyx and *salverform corolla,* the tube of which equals the calyx; the rotate limb purplish and about 3 lines wide: seeds 2—5 in each cell.—Sonoma Co.

5. **L. ambiguus** (Rattan), Greene. Stouter than the last; corolla much larger (½ in. long), not strictly salverform, the *slender tube little*

exserted, the obconic dark purple throat about as long as the rotate-spreading purplish lobes: seeds 2 in each cell.—At Oak Hill, four miles south of San Jose, *Rattan.*

     * * *Stem not dichotomous; flowers corymbose-capitate.*

     +—*Corolla salverform, the slender tube long-exserted.*

6. **L. androsaceus** (Benth.), Greene. *Stoutish,* 6—15 in. high: lowest leaves 3-, uppermost 5—7-parted, the divisions oblanceolate, those of the floral subulate-lanceolate, all acute, rather strongly hispid-ciliate: corolla more than 1 in. long, the slender purple tube 9—10 lines, the short turbinate *throat* about a line long, *very dark purple,* with a yellow border, the broad rounded or somewhat cuspidate segments lilac-purple (occasionally white), 3—4 lines long; style and filaments little surpassing the throat of the corolla.—Abundant on half shaded hillsides. May.

7. **L. parviflorus** (Benth.), Greene. Much more slender than the last, and scarcely as tall; leaves with narrow segments: *tube of corolla very slender,* 9—10 lines long; *throat yellow;* segments oval, 2—3 lines long, mostly pale yellow or white, tinged with red or brown on the outside: style and filaments half or more than half as long as the corolla-limb.—Plentiful in open grassy lands. April, May.

8. **L. acicularis,** Greene. Only 3—6 in. high, very slender, more rigid and less pubescent than the last; *leaf-segments linear-acerose: corolla golden yellow throughout,* the very slenderly filiform tube about thrice the length of the limb: stamens two-thirds the length of the obovate-lanceolate lobes; style short.—With the last but less common.

9. **L. rosaceus,** Greene. Commonly branching from the base and the branches decumbent, 3—5 in. high, stoutish and with short inter-nodes, these 5—7, not twice the length of the leaves; *segments of the lowest leaves obovate-spatulate,* of the upper spatulate-linear, those of the floral bracts subulate, pungently acute, spinulose-serrulate above the middle, more softly ciliate toward the base: corolla 1 in. long; *tube and limb rose-red, the ample throat orange.*—Only on sandy hills at San Francisco and southward. May, June.

10. **L. bicolor** (Nutt.), Greene. Very near the last, but dwarf (2—3 in. high); flowers rose-purple, the elongated corolla-tube ½—¾ inch long and less slender than in *L. parviflorus* but the *limb* much smaller, *only 2—3 lines broad.*—Low hills, and on grassy plains along the Mt. Diablo Range and far northward. April, May.

11. **L. ciliatus** (Benth.), Greene. *Rigid, strict,* ½—1 ft. high, *scabrous-pubescent:* internodes long: leaves with 5—9 linear rigidly ciliate segments: corolla rose-color, very small and slender, little longer than the floral leaves, the rotate limb only 2 lines broad.—Range of the last, but at greater elevations in the hills; usually under trees. May.

+ + *Corolla broadly funnelform above the short tube.*

12. **L. grandiflorus** (Benth.), Greene. Very rigid and strict, $\frac{1}{2}$—$1\frac{1}{2}$ feet high, the rigid linear-subulate leaf-segments 5—11, ascending, spinulose-serrulate on the margin and toward the base somewhat ciliate: corolla lilac; tube little exserted; limb more than $\frac{1}{4}$ inch broad.—At Alameda, San Francisco, and southward. June, July.

6. **PHLOX,** *Linn.* Opposite-leaved herbs with usually cymose and terminal inflorescence; differing from the last genus in that the stamens are unequal and unequally inserted on the upper part of the tube of the salverform corolla, and the leaves not rigid nor palmatisect. Our only species belongs to a chiefly Texan group of annuals with a scattered inflorescence: the upper leaves being alternate.

1. **P. gracilis** (Dougl.), Greene. Only a few inches high, pubescent: lowest leaves obovate; upper lanceolate: corolla less than $\frac{1}{2}$ in. long, rose-purple; tube hardly exceeding the linear calyx-lobes; limb only 2 lines broad.—Not rare in the more hilly districts. April, May.

ORDER LXXI. HYDROPHYLLACEÆ. .

Ours (with the exception of one gummy coriaceous-leaved shrub) herbs remarkable for much of the hispid and rough in pubescence, coincident with a thinnish and juicy herbage; the juice watery and without special properties. Leaves exstipulate, often pinnately lobed or dissected. Flowers complete, 5-merous, very often in unilateral and scorpioid false racemes or spikes. Calyx persistent, of 5 deep lobes or distinct sepals. Stamens borne on the tube of the corolla, alternate with its lobes. Ovary not lobed or divided, 1- or 2-celled; the 2 styles distinct at apex if not throughout (except in *Romanzoffia*). Fruit capsular; the 2 placentæ parietal, or borne on the half-partitions.

* *Ovary 1-celled, the placentæ forming a sac-like lining to the ovary-walls.*

Seeds uniform, globular or ovoid..NEMOPHILA 1
  " not uniform within the capsule................................EUCRYPTA 2

* * *Placentæ narrow, not lining the cell, sometimes almost partitioning it into two.*

Style-cleft at apex, or to the middle;
  Corolla blue or white, deciduous................................PHACELIA 3
    " yellow, persistent..EMMENANTHE 4
Style and stigma entire; flowers white............................ROMANZOFFIA 5
Styles 2, distinct to the base...ERIODICTYON 6

1. **NEMOPHILA,** *Nutt.* Diffuse annuals with tender herbage, pinnately lobed or divided leaves, and axillary peduncles usually 1-flowered. Calyx 5-parted, usually with a supplementary reflexed lobe at each sinus. Corolla deeply 5-lobed, broadly campanulate to nearly rotate, convolute

in bud; the throat appendaged within with 10 scales or plaits. Stamens and style mostly shorter than the corolla; filaments naked; anthers sagittate. Ovary tardily dehiscent by 2 valves, 1—16-seeded. Seeds globular or nearly so.

      * *Leaves alternate; stems beset with short retrorse prickles.*

1. **N. aurita,** Lindl. Stems both fleshy and very brittle, 1—3 ft. long, reclining or trailing by the prickly angles: leaves deeply pinnatifid above into mostly retrorse lobes, the petiole broadly winged and the wing auriculate-dilated at the base: lower peduncles 1 flowered, upper often 3-flowered: appendages of the calyx very small: *corolla 1 in. broad, dark violet,* its appendages broad, partly free: seeds globose, reticulate.—Shady slopes of hills, in San Mateo Co., and on Angel Island.

2. **N. membranacea** (Benth.). More slender, paler, glaucescent, prickly-angled and procumbent, 1—1½ ft. long: leaves pinnately divided into 3—9 linear obtuse nearly entire divisions; the petiole winged: fl. few or many on the peduncles, very small; *calyx and corolla without appendages, the latter white:* seeds globose, reticulated.—Shady hillsides from the Livermore Valley southward. Thoroughly congeneric with *N. racemosa;* only empirically placed under "*Ellisia,*" notwithstanding the absence of calyx-bractlets. March, April.

      * * *Leaves often opposite; stems not prickly-angled.*

3. **N. insignis,** Dougl. At length very branching, the branches ascending, 6—10 in. high: leaves pinnately parted into 7—9 oblong, sometimes 3—5-lobed, small divisions: *corolla rotate-campanulate, 1 in. broad, clear blue;* its internal scales short and rounded, *partly free, short-hirsute:* seeds oval, somewhat corrugated or tuberculate.—Very common in rich fields and on low hills. April, May.

4. **N. Menziesii,** Hook. & Arn. Smaller than the preceding, the leaves less divided: *corolla* as large, rather more *nearly rotate, white or very pale blue, sprinkled with dark dots toward the centre,* the spots confluent into a purplish eye; its *scales narrow, one edge wholly adherent, the other free and densely ciliate:* seeds oval or oblong, either smooth or somewhat tuberculate.—Not as common as the last, though plentiful in some parts of Marin and Napa counties. April, May.

5. **N. pedunculata,** Benth. Every way much smaller than the last; whole plant 2—4 in. high: *corolla seldom 2 lines wide, white, with purple reinlets.*—Rocky shades, and very wet plains; not uncommon. May.

6. **N. parviflora,** Dougl. Slender, weak and procumbent, neither the leaves nor the stems notably succulent: leaves pinnately 5—9-parted, the divisions obovate or oblong, obtuse: corolla campanulate, 3—5 lines broad, white; its appendages oblong, wholly adherent by one edge, nearly or quite glabrous.—Shady mountain sides. April—June.

2. EUCRYPTA, *Nutt.* Erect, with opposite bi- or tripinnately dissected leaves, and axillary peduncled racemes of small whitish flowers. Calyx 5-parted, not accrescent, not appendaged, campanulate. Corolla small, campanulate, without internal appendages. Capsule globose, 8-seeded, each valve in dehiscence liberating 2 oblong-ovoid seeds, and retaining between the placenta and the wall of the ovary 2 flattened and lenticular or meniscoid ones.

1. **E. chrysanthemifolia** (Nutt.), Greene. Stoutish, freely branching though very erect, 1—3 ft. high, very leafy: racemes rather dense, little surpassing the ample pubescent and slightly clammy tripinnatifid leaves: calyx-lobes ovate, acutish: corolla white with a bluish tinge: free rounded seeds corrugated; the concealed and flattened ones smooth.— Common in the hilly districts. April—June.

3. PHACELIA, *Jussieu.* Annuals, or a few perennials, with alternate simple or compound leaves, and more or less scorpoid unilateral racemes or spikes of blue or bluish flowers. Calyx mostly, or always, completely divided into 5 sepals. Corolla deciduous; the tube commonly with 10 vertical folds or lamellæ within. Capsule with narrow placentæ and 1—∞ corrugated or reticulate or pitted seeds.

  * *Seeds 1 or 2 on each placenta; testa areolate or favose.*

  ← *Perennials or biennials; leaves pinnatifid or entire.*

1. **P. Californica,** Cham. Stout sparingly leafy flowering *stems* 1—2 ft. high, *from a stout* much branched depressed and very *leafy woody caudex: herbage canescent with a minute close pubescence, and setose-hispid* with scattered long white hairs: leaves mainly of a large elliptic-lanceolate terminal lobe; the pinnæ below few and reduced: racemes erect, short and dense, in a short paniculate cluster at and near the summit of the stem: sepals equal, lanceolate, 2 lines long in maturity, erect, reticulate-veiny and with a strong midvein: capsule ovate-oblong, acutish, 1½ lines long, 1-seeded: seed ovate-lanceolate, a line long, deeply alveolate.—Sandy hills and plains near the sea. Feb.—Sept.

2. **P. imbricata,** Greene. Stems erect, several from a stout perpendicular perennial root, densely leafy at base only: panicle of geminate racemes long and lax, the branches widely spreading: pubescence much as in the last, but more scanty: corolla bluish: *fruiting calyces compressed and closely imbricated; sepals unequal, the broad deltoid-ovate outer one larger than the others,* these ovate-oblong, acutish, all hispid-ciliate, none showing either distinct midvein or reticulation; seed 1 only.—Dry hills and valleys of the mountain districts back from the sea. June, July.

3. **P. nemoralis,** Greene. Apparently only biennial, stout, erect, rather widely branching, 2—4 ft. high; herbage *light green, destitute of pubescence but strongly hispid with stinging hairs:* leaves simple, or with

a pair of pinnæ, rarely 2 pairs, at base, petiolate, without any conspicuous parallel veins: racemes geminate or ternate, short but not lax:
corolla small, ochroleucous: *sepals oblanceolate:* capsule 2-seeded: seed
deeply pitted.—Shady ravines and along stream banks.   June—Aug.

+ + *Annuals; leaves entire, or with shallow lobes.*

4. **P. malvæfolia,** Cham. Annual, the herbage of a *light green,
pubescent, very hispid with spreading or reflexed stinging hairs:* leaves
round-cordate, slightly 5—9-lobed, sharply toothed, 1—3 in. long: spikes
solitary or geminate: corolla 3—4 lines long, yellowish white, longer
than the unequal linear and spatulate sepals: stamens exserted: capsule
4-seeded: seeds alveolate-scabrous.—Damp shades, from San Francisco
and Alameda counties southward. June—Sept.

5. **P. Rattani,** Gray. Smaller than the last, beset with more slender
bristles: leaves ovate or oval, with truncate or subcordate base, incisely
lobed or crenate: calyx of *4 spatulate sepals and one larger and obovate:*
corolla whitish, little more than 2 lines long: stamens and style included:
seeds small.—Northern part of Sonoma Co., on Russian River, etc.

6. **P. Breweri,** Gray. Slender, diffusely branched, the stems 6—10
in. long, *canescently pubescent* and hispidulous: racemes slender, long
and lax: sepals linear: *corolla open-campanulate, violet,* twice as long as
the calyx: filaments not exserted: capsule ovate, acute, mostly 1-seeded:
seed favose.—Mt. Diablo, Mt. Hamilton and southward. May, June.

+ + + *Annuals (except n. 9); leaves oblong or narrower, pinnately lobed,
toothed, or compound, and the lobes or divisions toothed or incised.*

7. **P. distans,** Benth. Usually rather slender, branching, 1—2½ feet
high, the branches when present decumbent: leaves ample, very finely
and compoundly dissected: *spikes mostly scattered, solitary or geminate:*
sepals unequal, linear to spatulate, 1 or 2 more dilated upwards: *corolla
dull yellowish white,* 3—4 lines long, rotate-campanulate; the internal
appendages broadly semiovate, with a free pointed tip: stamens little
exserted: capsule globular: seeds rugose-tuberculate.—Plentiful at
Alameda; also in Marin Co. and southward. May—July.

8. **P. tanacetifolia,** Benth. Stouter than the last, seldom branching,
erect: leaves less ample and less dissected: *spikes all clustered and terminal:* sepals equal, linear, beset with rigid bristles: *corolla open-
campanulate,* ¼ in. long, *lavender-color or lilac-purplish; inner appendages wholly adnate:* stamens well exserted.—Mostly in fields and open
plains of the interior. May, June.

9. **P. ramosissima,** Dougl. Stems clustered and decumbent or
ascending, from a *perennial root: herbage pubescent and viscid-glandular:*
leaves divided into oblong or narrower pinnatifid-incised divisions:

kes short, dense: corolla ochroleucous or bluish: stamens and style
ly moderately exserted: sepals linear-spatulate, twice the length of
ɔ capsule: seeds oblong.—Mt. Diablo Range. June, July.

10. **P. Arthuri,** Greene. Annual, decumbent or nearly prostrate,
ɔ stoutish branches 2 ft. long: herbage setose-pubescent, not viscid,
e inflorescence hispidulous: leaves rather remotely pinnate, or some
rate, the lobes crenate-toothed: spikes many, solitary in the axils,
ort-peduncled: fl. biserial and crowded: *sepals entire, very unequal,*
ur small, at length partly enfolded by the accrescent rhombic-obovate
ute fifth: corolla only a line broad, light blue: stamens not exserted.
Found by the side of a street in the western part of Oakland, in 1887,
ᵥ *Arthur Simonds;* the plant now extinct in that locality.

11. **P. ciliata,** Benth. Erect and simple, or with several ascending
ranches, 1—1½ ft. high; stems scabrous; other parts pubescent or
ᵥaringly hirsute: leaves pinnately parted or cleft, the divisions or lobes
ᵥlong, pinnately incised: spikes short, at length rather loose, the
edicels short or none; *sepals* lanceolate to ovate, *accrescent, in age some-*
hat chartaceous, reticulate, 4—5 lines long and *sparsely bristly-ciliate:*
ᵥrolla smallish, bluish; capsule ovate, mucronate: seeds oval, favose.
-Common, especially east of the Mt. Diablo Range, on the plains; but
lso to the southward of San Francisco. May.

12. **P. suaveolens,** Greene. Annual, stoutish, freely branching from
he base, the branches ascending, 1—2 ft. long: *herbage very sweet-scented,*
oft pubescent and glandular-viscid throughout: cauline leaves oval,
ᵥoarsely-toothed, 1 in. long, on slender petioles of nearly equal length;
he lower with some lyrate lobes at or below the base of the main blade:
ᵥacemes solitary or in pairs, elongated, dense: sepals spatulate, entire,
'₄ inch long, exceeding the 4-seeded capsules: *corolla* bright blue,
ᵥarrowly funnelform, ½ in. long, the limb one half as broad: seeds oval,
ᵥlack, deeply favose-pitted.—At the Petrified Forest, Sonoma Co.

* * *Seeds 6—12 or more on each placenta; testa not rugose,*
but areolate-reticulate.

13. **P. circinatiformis,** Gray. Erect, sparingly branching from the
base, 6—10 in. high, *puberulent and very hispid:* leaves *ovate to oblong-*
lanceolate, parallel-veined: racemes dense: sepals linear or linear-spatulate, enlarging in age, greatly exceeding the capsule: corolla dull-white,
very small, scarcely wider than funnelform-tubular, 2—3 lines long;
calyx when mature 5 lines or more: capsule ovate, acute or mucronate,
6—16-seeded: seed scrobiculate.—Plentiful near the summit of Mt.
Diablo; also reported from Mt. Hamilton, but a rare plant. June.

14. **P. divaricata** (Benth.), Gray. Diffusely branching from the
base, the branches 6—18 in. long, more or less pubescent and hirsute:

leaves ovate or oblong, commonly entire, exceeding the petiole, the veins curving upwards: racemes loose; pedicels shorter than the calyx: corolla rotate-campanulate, ¾ in. broad, blue: seeds very many.—Common on dry hills about the Bay, preferring stony and poor soil. April—June.

15. **P. Douglasii** (Benth.), Torr. Diffusely branching, pubescent and hirsute: *leaves oblong-linear, pinnatifid into many pairs of lobes or segments; the terminal lobe not larger*, nor parallel-veined: *racemes long and lax:* sepals spatulate: blue corolla open-campanulate, ½—¾ in. wide; its appendages semioblanceolate: capsule ovate, mucronate, many-seeded: seeds roundish-oval, scrobiculate.—Near Lake Merced, thence southward. May.

4. **EMMENANTHE,** *Benth.* Herb (the genus in our view monotypical) with more the habit of *Eucrypta* than *Phacelia*, erect, the straight ascending racemes panicled at summit of the stem. Calyx deeply 5-parted, the divisions not widening upwards. Corolla campanulate, light yellow, persistent. Capsule as in *Phacelia*, and seeds pitted.

1. **E. penduliflora**, Benth. Villous-pubescent, slightly viscid, 1 ft. high more or less: lobes of the pinnatifid leaves very many, short, somewhat toothed or incised: corollas nearly ½ in. long, in age pendulous on filiform pedicels of their own length: filaments nearly free from the corolla: seeds several, 1 line long.—Summits of the higher Coast and Mt. Diablo ranges. June—Aug.

5. .**ROMANZOFFIA,** *Chamisso.* Delicate and pale herbs, of Saxifrageous aspect. Leaves mostly radical, alternate, round-cordate, crenately lobed, long-petioled. Stems loosely racemose; the flowers white. Calyx deeply 5-parted. Corolla funnelform, without appendages, deciduous. Stamens on the base of the corolla-tube, unequal. Filiform style and small stigma both entire. Capsule retuse, nearly or quite 2-celled. Pitted-reticulate seeds small, numerous, on linear placentæ.

1. **R. Sitchensis,** Bong. Filiform rootstocks tuberiferous: stems slender, erect, 4—8 in. high; pedicels spreading, longer than the flowers: calyx-lobes oblong-linear or lanceolate, not equalling the corolla, little shorter than the capsule: style long, slender.—Shady wet rocky places among the higher coast mountains. April—June.

6. **ERIODICTYON,** *Benth.* Shrubs, with coriaceous leaves balsamic and fragrant, white-woolly beneath. Inflorescence a terminal panicle of scorpioid cymes. Calyx deeply 5-parted, the segments not widening upwards. Corolla funnelform approaching the salverform. Styles 2, distinct to the base; stigmas clavate-capitate. Capsule crustaceous, small, globose, ovate, acute, 2-celled, 4-valved. Seeds few.

1. **E. Californicum** (H. & A.), Greene. Shrub 3—5 ft. high: lanceolate irregularly serrate leaves 3—8 in. long, resinous-glutinous above,

beneath very closely white-woolly between the reticulate veins: panicle of cymes naked: corolla pale blue, tubular-funnelform, thrice the length of the sparsely hairy calyx.—Dry hills in the sparsely wooded mountain districts. May—July.

Order LXXII. ASPERIFOLIÆ.

Plants intimately connected with *Hydrophyllaceæ*, the herbage in general still more pronouncedly rough-hairy or hispid, but less succulent. Inflorescence equally unilateral and scorpioid. Differing in that the single style is surrounded at base by a 4-lobed ovary, which ripens into 4 (or by abortion 2) usually distinct and very seed-like nutlets. The 5-cleft or -parted calyx mostly persistent; corolla deciduous, in ours mostly short-salverform and the throat closed by folds.

* *Corolla imbricate or quincuncial in æstivation.*

Nutlets depressed and horizontally extended;
  Nutlets rounded, the whole surface rough....................CYNOGLOSSUM 1
    " elongated, the margins prickly......................PECTOCARYA 2
Nutlets vertically elongated, but often incurved;
  Insertion basal;
    Nutlets carinate, transversely rugose......................ALLOCARYA 3
    " not carinate, irregularly roughened.............. LITHOSPEMUM 7
  Insertion nearly basal;
    Flowers white.... ....ALLOCARYA 3
    " yellow...AMSINCKIA 6
  Insertion lateral;
    Scar rounded, usually hollow.........................PLAGIOBOTHRYS 4
    " linear, forked at base..............................CRYPTANTHE 5
    " ovate or lanceolate...............................AMSINCKIA 6

* * *Corolla plicate in æstivation.*

Fruit nearly 4-lobed, at length separating into 4 nutlets...........HELIOTROPIUM 8

1. **CYNOGLOSSUM**, *Diosc.* Calyx 5-parted, persistent, open in fruit. Corolla short-salverform, with conspicuous arching crests at the throat concealing the short stamens and pistil. Nutlets 4, broad, depressed, the whole back covered with short stout prickles having minutely barbed tips, oblique or horizontal, separating at maturity from below upwards, and for a time suspended on a process connected with the style.

1. **C. grande**, Dougl. Perennial, stout, 2 ft. high, with long-petioled ovate-oblong leaves often a foot long, usually rounded at base, pubescent with soft slender hairs: panicle of short racemes small, on a long naked peduncle terminating the short leafy stem: corolla ⅓ in. broad, deep blue, with pinkish central folds.—In shades along streams, and on northward slopes. Jan.—April.

2. **PECTOCARYA**, *DeCandolle.* Low slender annuals, with strigose-hirsute pubescence, small narrow leaves, and small flowers near their

258   ASPERIFOLIÆ.

axils. Corolla minute, white. Nutlets flattish depressed and laciniate-bordered, or pectinately setose around the margin, the bristles or prickles uncinate at tip.

1. **P. penicillata** (H. & A.), A. DC. Diffusely spreading branches only a few inches long: *nutlets divergent in pairs, oblong*, surrounded by an undulate or pandurate wing which at the apex of the nutlet is thickly beset with uncinate bristles.—Common in the interior of Calif., especially southward; rare in our district, but found in Napa Valley.

2. **P. pusilla** (A. DC.), Gray. Erect, somewhat flexuous, 2—4 in. high: *nutlets equally divergent, cuneate-obovate*, wingless, and with a carinate midnerve, the acute margin beset with a row of slender uncinate bristles.—Napa Valley and northward.

3. **ALLOCARYA**, *Greene.* Low herbs, ours annual, with linear entire leaves, the lowest always opposite and connate-perfoliate: branches numerous and commonly depressed, racemose almost throughout. Plants vernal in their flowering, confined to low moist grounds, herbage usually light green and somewhat succulent, more or less hirsute. Pedicels turbinate-thickened and more or less distinctly 5-angled under the calyx, persistent, somewhat indurated in age. Calyx 5-parted to the base; segments spreading. Corolla salverform with short tube, yellow throat and white limb. Nutlets ovate or lanceolate, crustaceous, opaque or vitreous-shining, smooth or variously tuberculate and rugose, muriculate or even strongly glochidiate, often carinate on one or both sides, attached by an infra-medial or basal, concave, but sometimes raised and stipitate scar.

1. **A. stipitata**, Greene. *Erect, simple, or with ascending branches from the base*, 10—18 in. high: herbage light green, apparently glabrous, yet roughish, slightly, with sparse and short setæ: calyx nearly sessile; segments spreading, foliaceous and accrescent, in fruit often ½ in. long: corolla short-funnelform, ¼—½ in. broad; nutlets ovate-lanceolate, carinate for the whole length of the ventral face, and a little past the apex, the back covered with blunt tuberculations and interrupted transverse rugæ; *scar exactly basal*, roundish and separated from the body of the nutlet by a short but *distinct stipe*.—Napa Valley, and plains of the lower Sacramento and San Joaquin. May.

2. **A. Californica** (F. & M.), Greene. Slender, sparingly setose, *diffusely branching*, the branches 6—15 in. long, weak and reclining: racemes with few bracts at base: calyx-segments slender, not accrescent, spreading in fruit: nutlet ovate, ¾ line long, keeled, rugulose and granulated as in the last: *scar roundish*, nearly basal, *not stipitate.*—Common in low fields. April—June.

3. **A. stricta**, Greene. Slender, *strictly erect* and somewhat succulent, simple, or with several scarcely divergent spicate branches above, barely

5—6 in. high; glabrous, or nearly so, all except the floral leaves opposite: spikes dense, 1—2 in. long: flowers very small, white, with yellow centre: calyx segments closed over the immature fruit: *nutlets light gray, vitreous-shining, long-ovate,* about ½ line long, *with numerous close transverse rugosities;* insertion supra-basal, the scar linear with dilated base and about one-third the length of the nutlet.—Near the warm springs at Calistoga; plentiful, and probably local. April, May.

4. **A. diffusa,** Greene. Pubescence light, closely appressed: branches procumbent, a foot or less in length, *loosely racemose from the base,* the raceme leafy to the middle at least; lowest pedicel ½ in. long, the others hardly a line: calyx widely spreading, corolla small: *nutlets dark brown, broadly ovate, incurved,* ¾ line long, ventrally carinate down to the supra-basal oblong-lanceolate scar, the back with rather sharp granulations and rugæ, the latter favosely confluent.—Open hills, only toward the sea. April, May.

5. **A. humistrata,** Greene. *Stout and succulent,* the branches mostly *prostrate,* a foot long, racemose throughout: pedicels short and stout, commonly deflexed: calyx-lobes linear-spatulate, *in fruit greatly enlarged* (4—6 lines long) and *turned to one side, standing vertically* in a row: nutlets ovate-lanceolate, ¾ line long, straight, carinate ventrally down to the nearly or quite basal rounded scar, the back with minute muriculations and sharp-edged transverse rugulæ which are commonly tipped with a tuft of minute penicillate bristles.—Low subsaline plains of the interior, on the lower San Joaquin, etc. April, May.

6. **A. trachycarpa** (Gray), Greene. Size and habit of the last, but more branching and decumbent rather than procumbent, commonly more slender, rough with a coarser and somewhat spreading pubescence: racemes less open, leafy almost throughout: *segments of calyx linear, widely spreading:* corolla very small: nutlets ovate, straight, carinate on both sides, the *dorsal keel and nearly straight transverse rugæ dentate-interrupted;* scar suborbicular, nearly basal.—Habitat of the last.

7. **A. Greenei** (Gray). Habit, pubescence and inflorescence of the last, but a coarser, larger plant; nutlets a line long, ovate, straight, carinate ventrally down to the nearly basal ovate scar, the *back covered with coarse granulations and stout barbed prickles* ¼—½ line high, these distinct at base or more or less confluent into walled reticulations, the latter sometimes strongly developed and the prickles themselves correspondingly reduced or even nearly obsolete.—Very common on low plains of the interior. May.

8. **A. Chorisiana** (Cham.), Greene. Freely branching, the branches 10—18 in. long, at length reclining: racemes elongated, loose, leafy below; *pedicels filiform, 4—8 lines long:* calyx little accrescent, the spreading segments about a line long: corolla 3—5 lines wide: *nutlets*

ovate, little more than ½ line long, brownish and dull, carinate ventrally only, the *keel and scar closely approached,* but not covered *by the lateral angles,* the obtuse rugæ of the back running into more or less favose meshes among the numerous minute granulations: scar linear, short.— Common in wet grassy land among the Mission Hills, and in Marin Co.

4. PLAGIOBOTHRYS, *Fisch. & Mey.* Rather large but slender annuals with most of their leaves in a close radical tuft. Racemes spike-like, elongated, loose, naked or leafy-bracted; pedicels very short, filiform, persistent. Calyx 5-cleft or -parted, closed or campanulate, more or less accrescent in fruit, and when not too deeply cleft irregularly circumscissile near the base. Nutlets ovate or indistinctly cruciform in outline, carinate on both sides toward the apex, usually with well defined lateral margins, the back very regularly transversely rugose, smooth or roughened between the rugæ; insertion almost medial on a depressed gynobase; areola or scar rounded, hollow or solid.

* *Stems branched from the base, the branches prostrate.*

1. - P. canescens, Benth. Canescent with a pale soft-villous pubescence, the branches ½—1½ ft. long, leafy and floriferous throughout: *calyx thinnish, cleft to the middle,* the tube slightly inflated in age, and the *segments closed over the incurred-connivent nutlets;* these with strong transverse rugæ and minute intervening granulation.—Plains and hills of Alameda and Santa Clara counties, and thence both northward and southward, in the interior. April, May.

* * *Stems erect from the base, loosely branching above.*

2. P. nothofulvus, Gray. Stem mostly solitary from the depressed leaf-tuft, 1—1½ ft. high including the widely spreading flowering branches; herbage canescent with a short and fine pubescence: *calyx small,* cleft hardly to the middle, its teeth closed over the fruit, but the whole calyx, except the very base, *at length deciduous by circumcision,* exposing and releasing the ripe nutlets; these cruciform-ovate, with rather prominent rugæ, and dot-like white granulations intervening.— Very common on hills. March—May.

3. P. tenellus (Nutt.), Gray. Small and slender, seldom 10 in. high, soft-hirsute and canescent, the calyx rusty-yellowish, the erect stems with one or more small leaves: spikes rather few-flowered: calyx deeply cleft, not circumscissile, loosely connivent over the shining *somewhat cruciform nutlets;* these a line long, *smooth and glassy, with very straight transverse rugæ* which are either smooth or quite strongly muricate.— Very common in our northerly hilly districts. April, May.

4. P. campestris, Greene. Stouter than any of the preceding, 1—2½ ft. high, scarcely canescent, but hirsute, the calyx with a brownish pubescence, *cleft nearly to the base, the segments wholly herbaceous, per-*

sistent: nutlets large (1½ lines long), nearly a line wide in the middle, *abruptly stout-beaked, the body sharply carinate and laterally margined*, with or without sharp transverse rugæ and intervening muriculations.— Plains of eastern Solano and Contra Costa counties. May.

5. CRYPTANTHE, *Lehmann.* Pilose-hispid slender mostly rather rigid and erect annuals, with bractless spicate flowers. Herbage and root imparting no stain. Leaves alternate, narrow and entire. Calyx 5-parted to the base, deciduous: segments erect, usually closely embracing the fruit, the often attenuate and elongated tips sometimes spreading above it, and hispid with straight or hooked bristles. Nutlets 4 (sometimes by abortion 2 only, or 1), smooth, tuberculate or muriculate, attached from the base upwards commonly to near the apex; linear groove and transverse scar open or closed.

* *Nutlets muricate-roughened.*

1. **C. muriculata** (A. DC.), Greene. Stoutish, ½—1 ft. high, very hispid: *spikes rather short and dense*, usually *in twos or threes at the ends of the branches: calyx 2 lines long*, the segments merely acute, not long-tipped: nutlets a line long, of deltoid-ovate outline, light gray, scabrous-muricate over the flattish rather than rounded back: ventral groove and its basal fork mostly closed.—Mt. Diablo and southward, at considerable elevations. May, June.

2. **C. Jonesii** (Gray), Greene. With stouter stem than the last, but more slender and quite *distinctly panicled spicate branches: calyx only a line long*, and nutlets little more than ½ line.—Mt. Tamalpais, thence southward along the seaboard hills.

3. **C. micromeres** (Gray), Greene. Very slender, diffusely branched, 6—10 in. high, hispid, but green, not canescent: *spikes filiform*, very numerous but single: *calyx only ½ line long: ovate-trigonous nutlets acute, 3 of them muriculate-scabrous*, the other nearly or quite smooth.— Coast Range hills: an obscure but not rare plant.

4. **C. ambigua** (Gray), Greene. Usually stout, low, with many short ascending branches spicate almost throughout: herbage canescently-hispid: calyx ¼ in. long or more and, with *linear-elongated segments twice the length of the narrow-ovate acuminate papillose-scabrous gray nutlets;* these with groove and its basal bifurcation nearly closed.— Inner coast ranges northward, on dry hills.

* * *Nutlets smooth and shining.*

← *All the four ovules maturing into nutlets.*

5. **C. leiocarpa** (F. & M.), Greene. Usually diffusely branched from the base, the branches ½—1 ft. long; whole plant canescent with an appressed pubescence and some pilose-hispid hairs: inflorescence short-

spicate or somewhat glomerate and leafy-bracted: *calyx a line long, the segments scarcely attenuate* or prolonged above the nutlets; *groove of small ovate acute nutlet not forked at base.*—Sandy lands along the seaboard. April—Aug.

6. **C. Torreyana** (Gray), Greene. Erect, sparingly paniculate-branched, 1—1½ ft. high, hirsute-hispid, the calyx with rigid stinging bristles, its *sepals elongated and attenuate upward:* nutlets ovate, acute, the groove forked at base.—Very common among the hills and mountains, in clayey soil. May—July.

+-+ *Nutlet solitary, three ovules being abortive.*

7. **C. flaccida** (Dougl.), Greene. Rigidly erect, slender, usually simple up to the terminal set of rather strict spikes: minutely strigose-hispid: fruiting calyx erect, appressed to the rachis, narrow and slender, the filiform-linear sepals very hispid below with deflexed strong bristles: nutlet subterete, ovate-lanceolate, rostellate-acuminate, shorter than the sepals; groove enlarged below, but not furcate.—Very common on stony hillsides almost everywhere. May, June.

6. AMSINCKIA, *Lehmann.* Hispid annuals, with yellow flowers in elongated spikes. Calyx herbaceous; sepals 5, or 4 by the union of 2 into one broader one. Corolla salverform, the throat somewhat funnelform, the aperture often angular by folds, the lobes rounded, rarely somewhat unequal and the corolla therefore slightly bilabiate. Filaments short; anthers oblong or oblong-linear. Style filiform; stigma capitate, 2-lobed. Nutlets crustaceous, erect or incurved, smooth or rugose, commonly more or less ovate-triquetrous. Cotyledons 2-parted.

1. **A. lycopsoides,** Lehm. Herbage of a light *yellowish-green,* very hispid, the bristles from a pustulate base: *leaves* in a rosulate tuft, *lanceolate, slenderly acuminate:* stem at first erect, at length freely branching and the branches trailing, 1—2½ ft. long: spikes more or less leafy-bracted: sepals short: corolla pale yellow, very slender: nutlets ¾ line long, brown or blackish, rugulose and muriculate.—Sandy soils along the seaboard. March—July.

2. **A. intermedia,** Fisch. & Mey. Loosely branching above, 1—3 ft. high, less hispid than the last, *deep green: leaves linear, or the lowest oblanceolate:* sepals twice the length of the nutlets and half as long as the *deep yellow narrow corolla:* anthers oblong: nutlets very much incurved, carinate on the back, muricate-scabrous and obliquely rugose. —Very common in the Bay region, in alluvial soils. April—June.

3. **A. spectabilis,** Fisch. & Mey. Erect, slender, less hispid, 1—1½ ft. high: leaves linear: sepals narrowly linear-lanceolate, elongated, ferruginous-hispid: *corolla orange-yellow,* ½—¾ *in. long; lobes slightly unequal:* nutlets granulate-rugose, carinate on the back.—Hills of the Mt. Diablo Range. March—May.

4. **A. campestris.** Stoutish, 1—2 ft. high, the *short and rather dense spikes aggregated* at summit of the stem: pubescence strigose-hirsute rather than hispid: leaves all linear-oblanceolate: sepals short hardly twice as long as the nutlets, ovate-lanceolate, not ferruginous: *corolla inconspicuous*, its throat nearly closed by folds: nutlets very dark brown, irregularly-transverse rugose and more or less echinate-muricate. At Byron Springs, and near Bethany. March, April.

5. **A. echinata,** Gray. Erect, 1—2 ft. high, very hispid with white spreading bristles: leaves linear-lanceolate: sepals very narrow, yellow-hispid: corolla small and very slender: *nutlets almost prickly-muricate, the rugæ obsolete.*—Antioch and southward; perhaps here an immigrant from the Mohave district. March, April.

6. **A. tessellata,** Gray. Stout, rather loosely branching, very hispid: leaves oblong-lanceolate: spikes long and loose: calyx large and foliaceous, of 3 narrow sepals and 1 broad one, only loosely investing the nutlets and far surpassing them, bristly but hardly fulvous-hirsute: corolla small, orange-yellow, the *throat with folds: nutlets broadly ovate,* acute, not carinate but *flattish on the back, covered with warty granulations* running more or less distinctly into transverse ridges which, carried out to the edges, form a dentate border.—Plains of the lower San Joaquin and southward. March.

7. **A. collina.** Near the last, but of different habit, being slender and simple up to the few short dense terminal spikes: pubescence more hirsute and appressed: leaves narrowly linear-lanceolate, acute: calyx intensely fuscous, the sepals longer, less foliaceous: corolla with *no folds* in the throat: *nutlets* of ovate outline, flattened on the back, marked *with few and sharp interrupted transverse ridges, and intervening low tessellate granulations.*—Hills east of the Livermore Valley. March.

8. **A. grandiflora,** Kleeb. Stoutish, simple up to the short terminal spikes, hispid: lower leaves oblanceolate, upper lanceolate, all very acute, or even acuminate: sepals broad, often 4, or 3 only, very deeply fulvous-hirsute: *corolla 1 in. long, deep yellow;* stamens nearly sessile at the orifice of the very short proper tube, the funnelform-ampliate throat ½ in. long beyond this; limb more than ½ in. broad: *nutlets light gray, sharply triquetrous, perfectly smooth and shining,* the back concave rather than convex.—At Antioch, and on hills east of the Livermore Valley.

7. **LITHOSPERMUM,** *Diosc.* Herbs erect, with sessile leaves, and leafy-bracted flowers. Corolla salverform or funnelform. Stamens short, included. Nutlets ovoid, bony, sessile by the very base, the scar flat, not excavated.

1. **L. ARVENSE,** L. Annual, slightly canescent with minute appressed hairs, 1—2 ft. high: leaves linear or lanceolate, with prominent midrib:

corolla funnelform, ¼ in. long, whitish: the throat with puberulent lines: nutlets brownish, dull, coarsely wrinkled and pitted.—Occasional at San Francisco; native of Europe.

8. HELIOTROPIUM, *Theophr.* Genus differing from all others of this order, in our flora, by a corolla with plaited lobes, anthers connivent, and nutlets that are not seed-like in appearance, but resemble 4 separated closed cells of a capsular fruit.

1. **H. Curassavicum,** L. A very fleshy glabrous and glaucous depressed perennial: leaves obovate to much narrower, almost linear: spikes mostly in pairs, dense-flowered: corolla white, with yellow eye.— Common in low subsaline soils, chiefly in the interior.

ORDER LXXIII. CONVOLVULACEÆ.

Herbs or shrubs, with milky juice, the stems usually twining or trailing. Leaves alternate, petiolate, exstipulate. Peduncles axillary, 1-flowered, or cymosely several-flowered. Flowers regular, perfect, 5-merous. Sepals mostly distinct, persistent. Corolla mostly plaited and the plaits convolute. Stamens as many as the corolla-lobes and alternate with them. Ovary entire or lobed; usually maturing as a capsule with few and large seeds.

Stems twining or trailing; corolla plaited.............................CONVOLVULUS 1
  "   creeping; corolla not plaited.................................DICHONDRA 2
  "   erect or diffuse; corolla not plaited.........................CRESSA 3

1. CONVOLVULUS, *Pliny* (BINDWEED. MORNING GLORY). Corolla funnelform, plaited and the plaits dextrorsely convolute. Stamens not exserted. Style 1, cleft at apex; stigmas 2, linear to oblong or ovate. Capsule globose, thin-walled, 2-celled or imperfectly 4-celled, mostly 2—4-valved, with few and large seeds.

* *Species naturalized from Europe.*

1. C. PENTAPETALOIDES, L. *Annual,* slender, branching from the base, 6—15 in. high, pubescent: leaves spatulate-oblanceolate: peduncles 1-flowered, bibracteate toward the summit: *corolla small, purplish, deeply 5-lobed* the lobes ovate.—Common in fields along the eastern base of the Mt. Diablo Range. March, April.

2. C. ARVENSIS, L. *Perennial,* prostrate, the stems 1—3 ft. long: *leaves oblong-sagittate or hastate,* 1—2 in. long, the basal lobes short: pedicels 1—3-flowered, with a pair of subulate bracts near the base: *corolla white,* with a tinge of purple on the outside, *neither lobed nor angled.*—Very prevalent as a weed in fields and by waysides. May—Nov.

* * *Native species; calyx embraced by a pair of foliaceous bracts.*

3. **C. Soldanella,** L. Low, *glabrous, slightly succulent;* stems 10—15 in. long, prostrate: *leaves reniform, deep green and shining,* 1—2 in. long,

on stout petioles: corolla 1½ in. broad, pinkish: capsule 1-celled.—On sandy beaches only. May, June.

4. **C. subacaulis** (Hook. & Arn.). Fibrous-rooted low perennial; stem 1—18 in. long, erect, trailing, or in forms with longer stem somewhat twining: *leaves thin, sparingly pubescent,* oblong or ovoid or deltoid: with truncate or hastate base: bracts smallish, embracing but not enclosing the calyx: corolla campanulate-funnelform, angularly 5-lobed, 2 in. broad, cream-color, with purplish exterior.—Common and variable species of low hills along the seaboard.

5. **C. villosus** (Kell.), Gray. Stouter than the last, equally variable in size, form of leaves, etc., but stems more numerous, stouter, more leafy, and the whole herb *densely velvety-tomentose and white:* leaves rather sharply triangular-hastate as a rule, but variable: bracts narrow, but as long as the calyx: corolla smallish, funnelform, cream-color.— Mt. St. Helena, Mt. Diablo, and the higher inner range of hills generally.

6. **C. sepium,** L. Stems from a horizontal slender running rootstock, 2—3 ft. high, twining firmly: petioles, leaf-margins, etc., somewhat pubescent: leaves sagittate, very acute: bracts ovate-cordate, acute, large, completely enfolding the calyx: corolla pinkish, 2 in. long or more.—Plentiful in brackish marshes towards the mouth of Napa River and about Suisun Bay; its roots within reach of tide water; its stems twining upon rushes and sedges. June - Aug.

* * * *Native species; peduncles with small subulate bracts.*

7. **C. luteolus,** Gray. Shrubby, the herbaceous growing and flowering branches twining, the whole ascending shrubs and trees to the height of 20 ft. or more; the whole plant glabrous: leaves sagittate, 2 in. long: peduncles several-flowered, longer than the leaves; bracts linear-lanceolate, distant from the calyx about their own length: corolla rather open-funnelform, pale cream-color, 1½ in. long, the limb not lobed or angular. Var. **purpuratus.** Herbage distinctly glaucescent; corolla with broader shorter tube; color of limb from light to deep rose-purple. —Type very common on hills and along streams. The variety alone is found on islands in the Bay; occurring also on banks and ledges near the salt water on the mainland in Marin Co. May—Oct.

2. **DICHONDRA,** *Forster.* Prostrate creeping herbs, with round-reniform foliage, and small axillary flowers. Corolla campanulate, deeply 5-lobed. Ovary of 2 distinct carpels, each with a filiform style and capitate stigma, and maturing as an utricular 1-seeded fruit.

1. **D. repens,** Forst. Slender stems partly subterranean, rooting freely, when above ground pubescent: leaves glabrous, ½—1 in. wide, on long petioles: flowers short-peduncled: sepals obovate to spatulate,

obtuse, 1—2 lines long, rather exceeding the yellow corolla, and equalling the subglobose pubescent carpels.—Open hills about the Presidio, San Francisco, and above Mountain Lake.

3. CRESSA, Linn. Stems erect, branching from the base, leafy throughout, and bearing small white flowers in the upper axils. Calyx of 5 subequal sepals. Corolla with oblong tube and 5 oblong-ovate spreading lobes. Filaments filiform, exserted. Ovary 2-celled; capsule by abortion 1-seeded.

1. **C. cretica, L.** Slender stems and small oblong-ovate leaves canescently silky: fl. sessile or short-pedicelled in the upper axils, sometimes spicate-crowded as if in a leafy-bracted spike: white corolla 2—3 lines long, silky pubescent on the outside.—In hard subsaline clayey soils, usually near maritime salt marshes.

Order LXXIV. CUSCUTEÆ.

A single genus of parasitic herbs with twining filiform golden yellow leafless stems; floral organs much like those of some *Convolvulaceæ*.

1. CUSCUTA, Tragus. Flowers small, white, usually densely clustered. Calyx 5-cleft or -parted. Corolla campanulate, or subcylindric with spreading limb. Stamens usually with a fringed appendage below their insertion in the throat. Ovary globose, 2-celled, 4-ovuled. Styles in all our species 2. Capsule 1—4-seeded, dry and circumscissile or baccate. Embryo filiform, spirally coiled in fleshy albumen, destitute of cotyledons.

1. **C. Californica, Choisy.** Flowers 1—2 lines long, pedicelled in loose cymes: calyx-lobes acute: corolla-lobes lanceolate-subulate, as long as or longer than the shallow campanulate tube: *filaments nearly equalling the linear-oblong anthers;* appendages none, or represented by rudimentary inverted arches near the base of the tube: styles slender: *capsule depressed-globose.*—Attributed to the western part of California generally, and running into some marked varieties not known to us.

2. **C. salina, Engelm.** Stems very slender, densely clothing stems of *Salicornia* and other salt-marsh herbs: fl. 1½—2½ lines long, pedicelled in loose cymes: calyx-lobes ovate-lanceolate, acute, as long as the *denticulate corolla-lobes:* corolla-tube shallow-campanulate: *filaments about as long as the oval anthers;* fringed scales shorter than the corolla-tube, sometimes incomplete: styles as long as, or shorter than the acute ovary: capsule conical, surrounded by the withered corolla, usually 1-seeded.—Plentiful in salt marshes all around the Bay.

3. **C. subinclusa, Dur. & Hilg.** Stems coarse and few, ascending small shrubs to the height of a yard or more: fl. sessile or short-pedicelled, at length in clusters ½—1 in. thick, the individual fl. 2½—3½

lines long: calyx-lobes ovate-lanceolate, acutish, overlapping, much shorter than the *cylindric or urn-shaped corolla-tube:* corolla-lobes much shorter than the tube, ovate-lanceolate, acute, minutely crenulate or papillose: *anthers oval, subsessile:* scales narrow fringed, reaching to the middle of the tube: slender styles longer than the pointed ovary: capsule conical, capped by the withered corolla, usually 1-seeded.—Common in all the hilly districts, usually on shrubs or coarse herbs.

ORDER LXXV. S O L A N A C E Æ .

Herbs or shrubs, commonly rank-scented and with colorless narcotic juice. Leaves alternate, exstipulate. Flowers regular, 5-merous, 5-androus, their pedicels bractless. Corolla usually plaited in æstivation. Ovary 2-celled, with axial placentæ; style single. Fruit capsular or baccate, many-seeded. Seeds with curved embryo in fleshy albumen.

Corolla rotate or campanulate; fruit a berry.................................SOLANUM 1
Corolla elongated; fruit a capsule;
   Pericarp thick, prickly...DATURA 2
   Pericarp thin and smooth;
      Calyx merely toothed or lobed...............................NICOTIANA 3
      " 5-parted; segments foliaceous.............................PETUNIA 4

1. **SOLANUM,** *Pliny.* Calyx and corolla 5-parted or -cleft, the latter rotate or nearly so, and valvate in the bud. Filaments short: anthers distinct but connivent; the cells with a terminal opening. Style long: stigma entire. Fruit a berry, containing many flattened seeds.

1. S. VILLOSUM, Lam. Annual, stoutish, depressed, the branches 1 ft. long, somewhat flexuous, and with one or more slight angles, the whole plant villous-hirsute: leaves rhombic-ovate, 1 in. long or more, strongly sinuate-dentate: corolla white, minute: *berries of a. clear deep green* when ripe, and *half invested by the shallow-campanulate calyx.*—In fields and gardens.

2. S. ALATUM, Mœnch. Similar to the last, but with angular stem and *red berries.*—Said to have been found about San Francisco Bay; but we doubt the identification.

3. **S. Douglasii,** Dunal. Somewhat shrubby, widely branching, and even half climbing by the rough angles of the branchlets, 3—5 ft. high: leaves ovate with cuneate base, 1—2 in. long, nearly entire: *corolla bluish, 3—5 lines wide: berries black.*—Plentiful along stream banks and elsewhere in half shady places; flowering and fruiting in many instances almost throughout the year.

4. **S. umbelliferum,** Esch. Stoutish cinereous and tomentose-pubescent shrub, 3—4 ft. high: leaves thin, ovate or obovate, or oblong, obtuse entire, acute or cuneate at base, 1—2 in. long: umbels short-peduncled, few-flowered: *corolla about ¾ in. broad, deep blue, with greenish spots at*

base: berry 4—5 lines in diameter, *yellow* when ripe.—Common in moist thickets, or on more exposed bushy hillsides; flowering at all seasons: the leaves rarely somewhat pinnate.

2. DATURA, *Linn.* Coarse herbs of rank odor and narcotic-poisonous properties. Flowers large, solitary. Calyx tubular, deciduous by a transverse separation near the base. Corolla funnelform, strongly 5-plaited. Capsule thick, prickly.

1. **D. meteloides,** DC. Perennial; low and spreading stout flexuous branches often a yard long or more: leaves ovate or ovate-lanceolate, 5 —10 in. long, usually entire, pale with a soft whitish pubescence: calyx 3, and corolla 6—8 in. long, the latter pale blue or violet, the widely expanded limb with prominent slender-subulate long points: capsule nodding, 2 in. thick, subglobose, thickly armed with equal short weak prickles, bursting irregularly when ripe: seeds with thick margin.—Low sandy plains of the interior. July—Nov.

3. NICOTIANA, *Dalechamps* (TOBACCO). Strong-scented narcotic-poisonous herbs or shrubs, with entire leaves, and panicled narrowly funnelform flowers. Calyx oblong, 5-toothed or -lobed. Corolla-limb plaited and plaits convolute. Stamens not exserted: anthers short, opening lengthwise. Capsule many-seeded, 2-valved from the summit, the valves afterwards splitting into two.

1. **N. attenuata,** Torr. Viscid-pubescent, erect, 1—3 ft. high, *leaves* oblong-lanceolate, *acuminate, attenuate into a petiole:* fl. in loose terminal racemes: *calyx* ¼ in. long tubular-campanulate: *teeth short, equal,* triangular, acute: corolla white, 1 in. long or more, narrow-salverform, with short-lobed border ½ in. in diameter: capsule ovate, acute, exceeding the calyx.—Not known to occur quite within our limits; but found at Monterey, and at Stockton.

2. **N. Bigelovii,** Wats. Size of the last, and with the same clamminess: *leaves sessile: calyx with unequal linear-lanceolate lobes:* corolla tubular-funnelform, with deeply lobed limb 1 in. broad: the lobes broad-ovate, acute: capsule obtuse, shorter than the calyx.—Of the mountains chiefly, and originally; but now found at San Rafael and San Francisco, along the railroads and other public highways.

3. N. GLAUCA, Graham. Slender loosely branching soft-woody tree 12—20 ft. high; branches and foliage glabrous, glaucous: leaves ovate to oblong-ovate, subcordate, entire, long-petioled, rather thick and sub-coriaceous: fl. in loose terminal panicles; corolla 2 in. long, greenish-yellow, tubular, with constricted throat, and erect 5-crenate limb.— Native of Buenos Ayres; naturalized from Napa Valley and plains of the lower Sacramento southward.

4. PETUNIA, *Jussieu.* Viscid-pubescent herbs, with lateral rather than terminal flowers. Calyx 5-parted, the divisions foliaceous. Capsule 2-valved; valves entire.

1. **P. parviflora, Juss.** Small rather fleshy prostrate annual, freely branching, with narrow-spatulate ½ in. long leaves nearly sessile: fl. short-peduncled, 4 lines long: corolla purple, with a yellowish tube, its short retuse lobes slightly unequal: capsule ovoid.—Low moist lands at Alameda, thence eastward and southward.

Order LXXVI. SCROPHULARINEÆ.

Herbs or shrubs with a watery juice, not rarely a heavy narcotic odor. Leaves prevailingly opposite, though often alternate exstipulate. Flowers complete, mostly very irregular; the corolla usually bilabiate, imbricate in bud. Stamens mostly 2 or 4, with a rudimentary fifth. Style 1, undivided; stigma not rarely 2-lobed. Ovary 2-celled, the placentæ firmly united in the axis (except in *Diplacus*). Seeds ∞, small, often angular, occasionally winged.

1. VERBASCUM, *Pliny* (MULLEIN). Coarse biennials, the cauline leaves sessile, often decurrent; the spicate or racemose flowers ephemeral. Calyx 5-parted. Corolla nearly rotate. Stamens 5; anthers by confluence 1-celled; superior filaments woolly-bearded. Capsule 2-valved, ∞-seeded. Seeds rugose.—Our two species naturalized from Europe.

1. **V. THAPSUS, L.** *Densely woolly throughout:* stem simple, 3–6 ft. high, winged by the decurrent bases of the oblong entire crowded leaves: fl. yellow, in a dense spike a foot long or more, and 1–2 in. thick.— Upper part of Napa Valley, and elsewhere in mountainous districts.

2. **V. VIRGATUM,** Withering. *Green, but somewhat pubescent and glandular, rather slender,* 3–6 ft. high: leaves oblong, obtuse, crenate or sinuate, not decurrent: fl. yellow, in a long loose raceme; pedicels often in twos and threes and not longer than the calyx-lobes.—Not rare in fields and on roadsides.

2. LINARIA, *Brunfels.* Inflorescence simple, racemose. Corolla personate, spurred at base. Capsule many-seeded, opening by irregular perforations.

1. **L. Canadensis** (L.), Dumont. Glabrous annual, leafy chiefly as to the procumbent shoots at base of main stem, and the leaves of these opposite or whorled, oblong, 1–2 lines wide: stem 6–30 in. high, nearly naked, racemose at summit: pedicels erect, not longer than the filiform curved spur of the blue corolla.—Sandy soils.

3. ANTIRRHINUM, *Diosc.* Ours either glabrous, or glandular and oily-viscid herbs, with axillary or terminal and spicate-racemose flowers, their structure like those of *Linaria*, except that the corolla has a sac-like gibbosity or protuberance at base instead of a spur.

* *Stout perennials, with spicate-racemose flowers.*

1. **A. virga,** Gray. *Glabrous; stems strict and simple,* 3–5 ft. high: *leaves linear,* diminishing under the long spike-like raceme to subulate or setaceous bracts: purple flowers mostly secund, ¾ in. long: longer filaments with tips dilated to more than the width of the anthers.— Sonoma Co. and northward, among the higher hills. May, June.

2. **A. glandulosum,** Lindl. Size of the last, but more branching, very *glandular-pubescent and viscid* throughout: *leaves lanceolate:* corolla ¾ in. long. purplish, with yellow palate: capsule tipped with long and persistent style.—Mt. Hamilton and southward. June, July.

* * *Annuals, more or less slender and climbing.*

3. **A. vagans,** Gray. Diffusely branched, in age more or less *climbing by prehensile branchlets; sparsely setose-hirsute and more or less glandular* and viscid: leaves thickish, oblong-ovate to lanceolate, entire: oblong upper sepal equalling the tube of the corolla, the others linear and shorter: corolla light purple, ½ in. long: seeds tuberculate.—Very common in ravines among the hills and mountains. June—Oct.

4. **A. strictum** (H. & A.), Gray. Erect, very slender, nearly simple, 2 ft. high, glabrous, *climbing by the long filiform peduncles:* lowest leaves ovate-lanceolate; upper linear, the floral filiform, much shorter than the tortile peduncles: corolla violet, ½ in. long, with hairy palate and gibbous base; capsule crustaceous, tipped with a straight style of its own length.—Mt. Tamalpais and far southward. June—Aug.

4. **COLLINSIA,** *Nutt.* Annuals with opposite leaves, the lowest pairs of which are commonly ternately divided, the others merely toothed or entire. Flowers pedicellate, axillary and scattered, or in whorls forming a raceme. Calyx campanulate, deeply cleft. Corolla with very short proper tube, ventricose and gibbous or saccate throat, and bilabiate usually somewhat personate limb; the 2 lobes of the upper lip more or less recurved; middle lobe of the lower usually conduplicate, enclosing the stamens; these 4 in 2 pairs with long filaments; a gland at the base of the corolla on the upper side answering to the fifth stamen. Capsule ovate or globose; the rather few seeds somewhat peltate or meniscoid.

* *Corolla strongly bilabiate; the lowest lobe conduplicate.*

←*Flowers short-pedicelled, racemose at summit of stem.*

1. `C. tinctoria,` Hartw. Stoutish, 1 ft. high or less, viscid-hairy or glabrate: leaves more or less toothed, oblong or lanceolate, the upper sessile by a broad subcordate base: fl. nearly sessile; calyx-lobes linear or oblong-linear, obtuse: corolla yellowish or nearly white, marked with many purple lines and dots, or the purple prevailing; *throat so strongly saccate-ventricose* that its axis is at right angles with tube and limb; *upper lip very short.*—Mt. Diablo and northward through the higher hills; herbage imparting a stain. June.

2. **C. bicolor,** Benth. More slender, often 2 ft. high, usually glabrous or nearly so, seldom a little viscid and hairy: lowest leaves ternately compound, those seen at flowering all oblong-lanceolate: pedicels shorter than the acute calyx lobes: *lower lip of corolla purple, the upper little shorter,* paler or nearly white; saccate throat oblique to the tube.—Very common on open or shady hillsides, or, in an almost white-flowered smaller form, on open sandy plains. April—June.

3. **C. bartsiæfolia**, Benth. Puberulent and somewhat glandular, the calyx usually white-villous; stems rather stout, simple or branched, 6—10 in. high: leaves rather broader and more toothed than in the foregoing: *flower-whorls few: upper lip of corolla about the length of the curved-gibbous throat*, the whole nearly white; gland-like rudimentary stamen sessile, elongated, porrect.—Sandy soil along the seaboard; plentiful at San Francisco. April, May.

4. **C. Greenei**, Gray. Glandular-puberulent, slender, 4—8 in. high: leaves oblong-linear, tapering to the base, coarsely and sparsely toothed: fl. only 2—6 in the whorl. the pedicels as long as the calyx; lobes of calyx acutish: corolla rather slender, deep violet-purple; *upper lip short, crested near the base within* with a pair of callous teeth on each side, which are connected by a transverse ridge; lateral lobes of the lower lip small: gland small, sessile.—Rocky ledges along streams in the higher mountains of Sonoma Co., toward the Geysers. May, June.

+ + *Pedicels longer and fewer; the flowers solitary, or umbellate-whorled.*

5. **C. Franciscana**, Bioletti. *Slender and with thinnish foliage*, ½—2 ft. high, puberulent above, otherwise glabrous: leaves ovate or ovate-lanceolate, the upper sessile: *pedicels 1—6 in the axils of the uppermost leaves* or bracts, from slightly shorter to twice or thrice longer than the calyx: calyx-lobes acute: corolla ¾ in. long, the limb ½ in. broad, bluish, the upper lip pale, dotted with purple; throat a fourth longer than wide, closed at the mouth: gland subulate, bearing the yellowish rudiment of an anther.—San Francisco and San Mateo counties, on shady hillsides. May, June.

6. **C. arvensis**, Greene. Erect, simple or with several nearly erect branches from the base, 10—18 in. high, *glabrous except the* very sparsely setulose-hairy *leaf-margins:* lowest leaves oval or oblong, ½ in. long, on petioles of equal length, coarsely toothed or somewhat lobed; cauline lanceolate to linear, sessile, revolute: *fl. loosely racemose* (1, 2 or rarely 3 at each upper node), *deep violet-purple, ¾ inch long:* calyx-teeth lanceolate-subulate, twice the length of the tube; corolla with compressed saccate tube as broad as long (¼ in.); upper lip half the length of the lower, and paler: filaments very sparsely hirsute below.—Hills and open valleys of the Coast Range; plentiful in grain fields. April, May.

7. **C. sparsiflora**, Fisch. & Mey. *Puberulent throughout*, and the herbage reddish, 4—8 in. high, with few ascending branches: leaves mostly lanceolate, entire, the very lowest small, rounded, toothed and petiolate: pedicels usually 1 or 2 at node, three or four times the length of the calyx, the lobes of the latter lanceolate, twice as long as the tube: *corolla very small, hardly exceeding the calyx*, pale or deep violet-purple, with dots of deeper color.—Northward slopes and summits of the coast hills. April, May.

* Corolla bilabiate, but all the lobes, even the lowest, spreading.

8. **C. tenella,** Benth. Stems almost filiform, 5—10 in. long: lowest leaves round-ovate, toothed; middle cauline 3-lobed or -divided; uppermost and floral linear, entire: filiform pedicels solitary or in pairs at the nodes, longer than the leaves: corolla minute, little exceeding the calyx; its 5 lobes of nearly equal length, but the anterior one transversely oval, the other 4 oblong: capsule exceeding the calyx: cells 1-seeded.—Santa Clara and Sonoma counties, and far northward. April.

5. **SCROPHULARIA,** *Tourn.* Homely weed-like perennials, with opposite leaves, and loose cymes of small dull-reddish flowers in a narrow terminal panicle. Calyx deeply 5-cleft; lobes broad, rounded. Corolla short; tube ventricose; limb of 5 unequal lobes, 4 erect, the fifth and lowest recurved. Stamens 4, in 2 pairs; anthers transverse and by confluence 1-celled: a scale like rudimentary fifth stamen borne on the upper side of the throat of the corolla. Capsule ovate, septicidal, many-seeded. Seeds tuberculate-rugose.

1. **S. Californica,** Cham. Almost glabrous, dark green, or the stem purplish, 2—5 ft. high: leaves oblong-ovate with cordate base, coarsely and doubly toothed: inflorescence somewhat glandular-pubescent and viscid: branchlets of the panicle rather few-flowered and distinctly cymose. Var. **floribunda.** More slender; the herbage light green; leaves more sharply deeply and incisely toothed: branches of the panicle greatly elongated and remarkably flexuous as well as very many-flowered. —Type common at the seaboard; the variety belonging to hills of the interior.

6. **PENTSTEMON,** *Mitchell.* Herbaceous or shrubby, with opposite or verticillate leaves, and racemose-panicled mostly showy flowers; the peduncles from the axils of floral leaves or bracts. Calyx 5-parted. Corolla with ventricose usually elongated tube; limb more or less bilabiate; upper lip 2-lobed; lower 3-lobed. Stamens 4, declined at base, ascending above; anthers with cells mostly united or confluent at summit; the fifth stamen a conspicuous often more or less dilated and hairy filament. Capsule septicidal, ∞-seeded. Seeds angled.

* Anther cells divergent, or even divaricate.

1. **P. centranthifolius,** Benth. Herbaceous, *glaucous,* the erect and strict stems, 2—4 ft. high: cauline leaves ovate-lanceolate, sessile by a broadish clasping base; panicle narrow, 1 ft. long or more: *corolla narrow-tubular,* 1 in. long, obscurely bilabiate, *bright vermilion-red;* the short-oblong lobes alike save that the two posterior are united higher: anthers opening widely; sterile filament slender, naked.—Near Vacaville and Niles, and southward. May, June.

2. **P. Sonomensis**, Greene. Suffrutescent, evergreen, very leafy, 5—10 in. high, *slightly puberulent: leaves rather light green, coriaceous,* denticulate, ½—¾ in. long and nearly as broad, short-petiolate, the *uppermost round-ovate, acutish;* lower nearly orbicular and retuse: raceme terminal, sessile: calyx-lobes lanceolate, acuminate: corolla 1 in. long, deep red; segments nearly equal, not widely spreading: anthers slightly exserted: white-woolly.—Among rocks at the summit of Hood's Peak, Sonoma Co., and on Mt. St. Helena.

3. **P. corymbosus**, Benth. Suffrutescent, much branched, ascending 1—2 ft. high, *cinereous-pubescent:* branches leafy up to the corymbiform cyme: *leaves oblong or oval,* denticulate, ½—2 in. long: sepals lanceolate: corolla scarlet, narrow-tubular 1 in. long with abruptly spreading limb: anthers glabrous.—Summits of Mt. Hamilton, Mt. Diablo, and northward; growing on rocks. July, Aug.

4. **P. Lemmoni,** Gray. Slender shrub 2—5 ft. high, *bright green, glabrous:* leaves ovate-lanceolate, 1—1½ in. long, sharply though remotely serrate-toothed: panicle loose, virgate and racemose: peduncles longer than the subtending floral leaves, cymosely 2—7-flowered: *corolla small, dull yellowish and red,* with short tube and small unequal segments: sterile filaments strongly yellow-bearded on one side of the curved apex. —Mountains of Solano Co.

* * *Anthers horseshoe-shaped or reniform, ciliate.*

5. **P. heterophyllus,** Lindl. Tufted and bushy but only 1—1½ ft. high and somewhat woody at base: herbage glabrous, glaucescent, the stem commonly reddish: leaves linear-lanceolate and linear, subcoriaceous, entire: peduncles short, erect, 1—3-flowered: sepals lanceolate, acuminate: corolla glabrous, 1½ in. long, with ample ventricose tube and rather short bilabiate limb, the whole either red-purple or bright blue.—On the higher parts of both Coast and Mt. Diablo ranges.

7. **DIPLACUS,** *Nutt.* Evergreen glutinous low shrubs with opposite mostly serrate-toothed leaves revolute in vernation. Flowers showy, pedicellate, solitary in the axils. Calyx 5-angled, 5-toothed, persistent, finally ruptured by the splitting of the 2-valved subligneous capsule. Corolla with funnelform tube and bilabiate limb, all the segments spreading; the lower larger than the upper. Capsule linear-oblong, with a tubercular enlargement at the pointed apex at base of the style, firm-coriaceous, dehiscent by the upper suture, each valve bearing its own placenta which is lamellar and conceals the many minute seeds.

1. **D. glutinosus** (Wendl.), Nutt. Shrub 3—6 ft. high; branches puberulent: leaves linear-lanceolate, acutish, denticulate, in age loosely revolute, pubescent beneath with branching hairs: corolla 1 in. long, buff or salmon color; throat narrow-funnelform; lobes emarginate or

retuse and with other scarcely definable irregularities of margin.—Banks and hillsides, usually near streams. May—Dec.

8. EUNANUS, *Benth.* Annuals, either dwarf and depressed, or erect and somewhat strict. Herbage glandular-hairy and somewhat viscid. Flowers axillary and solitary. Calyx prismatic, 5-angled, 5-toothed. Corolla from slender funnelform and strongly bilabiate to almost salver-form and with slight inequality of lobes. Stamens 4, in unequal pairs. Style filiform; stigma bilamellar or somewhat peltate-funnelform. Capsule from cartilaginous to almost membranaceous, but dehiscent on one side only, or from that to a little below the apex on the other side, obtuse and shorter than the calyx, or acuminate and surpassing it. Seeds many, often muriculate.

* *Subacaulescent dwarfs, with greatly elongated corolla-tube: capsule short, cartilaginous, nearly or quite indehiscent.*

1. **E. subuniflorus** (Hook. & Arn.). Stem ¼—1½ in. high, very leafy: leaves rhombic-ovate to ovate-lanceolate, ½—1 in. long, entire or crenate-toothed: corolla dark red-purple, 1—1¼ in. long, its slender tube thrice the length of the calyx; *throat oblong-urceolate; upper lip* of limb consisting *of two broad erect lobes, the lower of a single small triangular tooth* (the lateral lobes obsolete): capsule ½ in. long, semi-translucent, obtuse, very gibbous, the acutely carinate posterior part twice as long as the obtuse anterior part: seeds ovoid, acute, granular-muriculate.— Damp hillsides and plains. March, April

2. **E. angustatus,** Greene. Nearly or quite acaulescent: leaves linear and linear-spatulate, 1 in. long, entire: calyx with constricted throat and ample foliaceous segments: corolla with *filiform tube at least 4 times the length of the funnelform throat: upper lip exceeding the lower:* capsule broadly ovate, acute, scarcely compressed: seeds few, large, favose-pitted. —Napa and Sonoma counties: an early vernal species like the last.

3. **E. tricolor** (Lindl.), Greene. Low, but with short decumbent leafy branches 1—3 in. long: leaves oblong-lanceolate, entire, or remotely serrate-toothed, 1—2 in. long: *corolla 2 in. long, with short narrow tube, ample throat and wide not very irregular limb,* the 3-lobed lower lip about as long as the 2-lobed upper, the whole corolla rose-purple, with markings of deep crimson and of yellow in the throat: capsule oblong-ovate, obtuse, slightly gibbous, compressed, both edges acute: seeds obovoid. —Plains of Solano and Contra Costa counties east of the Mt. Diablo Range. April, May.

* * *Less diminutive plants, and erect: capsules more elongated, thinner, definitely dehiscent.*

4. **E. Douglasii,** Benth. Rigidly *erect, 6—10 in. high,* the internodes longer than the leaves; these ovate to obovate and spatulate-lanceolate:

corolla dark red-purple; tube 1 in. long, widening into a short-funnelform throat and bilabiate limb; *lobes of the lower lip only half as long as the upper*, and more spreading: capsule 4—5 lines long, narrow, obtuse, subterete, unequal-sided: seeds few, large, obovate, minutely granular. Var. **parviflorus.** All parts of the plant thrice smaller; the corolla scarcely ½ in. long, its tube little exserted.—The type quite common on hillsides near the coast; the variety in the Vaca Mountains. April, May.

5. **E. Bolanderi** (Gray), Greene. Glandular-pubescent and viscid, ½—2 ft. high, simple or branched: leaves ovate and oblong 1—2 in. long: calyx-teeth unequal, acuminate: corolla red-purple, ½—1 in. long, *the tube rather abruptly widening to the not very irregular spreading limb: capsule fusiform-subulate*, as long as the calyx.—Mt. Hamilton, Mt. Diablo, etc. June—Aug.

9. **MIMULUS,** *Linn.* Herbaceous light green and flaccid plants, glabrous or slightly pubescent, some albuminous-viscid, others musk-scented, none resinous-glandular. Flowers peduncled, axillary and solitary, or becoming racemose by reduction of the upper leaves to bracts. Calyx 5-angled, commonly short and bilabiate, the uppermost lobe largest. Corolla in most of ours yellow, often personate, seldom otherwise than strongly bilabiate. Stamens 4; stigma bilamellar. Capsule obtuse, enclosed within the calyx; the membranaceous walls (apparently not valvate) tardily separating from the central and conjoined placentæ.

1. **M. cardinalis,** Dougl. Stout, viscid-pubescent, rather strong-scented when bruised, 2—4 ft. high: leaves ovate and ovate-lanceolate, erose-dentate, parallel-veined, sessile, 2 in. long: *corolla scarlet, 2 in. long, very irregular;* 3-lobed lower lip reflexed; upper less deeply divided, the undivided part erect, the lobes reflexed: stamens exserted.—Mountain streams, and on wet rocky hillsides. June—Sept.

2. **M. moschatus,** Dougl. Perennial, slender, *soft-villous, slimy, musk-scented:* stems 1 ft. long, decumbent, rooting at the joints: leaves oblong-ovate, 1 in. long, petiolate; corolla light yellow, ½ in. long. Var. **sessilifolius,** Gray. Leaves 2 in. long, sessile by a broad base: corolla 1 in. long, the ample spreading limb nearly regular: herbage only faintly or not at all musk-scented.—Swampy places in the Coast Range; probably the variety only, with us. May—Sept.

3. **M. floribundus,** Dougl. *Annual,* diffuse, slender, villous, *very slimy,* 6—10 in. high, leafy and flowering from the axils throughout: leaves ovate, 1 in. long or less, the upper shorter than the pedicels: calyx short-campanulate, ovate or oblong in fruit, and with nearly equal short triangular teeth: *corolla ¼—½ in. long, only twice the length of the calyx,* light yellow: capsule globose-ovate, obtuse.—Common in the mountains far north and south; not known in those about the Bay, but found in Lake Co., and to be expected in Sonoma and Marin.

4. **M. guttatus,** DC. An Alaskan species, whose type is in the Sierra Nevada, and eastward. We have but some remarkable varieties, or subspecies. Var. **grandis.** *Perennial, stout and fistulous, 2—5 ft. high;* stems terete, mostly simple above the decumbent and somewhat proliferous base: leaves orbicular to round-ovate, 1—3 in. long, the radical petiolate and sometimes lyrate by a few pinnæ along the petiole; floral reduced and connate-clasping: raceme pubescent, often faintly musky, not rarely 1½ ft. long: peduncles slender, 1 in. long or more: calyx ventricose-campanulate; teeth broadly triangular, the upper one largest: corolla 1—1½ in. long, strongly bilabiate, light yellow, the throat with many small red dots. Var. **insignis.** Apparently *annual,* simple or branching, 1—2 ft. high: calyx dotted with red, and *large corolla with very large dark red spots* on the palate and limb.—The first variety a conspicuous perennial of stream banks and some boggy places among the hills near the Bay. The second is of Napa and Sonoma counties, and the most beautiful *Mimulus* in our flora. April—July.

5. **M. glareosus,** Greene. Annual, slender, diffuse, the branches 10—12 in. long, these and the leaves *retrorsely pubescent and slimy:* leaves round-ovate, ¾ in. long, on slender petioles, irregularly toothed and slightly lobed, usually with a band of purple spots across the base: *pedicels very slender,* far exceeding the leaves: *calyx-teeth very unequal,* the uppermost one much the largest: corolla ½—¾ in. long, strongly bilabiate, yellow, with many purple dots.—Gravelly margins of mountain streams about Mt. St. Helena. July—Sept.

6. **M. arvensis,** Greene. Annual, erect, 1—2½ ft. high, stem somewhat 4-angled, sparingly leafy, loosely racemose from the middle: lower leaves long-petioled, roundish, coarsely toothed, hastate, and the whole becoming lyrate by the accession of several pairs of small leaflets below the main blade: *floral leaves soft-villous beneath, the plant otherwise glabrous: calyx 3—4 lines long, campanulate,* purple-dotted, the orifice commonly *almost truncate* and toothed rather than lobed and unequal, in fr. twice as large: capsule 2—3 lines long, compressed.—Plentiful in low fields. April—June.

7. **M. nasutus,** Greene. Annual, erect, ½—1½ ft. high, puberulent or nearly glabrous: *stem quadrangular and winged,* flowering from the base: leaves ovate to reniform-cordate, acute, coarsely toothed or lobed; the upper floral reduced to bracts; peduncles ascending, short: calyx broad-campanulate, the teeth or lobes very unequal; the upper one in maturity thrice the length of the others; the lower pair bent upwards and lying across and enfolding the other three: *corolla* comparatively small (½ in. long), *little surpassing the calyx, yellow, with a large dark-red blotch* on the lower lip.—Common about springy places in the hills almost everywhere, but especially in Marin Co. April, May.

8. **M. latidens.** Annual, *slender, usually much branched from the base*, 3—10 in. high, glandular-puberulent: leaves ovate to ovate-lanceolate, ½—1 in. long, 3—5-nerved, entire or sparingly toothed, short-petioled or sessile: pedicels as long as the fl., in fr. much longer: calyx oval, 3—6 lines long; teeth triangular-ovate: *corolla 3—5 lines long, nearly regular, but limb very small, white;* throat yellowish —Rich low plains of the lower Sacramento and southward. April, May.

10. **MIMETANTHE,** *Greene.* Villous and glandular ill-scented herb, with corolla of *Mimulus*; but calyx 5-cleft and slightly 5-sulcate, not angled. Style glabrous; stigma bilamellar. Capsule oblong-ovate, acuminate, chartaceous, dehiscent by the upper suture, and to just past the apex on the lower, the partial valves thus formed being reflexed. Seeds many, minute, favose-reticulate and glandular.

1. **M. pilosa** (Benth.), Greene. Annual, 6—15 in. high, much branched from the base, erect, leafy, villous with long viscid hairs, and with the odor of narcotic solanaceous plants: leaves lanceolate to oblong-ovate, entire, sessile: calyx slender-pedicelled, oblique, the upper segment longest and about equalling the tube: corolla light yellow, 3—4 lines long, little exceeding the calyx.—Gravelly banks and shores of mountain streams. June—Oct.

11. **HERPESTIS,** *Gærtn. f.* Ours a large herb creeping in mud, or a floating aquatic. Leaves opposite, entire. Flowers axillary. Calyx of 5 distinct very unequal sepals. Corolla campanulate. Stamens 4, in slightly unequal pairs; anther-cells parallel. Capsule thin, 2-valved, the valves 2-parted. Seeds many, elongated.

1. **H. Eiseni** (Kell.). Branches prostrate, rooting, ½—2 ft. long, stoutish and succulent: leaves oblong-obovate, 1 in. long, sessile, entire, several-nerved from the base: pedicels 1 or 2 in the axils, about as long as the leaves, pubescent: 3 outer sepals broad, the others narrow: corolla white, campanulate, 4—5 lines wide and as long, with a barely perceptible irregularity, the 2 upper lobes a trifle smaller and nearer together: anthers cordate-ovate, deep blue: seeds linear-oblong.—Stagnant pools along the lower San Joaquin and southward; certainly altogether distinct from *H. rotundifolia* to which it has been referred.

12. **GRATIOLA,** *Dodoens.* Low branching annuals, with opposite sessile leaves, and small solitary flowers on naked axillary peduncles. Calyx 5-parted; the divisions subequal. Small yellowish or white corolla with upper lip entire or 2-lobed; lower 3-cleft; all the lobes spreading. Stamens not exserted, only 2 fertile; their anthers with 2 transverse cells on a broad connective; anterior pair either wanting, or represented by sterile filaments. Style usually bent at tip; stigma of 2 flat lobes. Capsule many-seeded, 4-valved, the valves parting from the thick axial placenta.

1. **G. ebracteata,** Benth. Glabrous, obscurely viscid, erect, 2—5 in. high: leaves lanceolate, acute, mostly entire: calyx with no bractlets at base, the segments almost equalling the small yellow-throated white corolla: capsule globose.—Low wet places in fields; Sonoma Co., *Bioletti.*

13. **ILYSANTHES,** *Raf.* Slender glabrous low annuals, with opposite sessile leaves and filiform naked peduncles in their axils, the upper becoming racemose by the reduction of the subtending leaves to bracts. Calyx 5-parted; the divisions subequal. Upper lip of corolla short, 2-lobed, erect; lower larger, 3-cleft, spreading. Only 2 stamens fertile, these included, their anthers 2-celled; sterile anterior pair inserted high in the throat of the corolla, each of an . unequally 2-lobed filament; shorter lobe small, tooth-like, the longer glandular. Many-seeded small capsule 2-valved; edges of the valves separating from the partitions, these being left with the undivided placenta.

1. **I. gratioloides** (L.), Benth. Diffusely branching, 3—7 in. high: leaves ovate or oblong, sparingly toothed or entire: peduncles surpassing the leaves, in fruit divergent: corolla light purple or bluish, 3—4 lines long.—Muddy banks of the lower San Joaquin. July—Oct.

14. **PLANTAGINELLA,** *Ruppius.* Glabrous dwarf annual with running stems putting forth at their ends clusters of narrow entire fleshy leaves; 1-flowered scapes. Calyx and nearly regular corolla campanulate, 5-toothed and -cleft. Stamens 4, nearly equal; anthers with confluent cells. Style short, clavate; stigma thickish. Capsule globose, many-seeded, 2-valved; valves separating from the thin partitions; placentæ remaining central.

1. **P. aquatica** (L.), Mœnch. Leaves with petioles longer (when growing in water much longer) than the spatulate-oblong or oval blade, and exceeding the scapes, but the whole leaf usually only 1—3 in. long: corolla very small, white or purplish.—Margins of fresh water ponds and lakes on the San Francisco peninsula; usually terrestrial only.

15. **VERONICA,** *Fuchs.* (SPEEDWELL). Herbs, with at least the lowest leaves opposite. Flowers small, white or blue, in racemes, or solitary in the axils. Calyx and nearly rotate corolla each 4-parted; lower lobe of the latter, and sometimes the lateral ones, narrower than the upper. Stamens 2, one on each side of the uppermost corolla-lobe, exserted; anther-cells confluent at apex. Stigma somewhat capitate. Capsule compressed, 2-lobed, or at least emarginate, few—many-seeded; dehiscence loculicidal.

1. **V. Americana,** Schwein. *Perennial, glabrous,* the decumbent stems ½—2 ft. long; herbage rather fleshy: leaves oblong. serrate, truncate or slightly cordate at base, short-petioled: *racemes opposite,* slender peduncled, many-flowered: pedicels slender: *corolla blue,* with purple stripes: capsule turgid, many-seeded.—Shallow margins of streamlets, growing in sand or gravel.

2. **V. peregrina,** L. *Annual,* 3 –10 in. high, erect, sparingly branching, only the lower leaves opposite, linear-oblong, obtuse, entire or toothed; the upper gradually smaller: *fl. short-pedicelled in the axils: corolla very small, white:* capsule round-obcordate, many-seeded.—Common in wet fields. April, May.

3. V. ARVENSIS, L. Annual, simple and erect, or with few decumbent branches, 3 –6 in. high, pubescent: leaves ovate, crenate-serrate, the lowest petiolate, the bract-like floral ones linear-lanceolate: *spike-like raceme at length loose: corolla very small, deep blue:* capsule much shorter than the sepals, obcordate, the short style not exceeding the notch.—European weed, escaping from lawns and gardens.

4. V. BUXBAUMII, Ten. With many weak assurgent branches from the base; these 6—10 in. long: herbage somewhat pubescent, the hairs not gland-tipped: leaves round-ovate, petiolate (the floral alternate), coarsely serrate or crenate-serrate, ½ in long or less: *peduncles nearly 1 in. long: corolla bright blue, 4—5 lines wide:* capsule of 2 rhomboid-oval widely divergent lobes sparingly beset with small gland-tipped hairs and reticulate-venulose.—A pretty weed from Asia, plentiful in gardens at Berkeley, flowering from midwinter to early summer.

16. **WULFENIA,** *Jacquin.* Perennials, with flowers and capsules of *Veronica,* except that in our species the anther-cells are not confluent; but in habit quite different; all the leaves being radical and long-petioled. Flowers in a simple raceme or spike.

1. **W. cordata.** Scarcely rhizomatous, the tuft of rather coarse fibrous roots attached to the base of an elongated but upright leafy crown: herbage slightly rusty-hairy, especially the petioles and lower face of leaves; the latter ovate- to deltoid-cordate, 1—2½ in. long, obtuse, doubly crenate; petioles 2—3 in. long: scapes several, not equalling the leaves, few-flowered: fl. long-pedicellate, the pedicels and calyx with fairly abundant brownish deflexed hairs: corolla nearly rotate, light blue: capsule not known.—Northern slope of Mt. Tamalpais and in Mendocino Co. *(Synthyris reniformis cordata, Gray).*

17. **CASTILLEIA,** *Mutis* (PAINTED CUP). Herbs or suffrutescent plants, with alternate sessile leaves, the floral ones, or their tips, and the calyx-lobes, colored like petals. Flowers in terminal spikes. Calyx tubular, cleft in front or behind, or both; lobes 2 and lateral, or 4. Tubular corolla laterally compressed, especially the long conduplicate upper lip: lower lip short and minute, 3-toothed. Stamens 4; anthers 2-celled, the cells unequal. Capsule loculicidally 2-valved; valves bearing the placenta on their middle. Seeds ∞, with loose favose testa.

1. **C. affinis,** Hook. & Arn. *Annual, strict,* simple 1—3 ft. high: leaves linear-lanceolate, entire: lower fl. somewhat scattered; upper spi-

cate-crowded: calyx and upper bracts tipped with scarlet: corolla 1 in. long and more, surpassing the calyx, yellowish; lower lip short but prominent, its callous oblong teeth rather shorter than the keel beneath; upper lip nearly as long as the tube.—In wet places among the coast hills. April, May.

2. **C. latifolia,** Hook. & Arn. Soft-hirsute and viscid perennial 1–2 ft. high, branched from the base: *leaves round-obovate to oval,* 1 in. long, sometimes 3—5-lobed, especially the scarlet-tipped floral ones: the 2 calyx-lobes broad, notched or 2-lobed at the apex, surpassing the corolla-tube: *lower lip of corolla short, the teeth inflexed;* upper rather longer than the tube.—Moist slopes of hills near the sea.

3. **C. Douglasii,** Benth. Sparsely and rather stiffly hirsute, 1—2 ft. high: *lower leaves linear, obtuse, entire,* scarcely narrowed at base; upper usually broader and cleft or incised; floral oblong, with an ovate or oblong scarlet spot at tip: spike at length loose and interrupted: *calyx-lobes obovate,* scarlet at the obtuse tips, shortly and broadly 2-lobed, *scarcely exceeding the lower lip of the corolla.*—Northward slopes of the Coast Range; flowering almost all the year round.

4. **C. foliolosa,** Hook. & Arn. *Suffrutescent,* branching, 1—3 ft. high, *hoary or nearly white with matted wool* of intricately branched hairs: leaves short and numerous, many densely fascicled on short axillary branchlets, all linear and entire, or some with a pair of linear divaricate lobes; upper floral cleft, and with yellow or scarlet tipped segments: the 2 *calyx-lobes broad, retuse or merely notched,* nearly equalling the corolla; this with the lower lip very small.—On exposed southward slopes; almost always in flower.

18. **ORTHOCARPUS,** *Nutt.* Ours all spring annuals, with cleft foliage and often the colored-bracted spikes of *Castilleia;* floral structure much the same. Calyx 4-cleft, or cleft before and behind, and the divisions 2-cleft. Corolla with upper lip not greatly exceeding the lower in length, but the lower inflated and 3-plaited or 3-saccate. Stamens in some with but one anther-cell.

* *Lower lip only moderately inflated, plaited-3-saccate for its whole length: floral bracts more or less colored at the tips.*

1. **O. attenuatus,** Gray. Slender, strict, 1 ft. high or less, usually simple, pubescent: leaves linear-attenuate, entire, with few lobes: *spike very long, narrow and lax; bracts with slender barely white-tipped divisions:* corolla very narrow, ½ in. long, whitish, with purple spots on the lower lip, the narrow teeth of which nearly equal the upper.—Plains and low hills of the Napa and Sacramento valleys, etc.

2. **O. densiflorus,** Benth. Erect and simple, or branched from the base, 6—10 in. high, *soft-pubescent:* leaves with few slender lobes, or the

lower entire: *spike dense, at length cylindric;* bracts about equalling the
fl., the *linear lobes* with purplish and white tips: corolla 1 in. long or
less, purple and white, the teeth of the moderately inflated lower lip
shorter than the upper.—Common on hills or in lower lands.

3. **O. castilleioides,** Benth. More slender, with ampler foliage, the
broader denser *spikes soft-hirsute:* leaves narrowly lanceolate to oblong,
entire or laciniately cut into rather *short obtusish lobes;* bracts more
dilated and cuneate than the leaves, equalling the fl., herbaceous, white-
or yellowish-tipped: corolla 2 in. long, dull white, often purplish-tipped;
lower lip well inflated.—Common along the borders of salt marshes.

4. **O. purpurascens,** Benth. Erect, stoutish, 10—18 in. high, usually
with several ascending branches, *hirsute with copious white hairs:* leaves
once or twice pinnately parted, above the linear or lanceolate base, into
narrow-linear divisions: spikes dense, at length cylindric: bracts about
equalling the fl., their laciniate-lobed divisions and those of the calyx-
lobes tipped with crimson-purple: corolla 1 in. long; the tube whitish or
yellowish; lower lip with 3 sacs at its broad apex, these not larger than
the short rounded recumbent teeth or lobes; *upper lip long, obtuse and
hooked* at the apex, densely red-bearded.—Fields and hillsides.

* * *Lower lip of corolla conspicuously 3-saccate, much larger than the
slender upper one; floral bracts not petaloid-colored at tip.*

5. **O. pusillus,** Benth. Small very slender branching *red-purple herb
3—5 in. high:* leaves 1—2-pinnatifid, the floral 3—5-parted into filiform
segments, exceeding the *scattered inconspicuous dark red flowers:* corolla
glabrous, 2—3 lines long; upper lip longer than the 3-lobed lower one:
capsule globular.—Forming low dark red patches in moist grassy lands.

6. **O. floribundus,** Benth. Glabrous, 4—8 in. high, slender, somewhat corymbosely branching and erect: leaves pinnately parted at apex
into many linear-filiform divisions: spikes rather dense; bracts not
exceeding the calyx: *corolla very pale yellow;* lower lip strongly 3-sac-
cate, the teeth lanceolate, erect; *stamens at length well exserted* from the
upper lip.—Hilltops about San Francisco.

7. **O. faucibarbatus,** Gray. Stoutish and rather tall (1 ft. high),
fastigiately branching, *glabrous,* or the inflorescence with some hirsute
hairs; *herbage invariably green and slightly fleshy:* spikes long and dense;
corolla slender, 1 in. long, *even the sacs relatively small,* the folds within
villous-bearded, color either sulphur-yellow or white with a pinkish
tinge: stamens not exserted.—In moist meadows of the Coast Range;
the albino state more common in Santa Cruz Co.

8. **O. erianthus,** Benth. Smaller and more slender than the last,
but with as many fastigiate branches, soft-pubescent; *herbage,* or at least
the stem and bracts *dark reddish: corolla deep sulphur yellow,* the acute

slender falcate upper lip dark purple: tube very slender, but *sacs large and deep*, their folds villous within.—Very common, and on higher ground than the last.

9. **O. versicolor** (F. & M.). Slender as the last, and the herbage slightly reddish, seldom branching, or more than 6 in. high: leaves at apex filiform-cleft (as in all the group): corollas with shorter tube, and the broadly obovate *sacs almost twice as large as in the last;* color of the fl. *pure white, fading pinkish;* folds of the throat densely bearded. Var.. **roseus** (Gray). *Corolla deep rose-color* from the first (not white, then changing, as has been assumed).—Dry sandy hills, about San Francisco, and in Marin Co.; the flowers fragrant, as in no other species.

10. **O. lithospermoides**, Benth. *Stout, commonly simple*, or with 2 or 3 branches, 1—1½ ft. high, pubescent, *very leafy:* lower leaves lanceolate, entire; upper with a few slender lobes; floral with dilated base, and palmatifid tips nearly equalling the densely spicate flowers: calyx-lobes linear: corolla 1 in. long or more, strongly 3-saccate, the sacs ¼ in. wide, the whole of a *deeper than sulphur yellow, fading whitish.*— Moist plains and hillsides; later in flowering than the other species. May, June.

19. ADENOSTEGIA, *Benth.* Æstival and autumnal branching annuals, with alternate narrow leaves either entire or 3--5-parted. Flowers bracted, scattered, or spicate-crowded. Calyx spathaceous, of an anterior and a posterior leaf-like division, or the anterior one wanting. Corolla tubular, bilabiate; lips short and nearly equal; lower obtusely 3-toothed; upper compressed, its apex more or less uncinate-incurved. Stamens, with ciliate or bearded anther-cells. Capsule compressed. Seeds with a loose testa, pointed at one end.

* *Calyx 2-leaved; bracts and foliage gland-tipped.*

1. **A. rigida**, Benth. Puberulent and somewhat hispid, 1—2 ft. high, paniculately branched: leaves linear-filiform; lower entire; upper 3—5-parted, the floral with cuneate base and bristly-ciliate margins; divisions with dilated and retuse or notched gland-bearing tip: *fl.* crowded *in terminal heads:* corolla yellowish and purplish, ½—¾ in. long. Mt. Hamilton, and perhaps San Mateo Co. and southward. July—Sept.

2. **A. pilosa** (Gray), Greene. Soft-villous and somewhat hoary, 2—4 ft. high and loosely paniculate: leaves narrowly linear, entire; upper and floral often broader and 3-toothed: *fl. only 2 or 3 together at the ends of the branchlets:* corolla less than 1 in. long, white and purplish.—In the mountain districts generally. Aug.— Oct.

* * *Calyx 1-leaved; leaves and bracts not callous-tipped.*

3. **A. maritima** (Nutt.), Greene. Pale, glaucous and hoary-pubescent, corymbosely branched from the base, 1 ft. high or less: leaves and

bracts linear-lanceolate and lanceolate, acute: fl. spicate: corolla dull
purplish: *stamens 4*, in very unequal pairs; *anthers of the longer 2-celled,
of the shorter 1-celled*, only the base of the cell ciliolate.—Sandy salt
marshes from near San Francisco southward. Aug.—Nov.

4. **A. mollis** (Gray), Greene. Glaucous and villous-hirsute, 1—1½ ft.
high, fastigiately branched: leaves oblong-linear; lower entire, obtuse;
upper, and the bracts of the spike, with obtuse teeth or lobes: corolla
white and dull-purple: *stamens 2; anthers unequally 2-celled:* seeds
subreniform, with loose cellular-reticulate coat.—Brackish marshes
about Vallejo and Suisun. Aug.—Nov.

20. BELLARDIA, *Allioni.* Stoutish rather rigid erect annual, with
densely spicate terminal inflorescence. Calyx campanulate, 4-lobed, the
lobes toothed. Corolla strongly bilabiate, the lower and 3-lobed lip
equalling or exceeding the galeate upper. Stamens 4, in unequal pairs,
enclosed in the concavity of the galea; anther-cells mucronate. Capsule
ovate-globose, turgid; the thick placentæ bifid. Seeds minute and very
numerous, finely costate lengthwise.

1. B. TRIXAGO (L.), All. Simple or with a few branches, 1—1½ ft.
high: leaves lanceolate, crenate-serrate; the lower somewhat narrowed
at base, the upper broader below and subcordate-clasping: spike dense,
thick, tetragonal, several inches long: corolla ½—1 in. long, rose-color
and white.—Plentiful in an old field near Martinez, and escaping to the
uncultivated slopes adjacent. May.

21. PEDICULARIS, *Tourn.* Perennials, with alternate pinnately
divided leaves, and flowers in bracted spikes. Calyx irregular, 2—5-
toothed. Corolla strongly bilabiate: upper lip compressed and arched;
lower erect at base, 2-crested above, 3-lobed. Stamens 4, enclosed within
the galeate upper lip; anthers transverse, equally 2-celled. Capsule
ovate or lanceolate, oblique, compressed, loculicidal. Seeds ovoid.

1. **P. densiflora,** Benth. Herbage pubescent and dark reddish when
young, green and glabrate in age: stem 1 ft. high or less: leaves oblong-
lanceolate, 3—6 in. long, twice pinnatifid or parted, the divisions irregularly and sharply toothed: the upper reduced to bracts of the long
dense spike: calyx-teeth 5, lanceolate: corolla narrow, slightly clavate,
1 in. long, scarlet or crimson; lower lip only a fourth as long as the
upper: anther-cells with tapering or acute base.—Rocky hills in somewhat shady places. April—June.

ORDER LXXVII. OROBANCHACEÆ.

Plants without green herbage, and parasitic on roots of herbs and
shrubs. Floral structure that of *Scrophularineæ*, but capsules with
parietal placentæ. Seeds very small and numerous.

1. APHYLLON, *Mitchell.* Ours low viscid-pubescent plants, usually with peduncled flowers. Calyx regular, 5-cleft or -parted. Corolla more or less tubular and curved and not very strongly bilabiate. Stamens not exserted: anther-cells deeply separated from below upward, mucronate at base. Each valve of the capsule bearing a pair of contiguous placentæ.

* *Scapiform peduncles from a short crown or stem.*

1. **A. uniflorum** (L.), Gray. var. **occidentale.** *Scapes few, often 1 only,* slender, 1—5 in. high, *from an almost subterranean short crown:* lobes of the calyx subulate, longer than the tube: corolla $\frac{1}{2}$—$\frac{3}{4}$ in. long, deep blue-purple.—Wooded stony hills. April.

2. **A. fasciculatum** (Nutt.), Gray. *Scaly fleshy stem* rising several inches *above ground, bearing few or many fascicled peduncles* as long as the stem: calyx-lobes broader and shorter than in the last: corolla more than 1 in. long, sulphur-yellow, with reddish or purplish tints on the outside.—Sandy or gravelly hills. May.

* * *Inflorescence racemose or thyrsoid.*

3. **A. comosum** (Hook.), Gray. Branching at or near the surface of the ground: fl. on slender pedicels in a corymb or short raceme: calyx 5-parted, the lobes long and slender: corolla 1 in. long, rose-color or purplish; *upper lip 2-lobed or notched; lower 3-parted: anthers woolly.*— Open hills; parasitic on *Artemisia.* June.

4. **A. tuberosum,** Gray. Low, stout, minutely puberulent, the thickened base of the stem with imbricated scales: fl. in a dense thyrsoid cluster: *calyx unequally cleft, little shorter than the corolla; this with short sca cely spreading lobes:* anthers glabrous.—Mt. Hamilton. July.

ORDER LXXVIII. **LABIATÆ**.

Herbs (and a few shrubs) mostly keenly aromatic, with quadrangular stems, opposite simple exstipulate leaves and axillary solitary or cymose-congested flowers; but the inflorescence often more terminal and spicate or racemose. Calyx 3—5-toothed or cleft, regular or bilabiate. Corolla usually strongly bilabiate; upper lip entire or 2-lobed; lower 3-cleft or -parted. Stamens 2 or 4. Ovary 3-lobed, each lobe becoming a seed-like nutlet in the bottom of the persistent calyx. Seed erect from the base of the nutlet, usually exalbuminous.

* *Calyx and corolla both nearly regular.*

Stamens 4, long-exserted, curved...................................TRICHOSTEMA   1
   "   4, short, nearly equal.................................MENTHA   2
   "   2 only...LYCOPUS   3

* * *Calyx regular; corolla pronouncedly but not strongly bilabiate.*

Flowers densely glomerate terminally and in the upper axils;
   calyx short-tubular; corolla lobes short........KŒLLIA 4
Flowers in a simple terminal bracted flat-topped head...........MONARDELLA 5
Fowers 1 or 2 only in each axil;
 Stems slender, prostrate......................................MICROMERIA 6
 " shrubby, erect..SPHACELE 10

* * * *Calyx regular; corolla strongly bilabiate.*

Calyx with 10 spinescent teeth, some of them hooked............MARRUBIUM 16
Calyx with 5 more or less spinescent teeth;
 Nutlets with truncate summit................................LAMIUM 17
 " with rounded summit...................................STACHYS 18

* * * * *Both calyx and corolla bilabiate, the latter strongly so.*

Flowers few in each leaf-axil;
 Calyx oblong-campanulate, flattened on the back............MELISSA 7
 " closed in fruit, and with a casque-like projection on
   the back.............................SCUTELLARIA 14
Flowers densely verticillastrate;
 Verticillasters forming a close spike or thyrse;
  Upper calyx-lip truncate, 3-toothed......................BRUNELLA 15
  Calyx oblique at the throat; teeth subequal..............NEPETA 13
  " strongly ciliate, deeply and unequally 5-cleft......POGOGYNE 8
 Verticillasters more remote;
  Subtended by prickly-margined bracts....................ACANTHOMINTHA 9
  Anthers with long filament-like connective, 1 cell at each
   end, or 1 end with no cell;
   Upper lip of corolla manifest.........................SALVIA 11
   " " " obsolete.......................RAMONA 12

1. **TRICHOSTEMA,** *Gronov.* Herbs, ours annual, with a very keen somewhat acid smell, and blue flowers in axillary cymes. Calyx 5-cleft. Corolla with narrow tube and more or less oblique limb; the somewhat similar lobes oblong. Stamens with capillary filament very long exserted and curved, didynamous; anther-cells divaricate or divergent.

1. **T. laxum,** Gray. Diffusely branching, 1—2 ft. high, *minutely pubescent, sparsely leafy:* leaves lanceolate or oblong-lanceolate, acuminate but obtusish, 2—3 in. long, nervose-costate, rather slender-petioled: *cymes pedunculate, loose:* tube of corolla about 3 and limb 2 lines long.— From Sonoma Co. northward. July—Oct.

2. **T. lanceolatum,** Gray. Tall as the last, rather strict and simple, only a few ascending branches from the base: *herbage pale with a dense pubescence: leaves crowded,* costate-nerved, lanceolate or ovate-lanceolate, *tapering from near the broad base* to a very acute tip: cymes short-peduncled, dense: calyx villous: corolla somewhat pubescent.—Livermore Valley, and northward and southward, towards the interior. June—Oct.

2. **MENTHA,** *Pliny* (MINT). Fragrant perennial herbs, mostly spreading by slender creeping rootstocks. Calyx short, 5-toothed.

Corolla almost equally 4-lobed, small, hardly irregular, but upper lobe often broadest and emarginate. Stamens 4, similar, nearly equal, not declined.

* Old World species; inflorescence terminal.

1. M. VIRIDIS, L. Herbage green, nearly glabrous: stems erect, 2—4 ft. high: *leaves subsessile, oblong-lanceolate*, sparsely and sharply serrate, somewhat rugose-veiny: *flower-clusters crowded in narrow leafless terminal spikes:* calyx oblong-campanulate; teeth triangular-subulate, as long as the tube: corolla very pale purplish, glabrous.—Naturalized abundantly along streamlets and on moist banks.

2. M. PIPERITA, Huds. Tall as the preceding, but the herbage dark or reddish-green, with a very pungent flavor: *leaves* ovate-oblong. or oblong-lanceolate, acute, *distinctly petiolate;* spikes narrow but interrupted.—As common as the last, or even more so.

3. M. CITRATA, Ehrh. Aspect of the preceding, but leaves more rounded, thinner, with a comparatively delicate sweet ordor: flower-clusters a subglobose terminal head, with a few verticillasters beneath it.—By a streamlet near West Berkeley.

* * Flower-clusters in the leaf-axils; one species native.

4. M. PULEGIUM, L. Almost hoary with a short white-woolly pubescence: stems 1—2 ft. long, reclining or prostrate and rooting at the lower joints, the young sterile shoots above ground (not subterranean and shizomatous): leaves elliptic-ovate, short-petiolate, ½—·1 in. long, remotely denticulate-serrate; dense cymes globular, smaller (and the leaves smaller) towards the ends of the branches: calyx broad-funnelform, slightly bilabiate: teeth lanceolate-acuminate, half as long as the tube, 10-ribbed, the throat closed with hairs.—Wet banks of streams, above Santa Rosa, and islands of the lower San Joaquin.

5. M. Canadensis, L. From nearly glabrous to villous-hoary, 1—2½ ft. high: leaves oblong-ovate to oblong-lanceolate, acute, sharply serrate, tapering to the short petiole: calyx hairy: teeth triangular-subulate, half the length of the cylindraceous tube: corolla from white to pale lavender-color.—Frequent in marshes about the Bay, and along river banks.

3. LYCOPUS, *Tourn.* Herbs with the habit of the mints of our second or indigenous group; but herbage wholly scentless and rather intensely bitter. Leaves often deeply sinuate-toothed. Calyx campanulate, 4—5-toothed. Corolla with upper lip entire. Fertile stamens 2 only. Nutlets somewhat 3-sided, with thickened margins at the summit.

1. **L. lucidus,** Turcz. var. **Americanus,** Gray. Stoloniferous at base of the stem, 1—3 ft. high; stem sharply quadrangular toward the summit: leaves 2—4 in. long, lanceolate, acute or acuminate, sharply and coarsely serrate, almost sessile, glabrous or nearly so: calyx-teeth slen-

der-subulate, equalling the white corolla, not exceeding the nutlets:
rudimentary stamens slender, with thickened tips.—Suisun marshes.

4. KŒLLIA, *Mœnch.* Erect perennials, with densely crowded verticillasters at the upper nodes, subtended by a pair of reduced leaves.
Calyx with 5 short equal teeth; throat naked within. Corolla-tube
scarcely exceeding the calyx; the short limb bilabiate, the upper lip
nearly flat and almost entire; lower spreading and of 3 short obtuse
lobes. Stamens 4, straight, divergent, the interior pair slightly longer:
anther-cells parallel.

1. **K. Californica** (Torr.), O. Ktze. Herbage whitish with a very fine
and close soft pubescence: stem 2—3 ft. high, simple or with few terminal branches, all floriferous: leaves ovate to ovate-lanceolate, sessile
by an obtuse or subcordate base, entire or denticulate, 1—3 in. long:
heads terminal and in the axils of some of the uppermost leaves, com-
pacted with slender bracts: fl. white.—Contra Costa Co. hills.

5. MONARDELLA, *Benth.* Herbaceous or suffrutescent. Flowers
in large involucrate terminal heads. Calyx tubular, 10—13-nerved, 5-
toothed, the teeth short, straight, subequal: throat naked within. Corolla with slender tube; the bilabiate limb of 5 long and narrow linear or
oblong lobes. Stamens 4 exserted; anther-cells often divergent or
divaricate.

     * *Perennials, suffrutescent at the very base.*

1. **M. villosa,** Benth. Stems 1 ft. high, stoutish, not densely tufted:
leaves ovate, green and subglabrous above, villous beneath, ½—1 in. long,
conspicuously veiny, crenate-dentate: flowers deep purple.—Open grassy
hills near the sea.

2. **M. Sheltonii,** Torr. More slender and tufted, as well as more
shrubby at base: *leaves thinner, oblong,* entire or denticulate, *cinereous-
puberulent on both faces,* the upper ones subsessile; flowers of a lighter
purple approaching lavender-color.—Dry wooded hills from Napa and
Sonoma counties northward.

     * * *Annuals, branching above; leaves not toothed.*
      +— *Bracts of the head transparent between the veins.*

3. **M. Douglasii,** Benth. Pubescent: leaves lanceolate: *bracts ovate
to ovate-lanceolate,* cuspidate, somewhat hirsute, *fenestrate by greenish
veins running* through the hyaline surface from midrib *to stout marginal
nerves:* calyx-teeth rigid, subulate; tube hirsute.—Mt. Diablo.

4. **M. Breweri,** Gray. Puberulent: leaves ovate-oblong: *bracts
broadly ovate,* abruptly cuspidate, less translucent than in the last, the
veins more slender, destitute of strong marginal nerves: calyx-teeth triangular-subulate, acute.—Southern part of Contra Costa Co., Corral
Hollow, etc.

5. **M. undulata,** Benth. Minutely pubescent or glabrous, 8—15 in. high: leaves oblong-spatulate to nearly linear, obtuse, undulate-margined, petiolate: bracts and calyx villous; the former broadly ovate, obtuse or acutish, thin and somewhat scarious, parallel-veined: corolla rose-purple.—Hills toward the sea; Point Reyes, etc.

6. MICROMERIA, *Benth.* (YERBA BUENA). Very sweet-scented trailing evergreen undershrub, leafy, and with mostly solitary flowers in the axils. Calyx oblong-tubular, 13-striate, regularly 5-toothed. Corolla white, with erect emarginate upper lip, and spreading 3-parted lower one. Stamens 4; anthers 2-celled.

1. **M. Chamissonis** (Benth.). Stems 1—4 ft. long, often rooting at the ends: leaves round-ovate, thin-coriaceous, of a light yellowish green, 1 in. long, sparingly toothed: fl. on a filiform bibracteolate peduncle: calyx-teeth subulate: corolla pure white, 4 lines long, twice the length of the calyx.—Wooded northward slopes of hills near the coast.

7. MELISSA, *Tourn.* Sweet-scented perennial, with tufted stems, ovate rugose petiolate leaves, and white flowers clustered in the axils. Calyx campanulate-tubular, bilabiate, 13-striate; upper lip flattened, 3-toothed; lower 2-toothed, erect. Upper lip of corolla concave, emarginate; lower 3-lobed, the middle lobe largest. Stamens 4, converging under the upper lip; anther-cells divaricate, confluent.

1. M. OFFICINALIS, L. Herb 1—2 ft. high: leaves 1 in. long, ovate, truncate at base, deeply crenate-serrate: fl. few in the cluster, subsecund: corolla more than twice the length of the calyx.—By waysides, and along streamlets, at Temescal, Berkeley, etc.

8. POGOGYNE, *Benth.* Aromatic low annuals, with oblong-lanceolate leaves tapering to a petiole. Flowers crowded, the verticillasters mostly interrupted-spicate. Bracts and calyx hirsute-ciliate. Corolla blue or purple. Calyx deeply and unequally 5-cleft: teeth longer than the about 15-nerved tube. Corolla straight, tubular-funnelform, with short lips; the upper erect, entire; lower 3-lobed, spreading. Stamens 4, ascending; anthers 2-celled. Style bearded.

* *All 4 stamens antheriferous.*

1. **P. Douglasii,** Benth. Stout, 6—12 in. high: leaves oblong, spatulate, or oblanceolate: spikes dense; *bracts linear, acute:* lower segments of calyx twice or thrice the length of the tube, much longer and narrower than the others: corolla ¾ in. long, bluish-purple.—Common along ditches, and low places that become dry in summer. May—July.

2. **P. parviflora,** Benth. Not stout, 5—8 in. high: leaves narrower: *bracts obtuse:* calyx-segments *broader, less unequal,* the lower hardly longer, the upper shorter than the tube: corolla ½ in. long.—About San Francisco Bay, and northward.

* * *Flowers very small; upper stamens sterile.*

3. **P. serpylloides** (Torr.), Gray. Slender, almost diffuse, the branches 3—6 in. long: leaves obovate-oval or spatulate, 3—4 lines long, most of the lower distant with but few flowers in the axils, the upper approximate and the verticillasters dense and spicate-crowded: calyx hirsute, its lobes unequal but all much longer than the tube, the longer fully equalling the very small blue corolla: sterile filaments with capitellate rudiments of anthers: style with few and coarse hairs.—Plentiful in moist shades; Oakland Hills, etc.

9. **ACANTHOMINTHA,** *A. Gray.* General aspect of the preceding, and floral structure similar, but flowers less crowded; each verticillaster subtended not only by its leaves, but also by a pair of almost stipule-like coriaceous callous- and prickly-margined broad bracts.

1. **A. lanceolata,** Curran. Stoutish, ½—1 ft. high, pubescent, oily and ill-scented: leaves oblanceolate or oblong, sparingly spinulose-dentate: teeth of bracts and calyx long-aristate: middle tooth of upper calyx-lip erect, the lateral bent forward in fruit; lower lip 2-parted, its segments lanceolate and awns short: upper lip of corolla somewhat falcate-incurved, cleft at apex; lower with oblong entire lobes, the middle one longer and narrower: style sparsely pilose.—Mountains near Niles; also on Mt. Hamilton and southward to Monterey Co. June—Aug.

10. **SPHACELE,** *Benth.* Strong-scented shrubs, with ample leaves; the floral reduced in size, and the large flowers solitary in their axils. Calyx campanulate, deeply and subequally 5-toothed. Corolla with broad tube, and 4 short spreading lobes; the 5th and lowest one much longer, laid in a fold and erect. Stamens 4, distant; anther-cells diverging.

1. **S. calycina,** Benth. Bushy-branching, 2—5 ft. high, tomentulose-villous or glabrate: branches with ample thinnish and soft rugose ovate-oblong crenate leaves 2—4 in. long, short-petiolate; the uppermost and floral sessile: calyx-lobes triangular-lanceolate, little surpassed by the white or merely flesh-tinted corolla: anthers short.—Rather dry hills of the coast ranges. June—Aug.

11. **SALVIA,** *Linn.* Aromatic herbs and shrubs of various aspect; inflorescence in ours densely verticillastrate and interrupted. Calyx bilabiate, with upper lip 3-toothed or entire; lower 2-cleft. Corolla deeply bilabiate; upper lip erect; lower spreading or drooping, its middle lobe commonly large, notched or obcordate. Stamens 2, in the throat of the corolla; filaments short, the long curved connective appearing like a fork to the proper filament, its posterior portion ascending and bearing a single anther-cell, its opposite arm bearing a smaller or merely rudimentary anther-cell, or this arm of the connective itself almost obsolete.

1. **S. spathacea,** Greene. Very stout perennial, or the bases of the stems enduring, 1—3 ft. high, very villous and glandular, honey-scented, or when bruised more aromatic: lowest leaves hastate-lanceolate, obtuse, 3—8 in. long, on margined petioles; upper oblong, sessile, all very rugose, sinuate-crenate, white-tomentose beneath: fl. densely capitate-glomerate; heads large, interruptedly spicate: calyx ¾ in. long, spathe-like and closed by conduplication, the orifice oblique; 2 lower teeth very short: corolla crimson, 1½ in. long: anther-cell 1 only; rudimentary arm of the connective very short.—Hills near the coast, from San Francisco southward.

2. **S. carduacea,** Benth. Very stout erect annual, 1—2 ft. high: stem nearly naked, but with a cluster of ample sinuate-pinnatifid spinu-lose-toothed leaves at base, these and the whole plant, except the flower, white-woolly and thistle-like: head-like verticillasters 1—4, dense, 1 in. broad, equalled or surpassed by the ovate-lanceolate spinescently pec-tinate-toothed bracts: calyx long-woolly, many-nerved; its upper 3-toothed, the middle tooth large, the laterals smaller and distant: corolla 1 in. long, rather light blue; its tube exserted; upper lip erose dentic-ulate and cleft; lower with an excessively large flabelliform fimbriately many-cleft middle lobe: proper filaments very short; lower arm of the long filiform connective bearing a polliniferous anther-cell.—Dry Hills of Contra Costa Co. March, April.

3. **S. Columbariæ,** Benth. Slender annual, branching and leafy below, 8—18 in. high, naked and peduncle-like above, bearing few closely bracted rather large capitate clusters of small flowers: leaves rugulose, once or twice pinnatifid into toothed or incised divisions: upper lip of calyx large, arched, tipped with a pair of partly connate short-awned teeth, much exceeding the two small teeth of the lower lip: corolla small, hardly longer than the calyx, deep blue; upper lip small, notched; lower with small lateral lobes, and large unguiculate trans-versely oval 2-lobed middle one.—Throughout the coast ranges, on dry hills and sandy plains.

4. **S. mellifera,** Greene. Shrubby, 3—8 ft. high, with herbaceous flowering branches, leafy, cinereous-tomentose as to the growing parts, but glabrate in age: leaves oblong-lanceolate, petiolate, rugulose and crenate: dense capitate flower-clusters small, several at the end of each branchlet: calyx oblique, the teeth cuspitate and several-awned: corolla white or pinkish, small and little exserted.—Contra Costa and San Mateo counties, on hillsides.

12. RAMONA, *Greene.* Shrubs or undershrubs, with habit, foliage, inflorescence, and even the peculiar spathe-like oblique calyx of the Californian *Salvias;* but corolla with no proper upper lip; its throat inflated and horizontally split, the upper portion of this either obsolete, or vertically cleft and the segments divergent, exposing fully the stamens from their insertion on the bottom of the throat. Genitals all long-exserted. Stamens 2, consisting of a distinct filament articulated with the single arm of the connective present, this on the same plane with the proper filament, bearing the one anther-cell at its summit.

1. **R. humilis** (Benth.), Greene. Prostrate very leafy woody stems forming broad mats: leaves oblanceolate to spatulate-oblong, obtuse, petiolate, tomentose-canescent and finely rugulose: scape-like peduncles erect, a foot high more or less: verticillasters few-flowered: arched calyx with not very unequal lips, the upper pungently 3-toothed, the lower deeply 2-cleft: corolla, stamens and pistil deep violet; short tube of the corolla closed by hairs; sundered upper part of throat cleft into 2 lanceolate upturned and divergent segments; deeply concave and wide-open lower part showing stamens from their insertion, truncate at the upper end, the ampliate 3-lobed and crisped lower lip joined to this by a broad and short ligulate claw.—Common on the mountain sides near Mt. St. Helena, Hood's Peak, etc. April—June.

13. NEPETA, *Linn.* Calyx 15-ribbed, oblique at the constricted throat, unequally 5-toothed, 3 upper teeth longer than the 2 lower. Corolla with rather long and slender tube; upper lip erect, slightly concave, emarginate, or 2-lobed; lower spreading, 3-lobed. Stamens 4; anther-cells at length divaricate and confluent.—Both species Old World garden plants, not native here.

1. N. CATARIA, L. Perennial, *stoutish, erect,* 1—3 ft. high, with ascending branches; *hoary with a dense short pubescence: leaves triangular-ovate,* subcordate, coarsely serrate, about 2 in. long: verticillasters forming a dense terminal thyrsoid cluster: calyx-teeth lanceolate subulate: corolla ½ in. long, white, dotted with lilac.—Not rare in Marin, Sonoma and Solano counties.

2. N. HEDERACEA (L.). Stems *slender, procumbent,* rooting; flowering branches short, ascending: *leaves round-reniform,* crenate, glabrous or hispid-puberulent: flowers few: calyx ¼ in. long; teeth triangular-subulate: *corolla* ½ in. long or more, *blue.*—Moist banks by a roadside on Mt. St. Helena.

14. SCUTELLARIA, *Cortusi.* Perennial herbs not aromatic, with solitary flowers in the axils of leaves or bracts. Calyx without teeth, horizontally cleft into 2 lobes, the upper with a crest-like projection; the orifice closed in fruit, and the casque-like upper half of the calyx at length falling away from the mature fruit. Corolla with elongated and ventricose throat: upper lip erect, arched or galeate, the lateral lobes of the lower lip attached to it, the middle and lowest lobe of the corolla appearing to constitute the lower lip.

1. **S. Californica,** Gray. Slender, 8–20 in. high, *puberulent: leaves oblong, obtuse,* entire, short-petiolate, 1 in. long: *corolla 1 in. long, yellowish-white.*—Common in the mountain districts. May—July.

2. **S. tuberosa,** Benth. Slender, sparsely leafy, 2–4 in. high, from moniliform-tuberous rootstocks; *soft-pubescent or villous: leaves ovate,* coarsely and obtusely few-toothed, ¼–¾ in. long: *corolla* ½–¾ *in. long, deep blue:* nutlets strongly muricate.—Open hills, in clayey or stony soil. March—May.

15. BRUNELLA, *Tourn.* Low, perennial, with blue flowers in a terminal cylindraceous head or spike. Calyx oblong, 10-nerved, reticulate-veiny, bilabiate; lips flattened and closed in fruit; upper dilated, truncate, 3-toothed. the teeth broad and short; lower with 2 lanceolate teeth. Upper lip of corolla arched, entire; lower 3-lobed; its middle lobe drooping, rounded, concave, denticulate. Filaments 2-toothed at apex, the lower tooth bearing the anther, the cells divergent.

1. **A. vulgaris** (L.). Roughish-pubescent or glabrous, 6–10 in. high: leaves ovate or oblong, entire or toothed, petiolate: corolla violet or darker.—Moist borders of woods, and cool summits of hills.

16. MARRUBIUM, *Columna.* Bitter-aromatic whitish-woolly rugose-leaved perennial, with dense verticillasters of small white flowers in all the axils. Calyx 10-nerved, and with as many unequal teeth, some or all of these spinescent and the tips recurved. Corolla very short, white; upper lip 2-lobed; lower 3-cleft. Stamens 4, not exserted: anther-cells confluent.

1. **M. VULGARE,** L. (HOREHOUND). Stem stout, tufted, white-woolly, 2 ft. high: very rugose rounded and crenate leaves only hoary, rhombic-ovoid, 1–2 in. long, petiolate.—Very common wayside weed.

17. LAMIUM, *Pliny.* Decumbent slightly fleshy small herbs, with rounded toothed or lobed leaves, and pink or purple flowers in the axils. Calyx funnelform, 5-toothed. Corolla with rather long tube; upper lip vaulted, entire or notched; lower spreading, 3-lobed, the lateral lobes usually reduced to small teeth, the middle one emarginate. Stamens 4; anther-cells divaricate, confluent. Nutlets truncate at summit.

1. **L. AMPLEXICAULE,** L. Small annual, branching at base, 3–6 in. high, sparingly pubescent: lower leaves deltoid-ovate, petiolate; floral round-reniform, sessile and amplexicaul; all coarsely toothed: calyx-teeth triangular-acuminate, as long as the tube: corolla rose-purple, ½ in. long or more; the fl. rather few in the axils.—Fields of Sonoma Co., *Bioletti.*

18. STACHYS, *Diosc.* Ours all perennial, with upright rather sharply quadrangular leafy stems; flowers in the upper axils, or separate from the leafy part of the stem and in a terminal spike. Calyx narrow-campanulate or turbinate, 5—10-nerved, the 5 equal teeth often spiny-pointed. Corolla white or purple; tube not dilated into a throat; upper lip erect, concave, entire or emarginate; lower spreading, 3-lobed. Stamens 4, under the upper lip; anthers 2-celled.

    * *Corolla-tube little or not at all exceeding the calyx.*

      +— *Flowers whitish; herbage soft-hairy.*

    1. **S. ajugoides,** Benth. Villous with white hairs; stems *low and decumbent,* ½—1 ft. high: *leaves oblong, obtuse,* crenate-toothed, 1—3 in. long, often tapering to the petiole; the upper sessile: fl. about 3 in the axils of the upper ordinary leaves, but more numerous in those of the mere bracts above, and so becoming spicate: teeth of the short-campanulate calyx mucronate-acuminate.—In low moist fields. May—July.

    2. **S. albens,** Gray. Softly whitish-tomentose; stems *erect and rather strict, 3—5 ft. high: leaves* oblong-lanceolate to ovate, *subcordate at base,* obtuse, crenate, 2—4 in. long: virgate spike 3—9 in. long: teeth of the turbinate-campanulate calyx triangular, awn-pointed.—In marshy places only. June—Sept.

    3. **S. pycnantha,** Benth. Green, but hirsute or villous with soft spreading hairs; stems 1—3 ft. high: leaves oblong-ovate, cordate, obtuse, crenate, 2—4 in. long, all except the floral rather long-petioled: *fl. in a distinct bractless* (or nearly so) *dense cylindraceous terminal spike,* or the lowest verticillasters separated from the spike: calyx-teeth triangular, mucronate.—Only along streams, or about springy places among the hills. May—July.

    +— +— *Flowers purplish; pubescence coarser and shorter.*

    4. **S. Californica,** Benth. Stoutish, 2—5 ft. high, green and rather villous, the lower face of the ample ovate-oblong subcordate leaves resinous-viscid, the surface not very rugose, the whole plant heavily aromatic-scented: leaves gradually reduced to bracts of the interrupted-spicate inflorescence; fl. 6—10 in the whorl: teeth of the fruiting calyx ovate, mucronate-acute: corolla-tube somewhat exceeding the calyx; lateral lobes of the lower lip reflexed.—San Mateo Co. and southward, on banks of streams. June.

    5. **S. bullata,** Benth. More slender, 10—18 in. high, the sharp angles of the stem retrorsely hispid and the whole pubescence more harsh, the herbage not resiniferous or viscid: leaves oblong-ovate, coarsely crenate, strongly bullate-rugose: inflorescence as in the last, but fl. fewer (about 6 to the whorl): calyx-teeth pungent: corolla-tube not exserted; all the lobes of the lower lip eventually reflexed.—Very common in dry soils. March—June.

** *Corolla-tube far exceeding the calyx.*

6. **S. Chamissonis,** Benth. Very stout. 3—10 ft. high, mostly rough-hispid with retrorse bristles, especially on the angles of the stem: leaves 3—10 in. long, oblong-ovate, subcordate, crenate-serrate, villous or hirsute above, villous-tomentose beneath, rather long-petioled: calyx tubular-campanulate, its teeth cuspidate: corolla rose-red, 1 in. long, the tube exserted twice the length of the calyx.—Along the margins of mountain streams, in deep shades; of very rank growth in Bear Valley, beyond Olema, Marin Co.

ORDER LXXIX. VERBENACEÆ.

Square-stemmed opposite-leaved plants analogous to *Labiatæ;* but the inflorescence not verticillastrate; 5-toothed calyx not always persistent; the corolla less irregular. Ovary not deeply lobed; when mature splitting into 2 or 4 nutlets with lateral insertion, or covered with a pulp, and thus drupaceous.

1. **VERBENA,** *Pliny.* Herbs, with flowers in panicled spikes at summit of leafy stem or branches. Calyx prismatic, 5-angled and -toothed, at least in some species deciduous at the maturing of the fruit. Corolla salverform; limb unequally 5-lobed. Stamens 4, on the tube of the corolla, not exserted. Stigma of 2 dissimilar lobes. Ovary when mature splitting into 4 elongated laterally inserted nutlets.

1. **V. hastata,** L. Perennial, erect, 3—6 ft. high, *minutely pubescent: leaves oblong-lanceolate,* acuminate, coarsely and incisely serrate, some of the *lower hastately 3-lobed:* spikes many, dense, 2—4 in. long, in a close terminal panicle: corolla 2 lines long, deep blue; limb 2 lines wide.—Common on the banks of the lower Sacramento. Sept.

2. **V. prostrata,** R. Br. Stems rigidly erect and 2 ft. high, or more slender and ascending or decumbent, seldom or never prostrate: *herbage more or less hirsute-pubescent: leaves extremely variable,* obovate, spatulate-obovate, oblong or cuneate-oblong, tapering into a marginal petiole, sharply serrate, deeply incised, or pinnately 3—5-cleft: spikes long, slender, solitary or several, or not rarely many, short and dense, arranged in a terminal panicle: bracts subulate, not exceeding the calyx: corolla rather light blue, 2 lines long and broad.—Rather common in moist places along the Bay, and by streamlets among the hills.

3. **V. bracteosa,** Michx. Perennial, stoutish, rather rough-hirsute, much branched from the base and *nearly prostrate:* leaves cuneate-oblong or obovate, pinnately incised or cleft, or coarsely toothed; the lower narrowed into a short margined petiole, the uppermost passing into the bracts of the long dense *spikes* which are *squarrose with the rigid lanceolate sparsely hispid foliaceous bracts* that subtend the flowers: corolla very small and slender, blue.—Lower San Joaquin and southward.

2. LIPPIA, *Houston.* Ours low riparian herbs, bearing axillary peduncled and bracted capitate spikes of small whitish flowers in character much like those of *Verbena.* Ovary 2-celled, in fruit forming 2 one-seeded nutlets.

1. **L. nodiflora,** Michx. *Erect,* from creeping rootstocks, *herbaceous throughout* and rather slender: *leaves* oblanceolate or cuneate-spatulate, *serrate above:* peduncles exceeding the leaves.—Muddy banks of the lower Sacramento and San Joaquin.

2. **L. cuneifolia** (Torr.), Steud. Woody at base and diffusely branching, the branches often a yard long or more, rather rigid and coarse: leaves rigid, linear-cuneiform, incisely 2—6-toothed above the middle: peduncles usually shorter than the leaves; bracts rigid, broadly cuneate, abruptly acuminate.—River banks, and low subsaline plains of the interior.

DIVISION V.   APETALÆ AMENTIFERÆ.

Apetalous; mostly shrubs and trees with unisexual flowers; the staminate always (except in the first two orders), and often the pistillate also, in aments or catkins.

ORDER LXXX.   U R T I C A C E Æ .

Represented by very few species of two closely allied genera of herbaceous plants.

1. URTICA, *Pliny.* Perennials; the quadrangular stems and other parts bearing stinging bristly hairs. Leaves opposite, stipulate, serrate. Flowers monœcious or diœcious, green, clustered in ament-like axillary geminate racemes. Staminate fl. of 4 sepals, 4 stamens, and the cup-shaped rudiment of an ovary; pistillate with 4 sepals, 2 outer spreading, 2 inner erect, the latter becoming membranous and enclosing the ovate flattened achene. Stigma sessile, capitate, tufted.

     * *Annual; inflorescence of mingled fl. of both sexes.*

1. U. URENS, Ray (1660). Slender, erect or ascending, 1—2 ft. high, nearly glabrous: leaves thin, 1—2 in. long, ovate or ovate-oblong, very coarsely and deeply toothed; stipules small, free: flower-clusters mainly pistillate: fruiting sepals ovate, hispid on the margin, usually with one lateral bristle: achene 1 line long.—Very common in sandy soil about San Francisco, flowering all the year round.

     * * *Perennials; inflorescence unisexual.*

2. **U. Californica,** Greene. Stout but not tall, (2—3 ft. high), very hispid: stipules large, narrowly oblong: leaves broadly or somewhat

deltoidly ovate, acute, cordate at base, 3—5 in. long, ascending or spreading on stoutish petioles 1—1½ in. long: sepals broadly ovate, little exceeding the broadly ovate, minutely punctate achene which is little more than ½ line long.—Borders of thickets near streamlets on the seaward slope of the Coast Range in San Mateo Co., and no doubt all the middle Californian "*U. Lyallii*" is this. June, July.

3. **U. holosericea,** Nutt. Not only stinging-bristly, but also densely and finely hoary-tomentose, especially on the leaves beneath: stems very stout, 5—8 ft. high: leaves oblong· to ovate-lanceolate, acuminate, 2—4 in. long, often subcordate at base; stipules oblong, ½ in. long: inner sepals ovate, densely hispid, ½ line long, about equalling the broad-ovate achene.—Very common on banks of streams, and in other moist places.

2. **HESPEROCNIDE,** *Torr.* Very analogous to *Urtica urens*, but generically separated on account of the complete cohesion of the sepals of the pistillate flower into a membranous compressed oblong-ovate sac, with a minutely 2—4-toothed orifice.

1. **H. tenella,** Torr. Slender, 1—2 ft. high, hispid with branching hairs and bristly: leaves ovate, ½—1½ in. long, short-petioled, thin, obtusely serrate: flower-clusters lax, shorter than the petioles: perianth hispid with hooked hairs, ½—¾ line long in fruit: achene thin, minutely striate-tuberculate.—Shady banks in Napa and Contra Costa counties and southward. April, May.

Order LXXXI. PLATANACEÆ.

Represented by one species, of the only genus,

1. **PLATANUS,** *Theophr.* Monœcious trees, with exfoliating bark, ample palmately lobed leaves, sheathing deciduous stipules, and flowers in dense globose naked heads, without perianth, but subtended by clavate truncate minute hairy scales. Filaments very short; anthers clavate, with prolonged peltate connective. Style stigmatic on one side, persistent; ovules 1 or 2, pendulous. Fruit an obpyramidal achene.

1. **P. racemosa,** Nutt. Tree widely branching, often 60—80 ft. high: leaves broadly cordate, 3—5-lobed, densely rusty-tomentose when young; lobes acute or acuminate, sometimes toothed: fertile heads 2—7, in a moniliform spike; in fruit 1 in. in diameter.—Along all large streams.

Order LXXXII. BETULACEÆ.

Of *Betula*, the principal genus no species occurs within our limits; though *Alnus*, the only other genus of the order, is with us.

1. **ALNUS,** *Pliny* (ALDER). Ours monœcious deciduous trees, inhabiting stream banks or other wet places among the hills. Bark smooth.

Leaves alternate, simple, doubly toothed. Aments unisexual; staminate narrow-cylindric and pendulous; pistillate oblong or ovoid, erect. Bracts of staminate ament peltate, including 5 bractlets and about 3 flowers; perianth regular, 4-lobed. Stamens 2 or 4, opposite the lobes; anther-cells contiguous. Bracts of pistillate ament fleshy, imbricated, including 4 bractlets and 2 flowers, cuneate, slightly 4-lobed, in fruit persistent and woody. Nutlets seed-like and flattened.

1. **A. rhombifolia,** Nutt. Tree 30—50 ft. high; bark of branches dark brown; leaves ovate to ovate-oblong and obovate, slightly pubescent beneath, glabrous above, obtuse or acute, irregularly glandular-denticulate: *flowering in midwinter, long before the appearing of the leaves:* fruiting aments oblong, 6—8 lines long: *nutlets* a line long, very broadly obovate, *with a thickened margin.*—Not common. Some good trees formerly stood upon the site of the Judson Iron works near Shell Mound. Fl. Jan.

2. **A. rubra,** Bong. Bark of branches more or less dotted with white: leaves thick, rusty-pubescent beneath, ovate to elliptic, 2—8 in. long, acute, coarsely and rather obtusely toothed, the teeth crenate-toothed, the margin narrowly revolute: *flowers appearing in spring with the leaves:* fruiting aments roundish-ovate ½—1 in. long: *nutlets* 1½ lines long, orbicular or obovoid, *surrounded by a narrow thin wing.*— Very common. Fl. March, April.

Order LXXXIII. MYRICACEÆ.

Represented by one species of the genus

1. **GALE,** *Tourn.* Monœcious or diœcious shrubs or trees, with alternate often dotted and fragrant foliage. Flowers in sessile ovoid aments. Perianth none. Stamens several, monadelphous (in ours), axillary to a bract of the ament. Ovary 1-celled, 1-ovuled, with 2 sessile filiform stigmas. Fruit globular, nut-like, rough on the surface by an uneven deposit of wax.

1. **G. Californica** (Cham.). Evergreen tree, sometimes 30—40 ft. high, and trunk 1—2 ft. thick; the branches forming a round bushy or broader and somewhat depressed head: leaves thin-coriaceous, slightly tomentose below, oblanceolate, 2—4 in. long, acute, attenuate to a short petiole, serrate above the base: aments simple or somewhat compound, androgynous, 3—5 lines long; the small broadly ovate obtuse bracts more or less lacerately ciliate: fr. 2 lines in diameter; the waxy coat thin, granular-roughened, dull purplish.—Common in moist places among the hills.

ORDER LXXXIV. SALICACEÆ.

Trees or shrubs, with alternate simple stipulate leaves, and diœcious flowers in terminal aments, each flower subtended by a membranous bract, with no perianth. Stamens from two to several, central, or scattered on a glandular disk. Ovary 1-celled, bearing 2 subsessile stigmas. Fruit a 2-valved capsule, with many minute comose seeds.

1. **SALIX,** *Varro* (WILLOW). Branches terete. Buds covered by a single calyptriform scale. Bracts of the aments entire. Stigmas short. Ovary and capsule mostly slender-conical.

> * *Aments terminating leafy branchlets late in spring; stamens*
> *3—5; scales of fruiting ament deciduous.*

1. **S. nigra,** Marsh. Small tree with rounded head; trunk slender, with rough dark bark; *branches brittle at base:* leaves linear-lanceolate, slenderly acuminate from near the base, sometimes falcate, 4—6 in. long, closely serrate, green and glabrous on both sides, the midvein prominent; stipules semicordate or 0: fertile aments becoming loose, scales slightly toothed or entire, villous with crisp hairs: *capsules glabrous, ovate-conical,* brownish; styles very short: stigma notched.—The common river bank willow of the lower San Joaquin.

2. **S. lævigata,** Bebb. Taller and more shapely tree, with more elongated and symmetrical head: leaves larger, broader, lanceolate or oblong-lanceolate, acute or acuminate, 3—7 in. long, ¾—1½ in. wide, dark green glabrous and glossy above, paler beneath, minutely serrulate; *petioles downy-pubescent;* stipules semicordate or 0: aments rather dense, 2—4 in. long, somewhat flexuous; scales pallid, villous, dentate; in the staminate ament roundish-obovate, cucullate, in the pistillate narrower, truncate, with 2—4 teeth at apex; *capsule conical* from a thick base, acute, glabrous; style short or obsolete; stigmas emarginate.—Margins of mountain streams.

3. **S. lasiandra,** Benth. Shrub, or small tree with broad spreading head; branches yellow: leaves lanceolate, acute at base, the apex very slenderly acuminate, sharply and closely serrulate; pale beneath: *petioles glandular at the upper end;* stipules broadly semilunate, glandular-serrate: scales of cylindric staminate ament thin, yellowish, more or less toothed: stamens 5 or more: *capsules lanceolate;* stigmas bifid.—Low valleys and plains, in moist places.

> * * *Stamens 2, or 1 only.*

4. **S. longifolia,** Muhl. River-bank shrub, with slender stem and branches: leaves linear to lanceolate, long-acuminate, sessile or nearly so, usually 2—3 in. long, remotely mucronate-dentate or entire, seldom nearly glabrous, more commonly *from hoary to almost white with silky*

tomentum: aments on lateral leafy branches, and some terminal, often several together on a vigorous shoot, linear-cylindric: *scales* yellowish-villous, toothed, *deciduous:* stamens 2: capsules oblong-conic, obtuse, usually tomentose.—Common, but chiefly at some distance from the Bay.

5. **S. lasiolepis,** Benth.   Tree sometimes 30 or 40 ft. high, with smooth grayish trunk and broad rounded head: leaves oblanceolate or oblong-lanceolate, 3—5 in. long, obliquely acute or acuminate, more or less strongly pubescent when young; or even in age, coriaceous, unequally serrate, glaucous and somewhat ferruginous beneath; stipules mostly none: *aments subsessile, appearing long before the leaves,* cylindric, dense; *scales rounded, blackish, densely white-silky, persistent:* stamens thrice as long as the scale: filaments slightly united at base: capsules dark green, glabrous, acute.—The most abundant of willows, in the Bay region; found along all streams, and in all springy places.

6. **S. Bigelovii,** Torr.   Only a low tufted shrub, seldom 8 or 10 ft. high; branches very stout, pubescent: leaves obovate, or cuneate-oblong, obtuse, entire, silky-pubescent beneath: aments on short villous somewhat leafy peduncles, short and dense: ovary ovate, pedicellate.—Apparently confined to moist depressions among the sand hills near San Francisco; early-flowering like the last; in our opinion distinct from it.

7. **S. Sitchensis,** Sanson.   Shrub 6—15 ft. high, with straggling rather *slender recurring branches: these pubescent when young,* sometimes glaucous: leaves oblong-obovate to oblanceolate, acute, or the earliest obtuse and cuspidate-pointed, the base narrowed to a short petiole, lustrous-white and satiny-tomentose beneath, margin entire or obsoletely crenulate: aments erect, slender, dense: scales yellowish or tawny, sparsely villous: *stamen solitary:* capsule ovate-conic, acute, tomentose.—Banks of streams among the higher hills.

2. POPULUS, *Pliny* (POPLAR).   Trees with branchlets somewhat angular. Buds with several and imbricated scales. Bracts of the aments lacerately toothed or fringed. Stamens few or many, inserted on an obliquely truncate disk; filaments filiform; anthers purple. Capsule ovate-oblong to globose, 2—4-valved.

1. **P. trichocarpa,** Torr. & Gray.   Tree of somewhat conical outline, 30—50 ft. high: buds shining and viscid: *leaves broadly ovate to oblong-lanceolate, acuminate, 2—4 in. long, crenate,* finely puberulent when young, in maturity pale beneath; *petioles terete,* 1—2 in. long: staminate aments dense, 2 in. long; bracts slightly villous: rachis pubescent: disk broad, slightly pubescent; filaments as long as the anthers: pistillate aments 2 in. long (becoming 6 in.); rachis pubescent; bracts and dilated disk nearly glabrous; ovary densely pubescent; styles 3, broadly dilated and lobed: *capsules* almost sessile, *subglobose, pubescent,* 3-valved, ¼ in. thick: seed light-colored.—Along the larger mountain streams.

2. **P. Fremonti,** Wats. Large tree with broad rounded or depressed head, gray, cracked bark, and subterete branchlets: *leaves deltoid-ovate, sinuate-crenate; petioles flattened*, and, with the branchlets and leaf-margins, often pubescent when young: bracts and rachis of aments glabrous: fruiting aments 3—4 in. long: capsules ovate, 3—4 lines long, on stout pedicels 1—2 lines long; disk ¼ in. broad: valves of capsule 3: seeds white.—Valleys among the coast ranges, and also on plains of the interior.

Order LXXXV. JUGLANDEÆ.

Represented by a single species of the genus

1. JUGLANS, *Pliny* (WALNUT TREE). Trees with hard wood, and alternate exstipulate unequally pinnate somewhat resinous-aromatic leaves. Staminate flowers in long aments, 12—40 stamens to each of the 3-lobed green perianths; pistillate solitary, or few and spicate, their calyx adherent to the ovary, 4-toothed and bearing 4 small petals. Pistil 1; style short; stigmas 2, linear or clavate, fringed. Pericarp large, fleshy, indehiscent, enclosing a rugose nut. Seed without albumen; cotyledons fleshy, 2-lobed, rugose.

1. **J. Californica,** Wats. Tree 40—60 ft. high; trunk 2—4 ft. thick: leaflets 5—8 pairs, oblong-lanceolate, acute, 2—2½ in. long: aments loose, 4—8 in. long: fruit globose, little compressed, 1 in. thick: nut shallow-sulcate.—Frequent along streams, chiefly back from the seaboard.

Order LXXXVI. CUPULIFERÆ.

Monœcious trees or shrubs, with alternate simple pinnate-veined leaves, caducous stipules, and staminate flowers in cylindrical usually pendulous aments. Individual staminate fl. with a lobed or cleft perianth. Pistillate sessile in a cup-like involucre (1—5-flowered) covered with bract-like or spinescent appendages; perianth 6-lobed, adherent to the 2—6-celled and 4—12-ovuled ovary, this becoming a 1-celled 1-seeded nut inserted in or enclosed within an involucre. Seed without albumen; cotyledons large, fleshy.

1. QUERCUS, *Pliny* (OAK). Staminate flowers in slender usually pendulous aments; calyx 4—8-lobed or -parted; stamens 3—10; anthers 2-celled. Pistillate flowers single or in clusters; ovary 3-celled, 6-ovuled, bearing 3 styles or sessile stigmas, and enclosed by a scaly bud-like involucre which enlarges into a cup around the base of the rounded or elongated nut *(acorn)*, the 5 undeveloped ovules remaining as rudiments at base or summit of the perfect seed.

* *Stigmas nearly or quite sessile; bark usually light-colored; wood nearly white; foliage of a dull or pale green.*—WHITE OAKS.

    ← *Acorns ripening the first year; leaves deciduous.*

1. **Q. lobata,** Née. Stately tree, with slender, often long and pendulous branches: leaves oblong or obovate, 2½—5 in. long, deeply lobed or pinnatifid: acorns subsessile; *nut long-conical,* 1¼—2½ *in. long,* usually pointed: *cup deep-hemispherical, strongly tuberculate.*—Plains of the interior chiefly, but plentiful and of fine development in Napa Valley as well as elsewhere in valleys of the Coast Range.

2. **Q. Garryana,** Dougl. Not as large as the last; branches not drooping, but branchlets rigid, tomentose or pubescent: leaves more coarsely lobed; lobes broad, obtuse; acorns sessile or short-peduncled; *nut oval, often ventricose,* 1¼—1½ *in. long; cup small and shallow, with small lanceolate* slightly pubescent *scales.*—A tree of the hilly districts from Marin Co. northward.

3. **Q. Douglasii,** Hook. & Arn. Middle-sized tree, or larger, with rounded, or in age depressed head; branches and branchlets numerous; bark of trunk very light-colored: *leaves small* (2—3 in. long), *oblong,* sinuate or with shallow lobes, *bluish-green above, pubescent beneath:* acorn sessile or short-peduncled; nut elongated-oblong, ¾—1¼ in. long, mostly acutish; cup hemispherical, with ovate-lanceolate flat or sometimes tubercled scales—Mostly inhabiting the first and rather dry foothills of the inner ranges.

    ← ← *Acorns ripening the first year; leaves persistent.*

4. **Q. dumosa,** Nutt. *Shrub 4—8 ft. high,* the slender branches tomentose when young: *leaves coriaceous, 1 in. long or more, oblong, obtuse,* often acutish at base, sinuate or sinuate-toothed, on young shoots spinose-toothed, dark green above, pubescent beneath: acorns sessile, variable in size; nut oval, 1 in. long more or less; cup deep-hemispherical, 5—10 lines wide, usually strongly tuberculate, occasionally with flat scales. Var. **bullata,** Engelm. Leaves more rounded, convex above, usually hoary-tomentose on both sides.—A southern species, properly, but reaching our borders in San Mateo Co., and even southern Alameda; also recurring as far north as Lake Co., both type and variety.

    ← ← ← *Acorns biennial; trees evergreen.*

5. **Q. chrysolepis,** Liebm. Varying from the size of our largest oaks, to mere shrub of a few feet high: *leaves oblong,* acute or cuspidate, obtuse or subcordate at base, usually entire on old trees, sharply spinose-dentate on more vigorous young trees or shoots, *very pale and glaucous above, fulvous-tomentose beneath* when young, in age glabrate: acorns extremely variable in size: nut oval, obtuse, ½—1½ in. long; cup hemispherical, with scales almost hidden by fulvous tomentum, and

⅓—1 in. wide.—Scarcely other than small trees, or mere shrubs, of this species within our limits; but such forms common on the higher mountains, like Diablo and Tamalpais.

     * * *Stigmas on long styles; bark dark-colored; wood reddish;*
         foliage bright green.—BLACK OAKS.

6. **Q. agrifolia,** Née. Large tree, in our district mostly of low stature and widely spreading, the trunks or main branches often several from near the ground and ascending; tree not properly evergreen, but foliage persisting for a year: *leaves oval to oblong or oboroid,* 2—3 in. long, *sinuately spinose-dentate,* somewhat stellate-pubescent when young, in maturity mostly convex above, pale and glabrous beneath: *acorns annual* sessile or nearly so; nut rather narrow and tapering, 1—1½ in. long, 3—4 lines thick; cup turbinate, rather deep, with lanceolate imbricated slightly pubescent brown scales.—Common along streams, and in more open ground; at San Francisco only a large arborescent shrub.

7. **Q. Wislizeni,** A. DC. Resembling the last, but *evergreen;* leaves more coriaceous and lanceolate or oblong-lanceolate, commonly acute and less sinuately spinose-dentate: *acorns biennial:* cup turbinate, *very deep;* nut much as in the last.—Less frequent in our district; but small forms common on the higher hills and mountains.

8. **Q. Morehus,** Kell. Small tree (20—40 ft.), straight and symmetrical, or more lax and straggling: *leaves subpersistent and rather coriaceous, oblong-lanceolate, 3—4 in. long, acutish at base and long-petioled,* the margin coarsely sinuate-toothed and teeth subulate from a broad base: acorns biennial, solitary, on peduncles ½ in. long; cup hemispherical, with glabrous ovate scales ciliate below; nut oblong, obtuse, two-thirds exserted from the cup.—Marin and Contra Costa counties, and northward; nowhere plentiful; possibly a hybrid between the preceding and the next.

9. **Q. Kelloggii,** Newb. Middle-sized tree, not very symmetrical; bark of trunk rough, black: *leaves ample, deciduous,* broadly oval or obovoid, 4—6 in. long, *pinnately lobed,* the lobes tapering and entire, or broad and coarsely toothed, at first tomentose, at length glabrous: acorns biennial mostly on peduncles ½—1 in. long, often several together; cup hemispherical, rather deep, with ovate-lanceolate obtusish scales; nut oblong, obtuse, 1 in. long or more.—In the mountains chiefly, or along their bases in low valleys.

     * * * *Aments erect; flowers whitish; cup echinate.*

10. **Q. densiflora,** Hook. & Arn. Middle-sized or large symmetrical evergreen, with smooth bark and tomentose branchlets: *leaves oblong, acute, 2—6 in. long, coarsely and rather remotely spinulose-serrate, with strong feather-veins running from the midrib to each tooth,* tomentose

beneath, in age glabrate above: acorns solitary or in short-peduncled clusters; cup shallow, covered with linear or linear-subulate spreading scales and silky-tomentose within; nut very hard, oval or oblong, indistinctly trigonous at summit.—Moist woods of the Coast Range.

11. **Q. echinoides,** R. Br. Campst. Shrub bushy and low, 3—12 ft. high: *leaves* mostly 1—3 in. long, *oblong, obtuse, entire or crenate* and more or less revolute, the *feather veins obsolete toward the margin:* nut more elongated than in the last; cup with longer softer more or less tortuous linear appendages.—Hillsides of the inner coast ranges, mostly northward beyond our limits; but occurring at the Petrified Forest, Sonoma Co.

2. CASTANEA, *Brunfels.* Differing from the type of *Quercus* in that the staminate aments are erect, and sometimes panicled, and the pistillate in involucres situate at base of these. Nuts 1—several, enclosed in a pungently prickly at length dehiscent involucre.

1. **C. chrysophylla,** Dougl. Arborescent, erect, seldom 30 ft. high with us (much larger at the north), evergreen: leaves coriaceous, lanceolate, acuminate or acute, 1—4 in. long, densely yellowish-scurfy beneath: stout spines of involucre ½—1 in. long subverticillately branching: nut usually solitary, obtusely trigonous, ½ in. long.

ORDER LXXXVII. **CORYLACEÆ.**

Allied to the preceding order; represented here by one species of

1. **CORYLUS,** *Vergil.* Monœcious shrubs, with alternate simple leaves. Staminate flowers in pendulous aments appearing before the leaves. Stamens 4 to each obovate bract. Pistillate flowers in a short spike, 2 flowers to each bract, with small bractlets which become an involucre to the subglobose woody-shelled smooth nut.

1. **C. rostrata,** Ait. var. **Californica,** A. DC. Shrub seldom less than 8 ft., sometimes 20 ft. high, slender, with ascending and at last widely spreading branches: leaves broadly ovate or oval, 2—4 in. long, acute or abruptly acuminate, cordate or rounded at base, pubescent or even hirsute: fruiting involucre densely hispid, the bracts unitedly prolonged above the nut into a short broad tube; nut ½ in. thick.—By streams in the foothills and lower mountains, usually in the shade.

Subclass II. MONOCOTYLEDONOUS or ENDOGENOUS PLANTS.

Embryo with but one cotyledon. Leaves parallel-veined. Flowers having their parts mostly in threes.

Division I. CALYCEÆ PERIGYNÆ.

Perianth with tube adnate to the ovary, the stamens consequently perigynous, or else (as in the first order) gynandrous.

Order LXXXVIII. ORCHIDACEÆ.

Perennial herbs, the roots mostly few and very fleshy fibrous, or still more fleshy and tuberiform. Flowers usually inverted by a twist in the long ovary, the sepals and 2 lateral petals similar, the third and superior petal (apparently the lower) unlike the others and called the *lip*. Stamens coherent with the style and with it forming the *column*, with usually only the anther opposite the lower sepal perfect, and 2 rudimentary lateral ones; anthers 2-celled; pollen more or less coherent in 1—4 waxy masses. Stigma oblique, concave. Capsule dehiscing by 3 placentiferous valves. Seeds many, minute, resembling saw-dust.

Stem with ample leaves; flowers few, not small..................LIMODORUM 1
Stem-leaves few, small; fl. many, small, spicate;
    Lip undulate, not spurred at base............................ORCHIASTRUM 2
    " with a long spur at base...................................HABENARIA 3
Plants leafless and destitute of chlorphyll........................CORALLORHIZA 4

1. **LIMODORUM,** *Clusius.* Caulescent and leafy, erect, from creeping rootstocks. Flowers few, in a terminal conspicuously leafy-bracted raceme. Perianth spreading; sepals and petals subequal: lip free, deeply concave at base, narrowly constricted and somewhat jointed in the middle, the upper part petaloid-dilated. Anther 1, sessile behind the broad truncate stigma, 2-celled, obtuse; pollen-masses becoming attached above to the gland capping the small rounded beak of the stigma.

1. **L. giganteum** (Dougl.), O. Ktze. Stoutish, 1—4 ft. high, almost glabrous: leaves ovate below, lanceolate above, 3—8 in. long, acute or acuminate, somewhat scabrous on the veins beneath: fl. 3—10, greenish strongly veined with purple, on short pedicels; sepals ovate-lanceolate, 6—8 lines long, the upper concave; petals a little smaller; lip as long, the saccate base with erect wing-like margins, the dilated summit ovate-lanceolate, entire, somewhat wavy-crested.—Moist ground, Marin Co.

2. **ORCHIASTRUM,** *Micheli.* Stems leafy below, from fascicled tuberiform roots. Flowers small, in a 1—3-ranked spirally twisted spike. Lateral sepals somewhat decurrent; the upper and the petals coherent:

lip oblong, the dilated summit spreading and undulate, usually entire. Column short, oblique, ending in a stout terete stipe bearing the ovate stigma on the face.

1. **O. Romanzoffianum** (Cham.). Stout, 4—18 in. high, bracteate above: leaves oblong-lanceolate to linear: spike dense, 3-ranked, 1—4 in. long, conspicuously bracteate: perianth white, about 4 lines long; sepals and petals all connivent; lip recurved, ovate-oblong, contracted below the wavy-crenulate summit; *callosities obscure.*—Near the Presidio, San Francisco, *Bolander,* and in Marin Co. Sept.

2. **O. porrifolium** (Lindl.). Much like the last, but with smaller flowers, and consequently a narrower spike: *callosities at base of lip sharply prominent* and pointing downward.—Marin Co., *Behr.*

3. **HABENARIA,** *Willd.* Stems leafy-bracted or leafy, erect simple and solitary, from perpendicular fleshy-fibrous or tuberiform roots. Flowers small, green or white, in a terminal spike or raceme. Sepals and petals similar, convergent: lip flat, spreading entire (in ours), with a slender long spur at base on the outside.

1. **H. elegans** (Lindl.), Bolander. Stem rather slender, 1—2 ft. high, from an ovate or oblong tuberform root: leaves 2, radical, depressed, oblong, 3—5 in. long and 1½—2 in. broad, appearing in early spring, but dying and disappearing before the flowering period: *fl. small, light green, in a slender but dense long spike;* sepals and petals subequal, 2 lines long, obtuse; lip similar, with filiform spur 3—5 lines long: beak of stigma prominent, broad and rounded.—Wooded hillsides. June, July.

2. **H. maritima,** Greene. Robust, 6—16 in. high; at flowering time destitute of foliage, but the upper part of the stem bearing many lanceolate-subulate appressed and more or less imbricated green bracts ½ in. long or more: *spike* 1½—3 *in. long, 1 in. thick,* the flowers closely crowded, white, heavily honey-scented: sepals oblong, obtuse, 1¼ lines long, white, with a narrow and delicate deep green midvein; *petals* not quite equalling the sepals, oblong-lanceolate, the upper 2 plane, *deep green at base and well up the middle, otherwise white; the lip pure white* even to the prominently elevated and broad midvein: spur slender, longer than the ovary.—On dry hills near the sea at Point Lobos; leaves probably appearing in early spring and soon dying. Fl. Aug.—Oct.

3. **H. Michaeli,** Greene. Very robust, 8—12 in. high, *leafless, but the cylindric and apparently very fleshy stem bearing many* triangular or triangular-ovate acuminate *thin appressed bracts:* spike very dense, 3 in. long: fl. greenish; sepals and petals alike, ¼ in. long; lip broader, its spur a third longer than the ovary.—Open hills, under oaks, etc., from near Livermore southward.

4. **CORALLORHIZA,** *Haller.* Plants without green herbage, the solitary scapes from fleshy short jointed often coralline roots, and bearing

a few sheathing bracts in place of leaves, and a raceme at summit. Lateral sepals oblique at base, either decurrent into a short spur adnate to the side of the ovary, or forming a projecting gibbosity above it. Upper sepal and petals somewhat incurved: lip dilated, usually somewhat recurved, flat or concave, with a pair of longitudinal ridges near the base. Column bearing the caducous anther at summit. Pollen masses in 2 distinct pairs, sessile on a short oblong gland.

1. **C. multiflora,** Nutt. Scape 1—2 ft. high: sepals and petals 4—5 lines, buff, with reddish or purplish markings; spur a line long or more but quite adnate to the ovary; lip broadly ovate, subsessile, 3-lobed by a deep cleft on each side, the middle lobe rounded or emarginate, the lateral ones narrow and acutish, ridges of the body of the lip rather prominent: column stout, two-thirds the length of the petals.—Woods of the mountain ranges near the coast.

2. **C. Bigelovii,** Wats. Flowers fewer and smaller, altogether red-purple and pure white: lateral sepals oblique and, with the base of the column, strongly gibbous over the ovary: lip deeply concave, broad and somewhat auriculate, elliptical; spur none: column slender, broadly margined below.—More frequent than the last, and on both sides of the Bay; the flowers rather inconspicuous.

Order LXXXIX. IRIDACEÆ.

Stems in our species from creeping stout rootstocks, or a tuft of fibrous roots. Leaves 2-ranked, ensiform, sheathing. Flowers large, few or solitary, spathaceous-bracteate, regular, triandrous. Petal-like divisions of the superior perianth 6, in 2 series, convolute in bud. Stamens on the base of the outer series, or sepals; their anthers extrorse. Ovary 3-lobed, becoming a 3-celled capsule.

1. **IRIS,** *Theophr.* Stems from thick rootstocks, stout, terete. Flowers in the axils of spathaceous bracts along the flexuous upper part of the stem. Perianth-tube prolonged beyond the ovary; outer segments obovate above the claw, spreading or recurved; inner narrower, erect, at apex connivent. Stamens with linear anthers lying close beneath spreading or somewhat recurved large petaloid branches of the style. Capsule elongated, trigonous. Seeds flattened or turgid, horizontal, in 2 rows in each cell.

1. **I. macrosiphon,** Torr. Stem low and very slender, much shorter than the leaves (these 6—15 in. long), somewhat flattened: leaves narrow, acuminate: bracts linear-lanceolate, long-acuminate, 2½—4 in. long; fl. 1 or 2, short-pedicellate, with *filiform tube 1—3 in. long,* dark violet-purple: sepals 1½—2 in. long: capsule oblong-ovoid, abruptly acute at each end, 1 in. long: *seeds flattened and angular.*—Open hills about San Francisco, and elsewhere near the sea. March, April.

2. **I. Douglasiana,** Herbert. Stouter and taller than the last, ½—1½ ft. high: bracts broader, less acuminate: fl. 2 or 3, larger and on longer pedicels: *perianth-tube only* ½—1 *in. long:* capsule oblong, acutely triangular, 1¾ in. long: *seeds almost globular.*—Habitat of the preceding. April, May.

3. **I. longipetala,** Herbert. Stout, 1—2½ ft. high; leaves about as high: bracts large, acuminate, 3—4 in. long: *flowers* 3—5, pale blue, *on stout pedicels 1—2 in. long; tube funnelform,* ¼ in. long; *sepals* 2½—3 *in. long;* petals somewhat shorter, oblanceolate: capsule oblong, narrowed at each end, 2 in. long: seeds flattened.—Very plentiful on moist slopes of hills on the southern outskirts of San Francisco. May, June.

2. **BERMUDIANA,** *Tourn.* Stems and foliage from a tuft of coarse-fibrous perennial roots. Leaves narrow and grass-like. Stems flat. Flowers clustered within a pair of ensiform bracts, fugacious. Perianth 6-parted; segments only slightly differing in breadth, but of equal length, usually cuspidate, spreading. Stamens monadelphous; anthers oblong. Style short; stigmas filiform, involute. Capsule subglobose. Seeds several, nearly spherical.

1. **B. bella** (Wats.). Stems 1—1½ ft. high, glabrous or with scabrous margins, with 1—3 floriferous nodes at the summit: peduncles usually 2 at each node; spathes of 2 nearly equal bracts, scabrous on the keel, 4—7-flowered: *perianth deep blue-purple with yellow base,* expanding ¾ in. or more: *stamens monadelphous to near the summit;* anthers very small: capsule round-obovoid, ¼ in. high: seeds ⅔ line thick, obscurely pitted. —Abundant on moist slopes. March—June.

2. **B. Californica** (Ker). Scape broadly winged, 6—15 in. high, sur-passing the broad glaucescent leaves: *perianths* 3—7 to the spathe, *clear yellow,* ¾ in. broad: *filaments united at base only;* anthers linear-sagittate. —Boggy places on the San Francisco peninsula. Herbage turning black in drying, and staining paper dark purple.

Division II. CALYCEÆ HYPOGYNÆ.

Ovary superior; the stamens inserted at its base, either as distinct from the perianth, or as adherent to its tube.

Order XC. LILIACEÆ.

Ours all herbaceous plants; the stems usually from bulbs or corms or more or less fleshy rhizomes. Perianth regular, corolla-like, of 6 distinct or more or less united parts. Stamens opposite the segments, with 2-celled anthers (or confluently 1-celled). Ovary superior, 3-celled, becoming a capsular, or rarely a baccate fruit.

1. LILIUM, *Pliny.* Stem simple from a scaly bulb, leafy. Leaves narrow, sessile, scattered or whorled. Flowers large, showy, in a terminal raceme, or solitary, or subumbellate; pedicels not jointed. Perianth deciduous, funnelform or broader, with 6 distinct equal oblanceolate spreading or recurved segments, whitish or red, often spotted with brown. Stamens 5, hypogynous, included; filaments long; anthers oblong or linear, versatile. Style long, clavate, deciduous; stigma 3-lobed. Capsule coriaceous, 6-angled. Seeds many, flat, horizontal, in 2 rows in each of the 3 cells.

1. **L. rubescens,** Wats. *Bulb slightly oblique and rhizomatous,* 2 in. thick: broadly lanceolate scales 1 in. long: stem 1—7 ft. high: leaves glabrous, glaucescent, flat or undulate, lower scattered, upper in 3—7 whorls, oblanceolate, acute or acutish, 1—4 in. long: pedicels usually several, 1—3 in. long; *fl. nearly white becoming rose-purple,* somewhat dotted with brown; segments 1½—2 in. long, the upper third revolute: ovary wing-angled, alternate downward, ½ in. long.—Wooded slopes from Marin Co. northward.

2. **L. maritimum,** Kell. *Bulb conical,* 1—1½ in. thick, the scales closely appressed: stem 1—3 ft. high, slender: leaves seldom at all whorled, linear or narrowly oblanceolate, obtuse, 1—5 in. long: *fl. 1—5, long-peduncled,* horizontal, *deep reddish orange,* spotted with purple; segments lanceolate, 1½ in. long, the upper third somewhat recurved: oblong anthers 2 lines long.—Moist meadows near the coast, from Marin Co. northward.

3. **L. pardalinum,** Kell. Stem (from a thick branching rhizomatous mass of oblate scaly bulbs) 3—7 ft. high: leaves usually in 3 or 4 whorls of 9—15, with some scattered ones both above and below the whorls, narrowly lanceolate, sharply acuminate, deep green, thin, faintly 3-nerved; fl. usually very numerous, on long spreading pedicels; *perianth-segments 2—3 in. long,* ½—¾ *in. wide, strongly revolute, bright orange-red, with large purple spots* on the lower half: anthers red, 4—5 lines long: capsule narrowly oblong, 1½ in. long, angles acutish-—The most common lily of the Bay district, occurring in open marshy places, or in shady places along streams. June, July.

2. **FRITILLARIA,** *Gesner.* Stem, as in *Lilium,* upright and leafy, from a scaly bulb, but this commonly depressed-globose, often corm-like, the scales little flattened; flowers usually more campanulate, but sometimes funnelform. Perianth-segments deciduous, concave, mostly with a nectariferous pit near the base, the coloring commonly dull, but apt to be more or less checkered in light and dark shades. Stamens 6, included; anthers oblong, versatile. Styles slender, united to the middle or throughout. Capsule often acutely 6-angled, or even winged.

* *Perianth checkered more or less distinctly in two colors.*

1. **F. coccinea,** Greene. Stem stoutish, 8—18 in. high: leaves 2—3 in. long, linear-lanceolate, mostly 4—12, in 2 or 3 whorls near the middle of the stem: fl. mostly few (1—4), 1 in. long; segments not recurved at tip, spreading to the degree of the *campanulate-funnelform, yellow and scarlet,* not indistinctly checkered: styles distinct above; stigmas linear: capsule rather obtusely angled.—Wooded mountains of Sonoma and Napa counties. May.

2. **F. mutica,** Lindl. Bulb broad and flat, above thickly beset with tuber-like scales: stem 1½—3 ft. high, mostly 6—12-flowered. deep green

and glaucescent: leaves linear-lanceolate, in 3—7 whorls: *perianth cam-panulate, checkered with green and dark brown; segments* oblong-lanceo-late, strongly arched, with large oblong nectary and more or less strongly *crisped or crenulate-undulate* margin: styles distinct above: capsule 1 in. long, broadly winged. March, April.

* * *Perianth not checkered.*

3. **F. biflora,** Lindl. Bulb of few and thick fleshy ovate scales: stem stout, 6—18 in. high, 1—3-flowered: leaves 2—6, mostly near the base of the stem, scattered or somewhat whorled, oblong-lanceolate to linear, 2 –4 in. long: perianth campanulate, of a very dark lurid purple: style stout, parted into three above: capsule ½—¾ in. long, 6-angled. Var. **agrestis.** Taller, commonly 3—7-flowered: perianth wholly of a light yellowish green.—Type common on stony hills near the sea. The variety is of rich fields in the Livermore Valley and eastward. March.

4. **F. pluriflora,** Torr. Bulb as in the last; and plant every way like the variety *agrestis* except that the *perianth is narrowed at base* so as to be funnelform-campanulate, or turbinate-campanulate, and *the whole of a reddish purple;* nectary obscure: stigma shortly 3-lobed: stamens unequal.—Along the eastern bases of the inner coast ranges, in Solano, and apparently in Contra Costa.

5. **F. liliacea,** Lindl. Bulb as in the last two, or the scales fewer and larger: stem low: leaves few, those nearest the base often approx-imate, or even verticillate: segments of the 1 –3 whitish perianths broadly oblanceolate, without nectary, the whole *perianth tapering to the base: anthers oblong, mucronate: capsule stipitate,* ½ in. long and broad, truncate at both ends.—Open grassy hills near the sea, or in the Coast Range. April.

3. **ERYTHRONIUM,** *Linn.* Stem low, simple, 1—several-flowered, from an elongated bulb-like corm. Leaves 2, broad, thin, radical. Perianth broad-funnelform, with segments mostly recurved, deciduous. Stamens 6; filaments slender; anthers basifixed. Capsule membrana-ceous, obtusely triangular, loculicidally 3-valved. Seeds oblong-obovate, ascending in 2 rows in each cell; brown rugulose testa loose at apex.

1. **E. giganteum,** Lindl. Leaves 6 –10 in. long, narrowed to a short margined petiole: scape 10—15 in. high, 1—6-flowered: *perianth cream-color tinged with pink, yellow at base;* segments broadly lanceolate, 1—1½ in. long, recurved: capsule oblong-obovate, 7—9 lines long, very obtuse or retuse at summit.—In woods on northward slopes of the higher mountains of Sonoma Co., etc.

2. **E. Hartwegi,** Wats. Smaller; leaves not rarely 3: *perianths 1—3, pale yellow, with orange base,* the segments 1—2 in. long, *scarcely recurved.*—Near Healdsburg, Sonoma Co., and northward.

4. CALOCHORTUS, *Pursh.* Stem branching, from a membranous-coated corm (or this rarely fibrous-coated). Leaves few, linear-lanceolate, the radical 1 or 2 much larger than those of the flexuous stem. Perianth deciduous, of 6 more or less concave segments; the 3 outer small, comparatively colorless and therefore true sepals. Petals broadly cuneate-obovate, usually with a glandular pit near the base, this apt to be hidden by long hairs. Stamens 6; anthers erect, basifixed. Ovary triquetrous, 3-celled; stigmas sessile, recurved. Capsule membranaceous, 3-angled or -winged. Seeds ∞, in 2 rows in each cell, flattened; testa thin.

    * *Flowers nodding, subglobose with concave petals; capsule broad, obtuse, nodding.*

    1. **C. albus,** Dougl. Glaucous, 1—2 ft. high, almost paniculately branching: radical leaves 1—2 in. wide: *petals pearly-white, with a leaden-lavender tinge outside,* round-ovate, 1—1¼ in. long, bearded above the gland with long white hairs; *gland lunate,* shallow with 4 transverse imbricated scales fringed with short hairs: anthers linear-oblong, obtuse, mucronate: capsule 1—2 in. long, ½—1 in. wide.—Shady banks in the Coast Ranges near the sea.

    2. **C. pulchellus,** Dougl. Half as tall as the last, more branching and floriferous: radical leaf equalling or exceeding the stem: sepals greenish-yellow, 8—12 lines long, nearly equalling *deep yellow oblong-ovate acutish petals,* these glandular-ciliate, and with scattered short erect yellow hairs; *gland a deep pit,* produced upward and covered by stiff appressed yellow hairs: anthers broad, obtuse or acutish: capsule elliptical.—Cool summits of the more elevated Coast Range hills and mountains, in shade of thickets.

    * * *Flowers less concave, erect: capsules broad, nodding.*

    3. **C. Maweanus,** Leichtl. Stem low (3—10 in. high), 3—6-flowered: glaucous leaves surpassing the stem: petals white to purplish-blue, exceeding the purplish' sepals, broadly obovate, acute, somewhat arched and pitted, *the broad naked claw with a transverse semicircular scale, the rest of the surface within covered with long white or purplish hairs:* anthers lanceolate, acuminate: capsule oblong-elliptic, acutish.—Open hills.

    4. **C. lilacinus,** Kell. Stem 4—8 in. high: leaves long and broad: *fl. 4—10, in 1—3 close umbels or corymbs on long flexuous pedicels:* petals broad, pale lilac, with purplish claw, somewhat *hairy below the middle; gland* very shallow, *ciliate,* and with a narrow scale: anthers oblong, obtuse: capsule elliptic, obtuse at each end.—Open hills.

    5. **C. uniflorus,** Hook. & Arn. *Scarcely branching,* ½—1 ft. high, 1—2-flowered: sepals purplish, 6—8 lines long; *petals 10—12 lines, the summit broad, denticulate,* lilac, with purple base; *gland* small, shallow, *purple, densely hairy;* lower half of the petal with scattered hairs: anthers obtuse.—In low moist lands; plentiful at Calistoga.

* * * *Flowers open campanulate, both these and the capsules erect.*

6. **C. luteus,** Dougl. Leaves all very narrow, not conspicuous: stem 10—18 in. high, 1—3-flowered; sepals about equalling the petals, narrowly lanceolate, acuminate, erect, greenish yellow and purplish, with a brown spot at base: *petals broadly fan-shaped, 1 in. long, greenish yellow*, marked with brown-purple and slightly hairy in the middle portion; gland broad, rounded or somewhat lunate, densely covered with ascending yellow hairs, and with some scattered spreading ones surrounding it: anthers yellow, linear, obtuse: capsule attenuate upward, 1—1½ in. long.—Open plains and hillsides. May, June.

7. **C. venustus,** Benth. Habit of the preceding; foliage the same: campanulate *perianth* twice as large, *white or pale cream-color suffused with lilac above*, and with a conspicuous *red-purple spot near the top, a brownish spot overarched with yellow in the centre*, and a brownish base; gland large, oblong, densely hairy and surrounded with scattered hairs. —From Napa Co. southward, in the inner ranges of mountains.

8. **C. splendens,** Dougl. Like the last in habit, though more slender, branching, and with more numerous flowers: *sepals strongly recurved: petals deep lilac-purple*, with scattered white hairs below the middle, and a round densely hairy gland: anthers purple.—Hills at the western base of Mt. Diablo near Danville, and southward. June.

5. **CAMASSIA,** *Lindl.* Scape and linear carinate leaves from a tunicated bulb. Flowers in a simple scarious-bracted raceme. Perianth of 6 distinct and similar oblanceolate 3—7-nerved segments, nearly rotate-spreading. Stamens 6 on the base of the perianth; filaments filiform-subulate; anthers linear-oblong, versatile. Style slender, trifid. Capsule thick-membranaceous, 3-lobed and -angled, loculicidally 3-valved. Seeds several in each cell, ovate, more or less compressed and angled, with thin black shining testa.

1. **C. Quamash** (Pursh). Scape 1—2 ft. high: raceme loosely 10—20-flowered; pedicels ¼—1 in. long, shorter than the narrow bracts: fl. dark blue to very pale, 7—15 lines long.—Point Reyes, *Bigelow*.

6. **CHLOROGALUM,** *Kunth.* Tuft of radical linear carinate leaves and paniculately branching stem from a fibrous-coated bulb. Branches of panicle loosely racemose. Perianth white, or with a purplish tinge, persistent, remaining twisted over the growing and mature ovary. Perianth-segments distinct, ligulate. Stamens 6, adnate to base of segments; anthers versatile. Ovary subglobose; ovules 2 in each cell; style filiform slightly 3-cleft. Capsule broadly turbinate, 3-lobed, loculicidal. Seeds obovate, with close thin somewhat rugose testa.

1. **C. pomeridianum** (Ker), Kunth. Bulb large, long, densely and coarsely fibrous-coated: stem and spreading panicle 2—5 ft. high: leaves

½—1½ ft. long, strongly undulate: perianth rotate, its segments 8—10 lines long, white, with purple veins.—Common on rocky banks and hills; flowers vespertine, opening in the middle of the afternoon. June, July.

7. SCOLIOPUS, *Torr.* Nearly stemless glabrous perennials, the fascicle of coarse-fibrous roots bearing, on a short subterranean erect stem, a pair of ample oblong leaves and the nearly obsolete peduncle of an umbel, the elongated and sharply angled pedicels of which, resembling 1-flowered scapes, alone appear above ground. Perianth of 3 ovate-lanceolate spreading sepals and as many linear erect petals. Stamens 3, opposite the sepals; filaments short, cylindric, with a bulbous-dilated base; anthers oblong, extrorse, attached above the base. Ovary triquetrous; style very deeply 3-parted, the branches linear, stout, widely spreading and with a downward curvature, stigmatic only at the rounded apex, this held near the open anther by the curve in the style-branch. Capsule thin, bursting irregularly. Seeds numerous, oblong, slightly curved, sulcate-striate.

1. **S. Bigelovii,** Torr. Leaves 4—15 in. long, mottled or without dark spots, acute or obtusish: pedicels 3—15, slender, 3—8 in. long, sharply 3—4-angled, in age reclining and strongly tortuous or somewhat coiled: sepals spreading at base, bent downward abruptly from about the middle, the ground-color green, but this obscured by many dark-red striæ: petals as long, strongly revolute and nearly erect.—Mt. Tamalpais, in low woods; flowers very fetid. Jan.—March.

8. TRILLIUM, *Miller.* Perennials, with short fleshy rootstocks and fibrous roots. Solitary erect stems scariously sheathed at base, at summit bearing a whorl of 3 ample rhombic-ovate subsessile netted-veined leaves and a solitary 6-merous flower. Sepals 3, lanceolate, green-herbaceous, persistent. Petals 3, white or purplish to dark maroon, exceeding the sepals, persistent. Stamens 6, on the base of the perianth; filaments short, stout; anthers long, basifixed. Ovary sessile; styles none; stigmas linear or subulate, recurved. Fruit an ovate angular fleshy capsule 1—3-celled. Seeds ovate.

1. **T. sessile,** L. *Leaves and flower closely sessile; petals erect.* Var. **giganteum,** Hook. & Arn. Stout, 1—2 ft. high: leaves broadly rhombic-ovate, 3—6 in. long, usually broader than long, acutish: petals oblanceolate, 2—4 in. long, dark-maroon to rose-purple and white. Var. **chloropetalum,** Torr. Still larger, the obovate-elliptic obtuse petals twice the length of the sepals, and uniformly light green.—Only the varieties occur; the first plentiful in the Oakland and Berkeley hills, the second in the redwood districts of Marin Co. March, April.

2. **T. ovatum,** Pursh. Slender, 8—18 in. high: leaves 2—6 in. long, acute or abruptly acuminate, *narrowed at base, slightly petiolate: peduncle erect, 1—3 in. long:* petals spreading, lanceolate, acute, 1—2 in. long,

white changing to rose-purple, little surpassing the sepals: stigmas slender, sessile: capsule ovate, winged, ½—¾ in. long.—Coast Range, in open woods. April.

9. ZYGADENUS, *Michx.* Scape-like racemose stem and narrow carinate leaves from a tunicated bulb. Flowers greenish white, ascending or spreading on slender pedicels. Perianth nearly rotate; segments ovate to oblong-lanceolate, with a more or less pronounced green glandular spot at the mostly short-unguiculate base of the segments. Stamens 5, on the base of the segments; filaments subulate; anthers cordate before opening, peltate afterwards. Capsule of 3 nearly distinct and follicle-like cells. Seeds brownish, angled.

1. **Z. speciosus,** Dougl. Bulb oblong, 1--2 in. long: leaves light green, often 1 in. wide toward the base: stem stout, 1—4 ft. high, the ample subpyramidal raceme single, or with two or more accessory ones; lower pedicels 1–3 in. long: perianth free from the ovary, rotate, ¾ in. broad; *outer segments not unguiculate;* inner abruptly narrowed to a broad claw, all oblong-ovate, obtusish, the greenish yellow gland toothed on its upper margin by running into the strong elevated nerves of the segment: *stamens* nearly free, *shorter than the segments:* capsule 1 in. long. Var. **minor** (H. & A.). Seldom 6 or 8 in. high, simple, few-flowered: perianth larger: segments more oblong.—Type plentiful on bushy hills; the variety only in wet open ground, on the San Francisco peninsula. Feb.—April.

2. **Z. venenosus,** Wats. Bulb oblong-ovate, small: stem slender, 1—2 ft. high: leaves narrowly linear, scabrous on the margin: raceme mostly simple, short; *perianth-segments* ¼ in. long, *all contracted abruptly to a short claw;* blade rounded or subcordate at base; gland ending above in a well defined irregular line: *stamens as long as the perianth.*— Moist banks, and borders of boggy places, in Marin Co. May—July.

10. XEROPHYLLUM, *Michx.* Erect and simple-stemmed perennials, with thick lignescent rootstock and numerous narrow hard and dry serrulate leaves. Flowers many, white, in a dense raceme. Perianth of 6 distinct spreading oblong-lanceolate nerved persistent segments. Stamens at base of segments; filaments subulate-filiform; anthers rounded, dehiscent laterally. Ovary ovate, 3-lobed; styles 3, distinct, stigmatic on the inside, reflexed or recoiled, persistent. Capsule chartaceous. Seeds oblong, with thin wrinkled light-colored testa.

1. **X. tenax** (Pursh), Nutt. Radical leaves 2—3 ft. long, 2 lines wide, rigid: stem 2—5 ft. high, its scattered ascending gradually reduced leaves dilated at base: raceme 1—2 ft. long; lower bracts foliaceous and toothed, upper scarious: pedicels 1—2 in. long: fl. fragrant; segments 5 lines long, stamens somewhat longer.—Toward the summit of Mt. Tamalpais.

11. VAGNERA, *Adanson.* Erect simple ample-leaved perennial, with fleshy elongated horizontal rootstock, fibrous roots, and a terminal simple or compound raceme of small white 6-merous flowers. Perianth rotate. Filaments subulate; anthers rounded or oblong, versatile, introrse. Ovary ovate; style short, thick, 3-lobed. Fruit a globose 1—3-seeded berry.

1. **V. amplexicaulis** (Nutt.). Rootstock simple, stout: stem 1½—3 ft. high, somewhat pubescent: leaves ovate-oblong or elliptical, 3—7 in. long, sessile, amplexicaul: panicle of short racemes dense, short-peduncled, 2 in. long: perianth segments oblong-lanceolate, less than a line long: broadly subulate filaments both longer and broader than the perianth-segments: berry light red and with dots of dark red, usually 1-seeded: seed whitish, 1½ lines thick.—Woods of the Coast Range.

2. **V. sessilifolia** (Nutt.). Rootstock slender, branching, the plants thus forming close colonies covering the ground extensively: stem 1—2 ft. high, flexuous: leaves bright green, glabrous, 2—6 in. long, acute or acuminate, sessile, clasping, more or less plicate and furrowed (strongly so when young): raceme loose, flexuous, the spreading pedicels 2—7 lines long: perianth segments 2—4 lines long; stamens half as long: berry globose, red, 1—3-seeded: seeds brown.—Plentiful on moist northward slopes in thickets.

12. UNIFOLIUM, *Brunfels.* More diminutive herbs, with only 1—3 cordate leaves. Raceme short, simple, but the flowers often 2 or 3 together at a node. Perianth segments and stamens 4. Ovary 2-celled; stigma 2-lobed. Otherwise like the preceding genus.

1. **U. dilatatum** (Nutt.). Glabrous ½—1 ft. high, herbage deep green, not glaucescent, the 2 or 3 leaves 2—5 in. long, ovate to subreniform-cordate, with deep sinus and rounded lobes, the petiole 2 in. long or more: perianth segments oblong-obovate, deflexed; stamens shorter than these.—Woods of Marin Co. and northward.

13. DISPORUM, *Salisb.* Rootstocks short, erect. Roots fibrous. Stems branching. Leaves alternate, sessile, thin, many-nerved. Flowers 1—3 or more at the ends of the leafy branches. Perianth campanulate, of 6 white or greenish distinct segments. Stamens 6; filaments distinct filiform; anthers oblong, attached within above the base. Ovary ablong or ovate; style slender, entire or with 3 short spreading stigmas. Fruit a red berry, 3—6-seeded.

1. **D. Menziesii** (Don.), Britton. Somewhat woolly-pubescent: stems 1- 3 ft. high: leaves ovate to ovate-lanceolate, narrowly acuminate, rounded or subcordate at base, 2—5 in. long: fl. 1—5; *perianth whitish, almost funnelform, the segments being erect,* nearly 1 in. long; style more

or less woolly above, slightly 3-cleft: *fruit oblong-obovate, attenuate above* into a short somewhat villous beak, about ½ in. long, bright salmon-color.—Woods of Marin Co. April, May.

2. **D. Hookeri** (Torr.), Britton. Roughish-pubescent, 1—2 ft. high: leaves ovate, deeply cordate, 1½—3 in. long, the uppermost oblique: *perianth green, the segments spreading* to the campanulate: *berry obovate, obtuse,* scarlet.—Wooded ravines in the Oakland Hills, etc. May, June.

14. CLINTONIA, *Raf.* Apparently acaulescent, the very short stem (from fibrous roots) bearing from beneath the ground, large oblong or oblanceolate leaves, and a scape with a solitary, or many and umbellate flowers. Perianth campanulate, of 6 distinct oblanceolate deciduous segments. Stamens 6; filaments filiform; anthers oblong or linear, versatile, attached on the inner side above the base. Ovary ovate-oblong; style slender. Fruit a smooth ovoid few—many-seeded berry.

1. **C. Andrewsiana,** Torr. Nearly glabrous, only the inflorescence notably pubescent: leaves ½--1 ft. long, 2—4 in. wide, oblong or oblanceolate, acute or abruptly acuminate; scapiform peduncle 1—2 ft. high, often with a foliaceous bract: fl. deep rose-purple, many, in a terminal umbel and one or more lateral umbellate fascicles; pedicels unequal, 1 in. long or less: perianth gibbous at base, 4—7 lines long: filaments pubescent; anthers a line long: berry 4—5 lines long, the cells 8—10-seeded.—Deep shades of the Coast Range forests.

15. DICHELOSTEMMA, *Kunth.* Leaves (about 2) fleshy, linear, concave above; these and the long tortuous or twining scape from a depressed fibrous-coated corm. Scape with a solitary umbel, this sub-tended by 3 or more thin spathaceous bracts. Perianth-tube thin, more or less inflated and angular or saccate; segments about equalling the tube. Stamens 6, on the throat of the perianth; filaments disappearing from the surface of the thin tube below, above the insertion developed into petaloid appendages, those opposite the sepaline segments with or without an anther, the others always antheriferous; anthers basifixed.

✱ Flowers blue or violet.

1. **D. congestum** (Sm.), Kunth. Scape 3—5 ft. high, flexuous, but apparently never twining: *fl. blue-purple, in a dense capitate raceme* (the pedicels united into a central axis): perianth 6—8 lines long; tube slightly constricted above, about as long as the rotate-spreading segments: *fertile stamens 3; staminodia bifid, spreading* and deeply colored like the perianth.—In fields and along borders of thickets. May, June.

2. **D. capitatum** (Benth.), Wood. Scape ½—1½ ft. high, very tortuous, not rarely twining: bracts of the umbel of a dark and lustrous purple, much darker than the flowers; pedicels of the umbel very

unequal: perianth-tube funnelform, not constricted, shorter than the segments: *stamens all antheriferous, their appendages* forming a corona that is *connivent.*—Rocky hills, in the open country. March, April.

* * *Flowers rose-purple or pinkish.*

3. **D. Californicum** (Torr.), Wood. Scape 4—10 ft. high, in smaller plants tortuous, in taller firmly twining by several abrupt turns, *perianth pink to rose-red*, 6—8 lines long; *tube* 3—4 lines long and broad, *hexagonal, the angles somewhat saccately produced* above the middle; segments rotate, the very tips recurved: fertile stamens 3; anthers linear-sagittate; appendages of these and the staminodia emarginate and ciliolate-scabrous.—Foothills of the coast ranges. May, June.

* * * *Flowers scarlet, but green-tipped.*

4. **D. Ida-Maia** (Wood). Scape not twining, but distinctly tortuous; *perianth-tube 1 in. long, scarlet*, broadly tubular, somewhat 6-saccate at the truncate base; segments short, spreading, chrome-green: fertile stamens 3; staminodia and appendages of perfect stamens yellow.—Coast Range woods from Marin Co. northward. May, June.

16. **HOOKERA,** *Salisb.* Scapes straight, firmly erect. Umbel few-flowered, the pedicels firm and perianth erect. Perianth-tube thick, opaque, turbinate; segments equalling the tube, spreading, or recurved at tip. Filaments 6, stout, angular, inserted at the throat but prominent down to the base of the tube; 3 antheriferous, the alternate 3 bearing white-petaloid lamellæ. Anthers basifixed.

1. **H. coronaria,** Salisb. Scape stout, 1 ft. high: pedicels 3—10, 1—3 in. long: perianth 1 in. long or more, purple: *anthers* 4—5 lines long, *exceeding the oblong-lanceolate staminodia.*—Very common and generally distributed. May, June.

2. **H. minor** (Wats.), Britten. Scape slender, 3—6 in. high: pedicels 2—6, mostly 1—3 in. long: perianth usually less than 1 in. long; limb rotate: *anthers* 2 lines long, *shorter than the retuse or emarginate staminodia.*—Plains of the interior, if at all within our limits.

3. **H. terrestris** (Kell.), Britten. Scape usually altogether subterranean, only the umbel above ground: pedicels 2—30, slender, 3—4 in. long: perianth less than 1 in.; limb rotate: anthers 1½ lines long, shorter than the yellowish *emarginate staminodia, the margins of which are revolute.*—Near the coast from San Francisco northward. June.

17. **CALLIPRORA,** *Lindl.* Slender scape and few broad-linear thinnish leaves from a depressed fibrous-coated corm. Flowers yellow, with dark brown lines. Perianth-tube short; segments rotate-spreading. Filaments, below their insertion (at throat of perianth) coalescent with the tube and not obvious, free and broadly appendaged above it, all (6) antheriferous; anthers versatile.

1. **C. ixioides** (Ait. f.). Scape a foot high: leaves 2: *filaments* alternately long and short, all *bifurcate at the winged summit*, the oblong anther on a cusp in the middle between the forks.—From below San Francisco northward, in the Coast Range. May.

2. **C. lugens** (Greene). Like the preceding in size and general appearance, but the broad *appendages of the filaments rounded, not forked*, at summit: perianth deep saffron-color within; exteriorly the entire *tube, and midvein of the segments, dark brown* approaching black.—Vaca Mountains; collected only by the author. May.

18. **TRITELEIA,** *Douglas.* Vegetative organs as in *Calliprora.* Flowers blue, or bluish. Perianth-tube funnelform or turbinate, not inflated or angular or saccate though thin. Stamens 8, all antheriferous; · filaments coalescent with the upper portion of the perianth-tube, reappearing near the base in the form of thin but prominent crests; anthers small, versatile. Ovary on a slender stipe.

1. **T. laxa,** Benth. Scape 1—2 ft. high: umbel 10—30-flowered: *perianth* 1½ in. long, *funnelform, pale to deep violet*, cleft nearly to the middle: filaments free for a line's length above the insertion; anthers ovate-lanceolate, 2-lobed at base, erect, though fixed near the middle: stipe of ovary ½ in. long.—Very common in fields and on open hillsides. May, June.

2. **T. peduncularis,** Lindl. Scape 1—3 ft. high: umbel 15—30-flowered; pedicels very long, often 6—10 in.: *perianth broad-funnelform, pale rose-purple to nearly white*, 1 in. long, cleft below the middle; segments widely spreading; anthers oblong-linear, retuse at summit.—Wet banks of streams and in springy places. June, July.

19. **HESPEROSCORDUM,** *Lindl.* Sufficiently differentiated, as a genus, from the preceding, by its rotate-campanulate perianth cleft to the middle, the segments taking no angle of divergence of their own (not spreading), and filaments deltoid-dilated and monadelphous below; the anthers basifixed and erect.

1. **H. lacteum,** Lindl. Scape slender, 1—2 ft. high; perianth 5—8 lines long and as broad, white, with green veins, or sometimes with a tinge of lilac: stamens all alike: capsule short-stipitate.—Plentiful in rather low rich soils. May, June.

20. **MUILLA,** *S. Wats.* Somewhat succulent scape and semiterete leaves from a coated corm. Umbel with several narrow scarious bracts. Perianth segments and stamens 6 each and distinct or nearly so, white or greenish. Genus differing from all the foregoing umbellate-flowered genera in that the pedicels are not jointed under the perianth, and from *Allium* in that the plants are destitute of alliaceous properties.

1. **M. maritima** (Benth.), Wats. Corm depressed, ½—¾ in. thick, fibrous-coated: leaves many, 4—8 in. long, scabrous-margined: scape hardly longer than the leaves, erect, straight: pedicels 5—15: perianth-segments 2—2½ lines long, the sepaline oblong, slightly revolute, the petaline oval, plane: capsule ¼ in. long, abruptly beaked by the stout style: seeds black, oblong, compressed and angular.—Borders of salt marshes about the Bay, and in subsaline soils of the interior; but also on the highest parts of the Mission Hills, in dry stony soil. March, April.

21. **ALLIUM,** *Pliny.* Scapes and leaves from a tunicated bulb, or rarely from a coated corm: the whole plant always with alliaceous odor and flower. Umbel with a 2-valved (rarely 3—5-valved) spathe; pedicels not jointed. Perianth of 6 distinct or nearly distinct, usually equal segments, often gibbous at base. Stamens 6, on the base of the segments; filaments often dilated below; anthers versatile. Ovary sessile, deeply 3-lobed, with very short axis. Capsule obtusely 3-lobed, obovate-globose, often crested. Seeds obovoid, wrinkled, black.

* *Scape flattened, 2-edged, and leaves 2, from a coated bulb.*

1. **A. falcifolium,** Hook. & Arn. Leaves 2, broadly linear, flat, falcate, 3—4 lines wide: bulb-coats with no reticulation: scape 2—3 in. high, 1—2 lines wide: fl. rose-colored, the segments lanceolate, attenuate and spreading at the tips, glandular-serrulate, 4—6 lines long; stamens and style little more than half as long.—Inner coast hills, from Solano and Sonoma counties northward. April.

2. **A. Breweri,** Wats. Rather smaller, but bulbs larger (½—¾ in. thick): scape 1—2 in. above ground: 2 leaves rather broader: segments of deep rose-colored perianth nearly erect, 5—6 lines long, only a third longer than the stamens: *ovary and capsule with a thick crest on each cell.*—Summit of Mt. Diablo.

* * *Scape terete, leaves several, from a coated bulb.*

3. **A. lacunosum,** Wats. Bulb-coats pale, *pitted by a transversely oblong or somewhat quadrate reticulation,* the outline of the cells very minutely sinuous: scape 3—6 in. high: fl. 5—20 on pedicels 3—5 lines long, the oblong-lanceolate acuminate segments 3—4 lines long, little exceeding the stamens: filaments narrowly deltoid at base: ovary-cells with an obtuse ridge toward the summit on each side.—Summits of the higher mountains from Mt. Diablo southward. June.

4. **A. serratum,** Wats. *Bulb-coats with a close horizontally serrate reticulation:* perianth deep rose-purple, 5 lines long; sepaline segments oblong, abruptly acute, somewhat spreading; petaline lanceolate-oblong, erect, with merely the tips spreading, their margins entire: filaments with narrow-deltoid base: crests of ovary central.—Grassy hills toward the sea. June.

5. **A. attenuifolium**, Kell. Bulb-coats white, with a delicate transversely sinuate or serrate reticulation, the vertical lines especially also minutely sinuous: *leaves several, very long and slender; scape 10—18 in. leafy below: bracts of the spathe 2, short, abruptly pointed:* pedicel 6—8 lines long: fl. white, the oblong-lanceolate acuminate segments 3—4 lines long, exceeding the stamens.—Mountains of Marin and Sonoma counties. June.

6. **A. monospermum**, Jepson, MSS. Size and habit of the last, nearly; but scapes 3 or 4 from the bulb, and this red: *bracts of the umbel 3, broad, acuminate;* pedicels 50—80: perianth pale-purplish: filaments with broadly deltoid and connate bases: *capsule* (by abortion) *1-celled, 1-seeded.*—Summits of Vaca Mountains, Solano Co., *Jepson.* June.

* * * *Scapes terete, lateral, from a coated corm.*

7. **Bolanderi**, Wats. *Corms small clustered, oblique, the coats with a delicate undulate-serrate reticulation:* scape from one end of the corm, only a few inches high, slender; pedicels 5—15: perianth rose-colored; segments acuminate, 4—5 lines long; stamens only half as long.—Mt. Hamilton, and again to the northward of our limits, in Humboldt Co.

8. **A. unifolium**, Kell. *Corm large* (8—12 lines long), the chartaceous *coat with a close contorted reticulation:* scape stout, *1—2 ft. high,* with 2 or more long leaves at its base and sheathing it: bracts 2, acuminate: pedicels 10—30: rotate-campanulate perianth rose-red or rose purplish, nearly 1 in. in diameter; segments acute or acuminate, a third longer than the stamens.—Common in rich rather moist lands. April, May.

CORRIGENDA.

Page 9. For **P. erassifolium** read **P. crassifolium.**

Heading of page 81. For Polygoneæ read Polygaleæ.

Page 93. For **L. Crassifolius** write **L. crassifolius.**

Page 101. For Var. **augustatum** read Var. **angustatum.**

Page 213. For **S. curycephalus** read **S. eurycephalus.**

INDEX.

GLOSSARY.

Abnormal.—Deviating from the normal or usual.

Abortion.—The suppression or imperfect development of any part.

Abortive.—Imperfectly developed.

Acaulescent.—Stemless or apparently so.

Acerose.—Needle-shaped, as a pine leaf.

Acicular.— Needle- or bristle - shaped; more slender than acerose.

Aculeate.— Having sharp points or prickles.

Acuminate.—Tapering to a point.

Acute.—Sharp at the end, or at the edge or margin.

Adnate.—United; used properly of the surfaces of different organs, as of calyx and ovary.

Æstivation.—Arrangement of the parts of the perianth in the bud.

Achene.—A dry, hard, indehiscent 1-celled and 1-seeded seed-like fruit.

Albumen.—The nutritive material of the seed, within its coats and exterior to the embryo.

Albuminous.—Having albumen.

Alliaceous.—With the odor of onions or garlic.

Alternate.— Following one another at intervals; not opposite, intermediate.

Alveolate.—Deeply and closely pitted.

Ament.—A unisexual, usually pendulous, spike with scaly bracts.

Amplexicaul.—Clasping the stem.

Androecium.—The aggregate of the pistils in a flower.

Androgynous.—Having both male and female flowers.

Annual.—Of not more than one year's duration.

Annular.—Having the form of a ring.

Anterior.—Equivalent to inferior or lower, in the sense of away from the axis and toward the bract.

Anther.—That part of the stamen which contains the pollen.

Antheriferous.—Bearing anthers.

Apetalous.—Having no corolla.

Apex. —The top or summit of a thing.

Apical.—At the apex.

Apiculate.—Abruptly ending in a short point or tip.

Appressed.—Lying close to the surface.

Aquatic.—Growing in water.

Arachnoid.—Resembling cobweb.

Arborescent.—Becoming a tree, or tree-like.

Arcuate.—Curved like a bow.

Areola.—A pit-like scar, as that left by the corolla in Compositæ.

Areolæ.—The spaces in any reticulated surface.

Areolation.—Any system of reticulated markings.

Aril.—An expanded appendage to the hilum, enveloping the seed.

Arillate.—Having an aril.

Arilliform.—Resembling an aril.

Aristate.—Having an awn.

Aristulate.—Having a very small awn.

Articulated.—Jointed.

Ascending.—Rising somewhat obliquely, not erect.

Attenuate.—Narrowing gradually; tapering.

Auricle.—A small ear-like lobe at the base of a leaf.

Auriculate.—Furnished with auricles.

Awn.—A bristle-like appendage.

Axil.—The angle formed by a leaf or branch with the stem.

Axile or *Axial.*—Situated in the axis or relating to it.

Axillary.—Situated in an axil.

Axis.—The central line of a body in the direction of its length.

Baccate.—Berry-like; pulpy.

Banner.—A name often applied to the uppermost petal of a papilionaceous flower.

Barb.—A sharply reflexed point upon an awn, etc., like the barb of a fish-hook.

Barbellate.—Beset with very short stiff hairs.

Barbellulate.—Sparsely beset with short very fine hairs.

Bark.—The outer covering or rind of a stem.

Basal.—At, from, or relating to the base.

Basifixed.—Attached by the lower end.

Beak.—A prolonged tip.

Berry.—A simple fruit of which the whole substance, excepting the seeds, is pulpy.

Bifid.—Two-cleft.

Bipinnate.—Twice pinnate.

Bladdery.—Thin and inflated.

Blade.—The expanded portion of a leaf or petal.

Bract.—A leaf or modification of a leaf subtending a flower or flower-cluster.

Bracteate.—Having bracts.

Bracteolate.—Having bractlets.

Bractlet.—A secondary bract upon the pedicel of a flower.

Branch.—A division of a stem.

Branchlet.—A secondary or ultimate division of a stem.

Bristle.—A stiff hair or bristle-like appendage.

Budscales.—The scales which form the outer coats of a leaf-bud.

Bulb.—A subterranean roundish body, formed of fleshy scales or coatings; essentially a rudimentary stem or leaf-bud, and at length developing a flowering stem and often leaves.

Bulbiferous.—Bulb-bearing.

Bulbilliferous.—Bearing many small bulblets.

Bulblet.—A small bulb formed in the axil of a leaf or bract.

Bulbous.—Producing bulbs; bulb-like.

Caducous.—Falling very early; not at all persistent.

Calcarate.—Spurred.

Calyculate.—Having an involucre resembling a second external calyx.

Calyptra.—In mosses, the hood which at first covers the capsule. Otherwise, the lid-like deciduous calyx of some flowers.

Calyx.—The outer envelope of a flower.

Campanulate.—Bell - shaped or cup-shaped, with broad base.

Canescent.—Hoary with a grayish pubescence or puberulence.

Capillary.—Very slender and hair-like.

Capitate.—Subglobose and terminal, like a head; collected in a head.

Capitellate.—Diminutive of capitate.

Capsular.—Relating to or like a capsule.

Capsule.—A dry dehiscent fruit formed from a compound pistil.

Carinate.—Keeled.

Carpel.—A simple pistil or one of the several parts of a compound one.

Cartilaginous.—Firm and tough like cartilage.

Caruncle.—An outgrowth at the base of a seed; sometimes applied to an enlargement of the rhaphe.

Catkin.—The scaly unisexual soft-silky ament of willows.

Caudate.—Having a tail or slender tail-like appendage.

Caudex.—The trunk of a palm or other arborescent endogen; or the persistent base of any herbaceous perennial.

Caulescent.—Having a manifest stem.

Cauline.—Belonging to the stem.

Centrifugal.—Developing from the centre outward, as in the cyme.

Centripetal.—Developing from the margin toward the center, or from below upward, as in the corymb, raceme, etc.

Cespitose.—Growing in tufts or turf-like; forming mats.

Chaff.—Small dry scales, usually membraneous or scarious.

Chartaceous.— Having the texture of parchment or writing-paper.

Ciliate.—Having the margin, or sometimes the nerves, fringed with hairs like eye-lashes.

Cinereous.—Ash-gray, the color of wood ashes.

Circinate.—Rolled up from the tip into a coil.

Circumscissile.—Dehiscing by a transverse circular line of division.

Clavate.—Club-shaped; enlarged gradually toward the summit.

Claw.—The elongated narrow base of a petal or sepal.

Cleft.—Cut somewhat deeply, usually about half way to the center or mid-rib.

Climbing.—Rising by the aid of some support.

Coalescent.—United; used properly in respect to similar parts, as the stamens in Malvaceæ.

Cohesion.—The sticking together of parts, or their more intimate coalescence or adnation.

Colored.—Of other color than green.

Column.—A body formed by the union of filaments (stamineal) or, in orchids, of the stamens and pistil.

Coma.—A tuft of hairs, especially upon a seed.

Commissure.—The surface by which two carpels cohere, as in Umbelliferæ.

Comose.—Having a coma.

Complicate.—Folded together.

Compound.— The opposite of simple; consisting of more than one; divided.

Compressed.—Flattened laterally.

Conduplicate.—Doubled together lengthwise, of leaves.

Confluent.—Blended or running together.

Conical.—Shaped like a cone; narrowing to a point from a circular base.

Connate.—United in one; growing together.

Connective.—The portion of the filament which connects (or separates) the cells of the anther.

Connivent.—Coming in contact; converging together.

Constricted.—Contracted or drawn together, as a bag by its string.

Continuous.—Not interrupted by joints or otherwise.

Contorted.—Twisted; in æstivation, an equal and uniform somewhat oblique overlapping and rolling up of the parts of the circle.

Contracted.—Reduced in width or length.

Convolute.—Rolled together from one edge. See contorted.

Cordate.—Heart-shaped, i. e., ovate with rounded lateral lobes projecting beyond the base and forming a sinus.

Coriaceous.—Of the stiffness and consistence of leather.

Corm.—A solid fleshy rounded or depressed subterranean body, at the base of a stem, and bulb-like in appearance.

Corneous.—Of the consistence of horn; horny.

Corolla.—The inner perianth, within the calyx, consisting of the petals.

Corymb.—A flat-topped or convex open inflorescence, with short axis, flowering from the margin inward; a depressed raceme.

Corymbose.—In corymbs, or resembling a corymb.

Costa.—A rib, mid-rib, or mid-nerve.

Costate.—Having one or more longitudinal ribs or nerves.

Cotyledons.—The seed-lobes or leaves of the embryo.

Creeping.—Running upon or under the ground and rooting.

Crenate.—Scalloped; having rounded teeth with shallow acute sinuses.

Crenulate.—Finely crenate.

Crested.—Having an elevated ridge or appendage like the crest of a helmet.

Cruciferous.—Belonging to the Cruciferæ, with cruciform or cross-shaped corolla.

Crustaceous.—Hard and brittle.

Cucullate.—Shaped like a hood or cowl, concave and somewhat arched, or like an ovate leaf with edges at base inrolled.

Culm.—The hollow jointed stem peculiar to grasses.

Cuneate or *Cuneiform.*—Wedge-shaped; triangular with the angle downward.

Cupule.—A cup-shaped involucre, enclosing a nut, as of an acorn.

Cusp.—A sharp rigid point.

Cuspidate.—Terminating in a cusp.

Cyathiform.—Cup-shaped with a somewhat flaring mouth.

Cylindrical.—In the form of a cylinder.

Cyme.—A broad and flattish inflorescence, flowering from the center outward.

Cymose.—In cymes or cyme-like.

Deciduous.—Falling off after a time; not persistent.

Declinate or *Declined.*—Bent or curved downward.

Decompound.—Repeatedly compound or divided.

Decumbent.—Reclining at base, the summit ascending.

Decurrent.—Running down the stem, applied to a leaf with blade prolonged below its insertion.

Decussate.—In pairs alternating at right angles, or similarly in threes.

Definite.—Of a constant number, not exceeding twenty; limited or determinate, as *definite inflorescence*, in which a flower terminates the axis.

Deflexed.—Bent or turned down abruptly.

Dehiscence.—The regular opening of a capsule or anther-cell at maturity.

Dehiscent.—Opening by valves, slits, etc.

Deltoid.—Having the shape of the Greek letter delta; broadly triangular.

Dendroid—Tree-shaped; branching in the form of a tree.

Dentate.—Toothed; having symmetrical teeth projecting straight outward.

Denticulate.—Minutely toothed.

Depauperate.— Impoverished; reduced in size by unfavorable surroundings.

Depressed. — Somewhat flattened from above.

Dextrorse.—Toward the right hand; applied to spirals as seen from without. It is frequently used as if the spiral were seen from within, in which case it indicates just the opposite direction.

Diadelphous.—In two sets or clusters.

Diandrous.—Having two stamens only.

Dichotomous. — Forking regularly by pairs.

Dicotyledonous.—Having an embryo with two cotyledons.

Didymous.—In pairs; twin.

Didynamous.—Having four stamens disposed in two unequal pairs.

Diffuse.—Widely spreading; widely and loosely branched.

Digitate.—Fingered; applied to a compound leaf having the leaflets all diverging from the top of the petiole.

Dimorphous.—Occurring in two forms.

Diœcious.—Unisexual, the flowers of different sexes borne by separate plants.

Diœcio-polygamous. — Diœcious with some perfect flowers intermingled.

Disciform.—In the shape of a disk, depressed and circular.

Discoid.—In Compositæ, having disk-flowers only, without rays.

Disk.—A dilation or development of the receptacle around the base of the pistil. In Compositæ the inner series of tubular flowers as distinct from the rays.

Dissected.—Deeply cut or divided into numerous segments.

Divaricate.—Widely divergent, nearly at right angles.

Divergent.—Receding from each other.

Divided.—Cleft to the base or to the midnerve.

Dorsal.—Upon or relating to the *dorsum* or back.

Drupaceous.—Resembling or of the nature of a drupe.

Drupe.—A stone-fruit; a fleshy or pulpy fruit with the seed or kernel inclosed in a hard or strong casing (*putamen*).

Drupelet.—A diminutive drupe, as each of the several parts of a blackberry.

Echinate.—Beset with prickles.

Elliptical.—In the form of an ellipse, oblong with both ends uniformly and somewhat gradually rounded.

Emarginate.—Notched at the extremity.

Embracing.—Clasping at base.

Endocarp.—The inner layer of the pericarp lying next to the seed.

Endogenous.—Growing from within, instead of by superficial increments, the growth ordinarily being general throughout the substance of the stem.

Endogens.—Plants with an endogenous structure.

Ensiform.—Sword-shaped, as the leaf of an *Iris.*

Entire.—With the margin uninterrupted, without teeth or division of any sort.

Ephemeral.—Lasting but a day, or for a very short time.

Epigynous.—At or upon the top of the ovary.

Equitant. — Astride; of conduplicate leaves which fold over each other in two ranks, as in *Iris.*

Erect.—Upright; perpendicular to the surface of attachment.

Evergreen.—Bearing its foliage through all the seasons.

Exalbuminous.—Destitute of albumen.

Explanate.—Opened out flat.

Exserted.— Projecting beyond an envelope, as stamens standing out of the corolla.

Exstipulate.—Without stipules.

Extrorse.—Directed outward.

Falcate or *Falciform.*— Sickle-shaped; manifestly curved and more or less flattened or folded.

Farinaceous. — Mealy; containing or yielding flour or starch.

Farinose.—Covered with a white mealy powder.

Fascicle.—A close bundle or cluster.

Fastigiate.—With branches erect, parallel and near together, as in the Lombardy poplar.

Faveolate, Favose.—Pitted or honeycombed.

Ferruginous.—Of the color of iron-rust.

Fertile.—Capable of producing fruit, as a pistillate flower; applied also to a pollen-bearing stamen.

Filament. — That part of the stamen which supports the anther; any thread-like body.

Filiform.—Thread-shaped; long, slender and terete.

Fimbriate.—Fringed with narrow processes; having the margin finely dissected.

Fistular, Fistulous.—Hollow and cylindrical.

Flabelliform.—Fan-shaped; dilated and rounded above, from a cuneate base.

Flexuous.—Bent or curving alternately in opposite directions.

Foliaceous.—Leaf-like in structure and appearance; leafy.

Follicle.—A pod formed from a single pistil, dehiscing along the ventral suture only.

Follicular.—Pertaining to a follicle or like it in structure.

Foveate.—Pitted; marked by deep depressions.

Foveolate.—Diminutive of the last.

Free.—Not adnate to other organs.

Fructification.—The bearing of fruit, or the organs concerned in the production of fruit.

Fruit.—The matured seed- or spore-vessel, of whatever kind, with its appendages and contents.

Frutescent.—Shrubby or somewhat so.

Fugacious.—Very soon falling; of extremely short continuance.

Fulvous.—Dull brownish or grayish yellow.

Funiculus.—The stalklet of an ovule or seed.

Funnelform.— Tubular, but expanding gradually from the narrow base to the spreading border or limb.

Fusiform.—Spindle-shaped, i. e., tapering toward each end from a thickened middle.

Galea.—A helmet; applied to the helmet-shaped upper lip of the corolla in Labiatæ, Aconitum, etc.; also to the upper lip of some Scrophularineæ. though not so shaped.

Galeate.—Having a galea.

Geminate.—In pairs; binate; twin.

Geniculate.—Bent abruptly at an angle, like the knee.

Gibbous.—Protuberant; swelling out at one side.

Glabrate.—Becoming glabrous.

Glabrous.—Without any kind of hairiness.

Gland.—Any secreting structure, depression or prominence on any part of a plant, or any structure having such an appearance.

Glandular.—Bearing glands, or gland-like.

Glaucescent.—Somewhat glaucous; becoming glaucous.

Glaucous.—Covered with a fine whitish bloom that is easily rubbed off; having a bluish-hoary appearance.

Globose, Globular.—Round; spherical, or nearly so.

Glochidiate.—Barbed, like a fish-hook.

Glomerate.—Closely clustered.

Glomerule.—A compact somewhat capitate cyme.

Glutinous.—Viscid; sticky; covered with gummy secretion.

Granular.—Composed of small grains or grain-like bodies; rough with grain-like prominences.

Gymnosperms.—Plants having naked seeds, or in which the typically naked ovule is fertilized directly by the pollen without the intervention of a stigma.

Gynandrous—Having the stamens adnate to the pistils and style, so as to be apparently borne at or upon its summit, as in Orchids.

Gynobase.—A short thick prolongation of the axis or receptacle upon which the pistil rests.

Gynœcium.—The aggregate of the pistils of a flower.

Habit.—The general form or mode of growth of a plant

Habitat.—The locality or geographical range of a plant.

Hamate.—Curved at the end into a hook.

Hastate.—Triangular or arrow-shaped with basal angles or lobes directed outward.

Head.—A cluster of flowers, which are sessile or nearly so upon a very short axis or receptacle; a shortened spike.

Herb.—A plant that has no persistent woody growth above the ground.

Herbaceous.—Having the character of an herb; not woody or shrubby.

Hilum.—The scar or place of attachment of the seed.

Hirsute.—Pubescent with rather coarse or stiff hairs.

Hirsutulous.—Diminutive of hirsute, i. e., sparingly and shortly hirsute

Hispid.—Beset with rigid or bristly hairs.

Hispidulous.—Minutely hispid.

Hoary.—Grayish-white with a fine close pubescence.

Hyaline.—Transparent; translucent.

Hybrid.—A cross between two species, produced by the fertilization of the flower of one species by the pollen of another

Imbricate.—Overlapping, like shingles on a building.

Incised.—Irregularly, sharply and deeply cut.

Included.—Enclosed by the surrounding organs; not exserted.

Incurved.—Curved inward.

Indigenous.—Native to the country.

Indument.—The hairy, silky, woolly scurfy or other such clothing of leaf or stem.

Induplicate.—With margins folded inward.

Inferior.—Lower; that part of a flower, etc., which is toward the bract; applied also to a calyx that is free from the ovary, and to an ovary that is adnate to the calyx.

Inflated.—Bladdery.

Inflorescence.—The flowering portion of a plant, and especially the mode of its arrangement.

Inserted.—Attached to or growing upon.

Insertion.—The place or mode of attachment of an organ.

Internode.—The part of a stem between two joints.

Introrse.—Turned inward toward the axis.

Involucel.—An inner or secondary involucre; that which surrounds an umbellet.

Involucrate.—Having an involucre.

Involucre—A circle or circles of scales, bracts or leaves, distinct or united, surrounding a flower or flower-cluster; in Umbelliferæ, the bracts subtending the umbel.

Involute.—Rolled inward.

Irregular.—With parts unlike in size or form, or both.

Keel.—A central dorsal ridge, resembling the keel of a boat. Also the lower pair of petals in a papilionaceous flower.

Labiate.—Lipped; applied to an irregular corolla or calyx which is unequally divided into two parts or lips.

Lacerate.—Torn; irregularly and deeply cleft.

Laciniate.—Cut into narrow slender teeth or lobes.

Lamella.—A thin plate or scale.

Lamellar.—In form of a plate or scale.

Lamina.—The blade or dilated portion of a leaf.

Lanate.—Covered with long curled hairs like wool.

Lanceolate.—Shaped like a lance-head; tapering upward from a narrowly ovate or subovate base.

Lanuginous.— Provided with wool; woolly.

Lateral.—At the side: attached to the side.

Lavender-color.—A pale grayish blue.

Leaf-blade.—The dilated portion of a leaf.

Leaflet.—A separate division of a compound leaf.

Legume.—A normally 1-celled capsule, formed from a single carpel, but dehiscing by two valves, as in the Pea, Bean, etc.

Lenticular.—Lens- or lentil-shaped; of the form of a double-convex lens.

Ligneous.—Woody.

Lignescent.—Becoming woody.

Ligule.—A small tongue-like or strap-shaped body. Applied to the corolla of ray flowers in Compositæ.

Ligulate.— Furnished with a ligule; strap-shaped.

Liliaceous.—Lily-like.

Limb.--The dilated and usually spreading portion of a perianth or petal, as distinct from the tubular part or claw.

Line.—The twelfth part of an inch, nearly equivalent to two millimeters.

Linear.—Narrow and elongated, with parallel margins.

Lineate.—Marked with lines.

Lip.—Either of the two divisions of a bilabiate corolla or calyx; in Orchids. the upper petal, usually very different from the others.

Lobate, Lobed.—Divided into or bearing lobes.

Lobe.—Any division of a leaf, corolla, etc., especially if rounded.

Loculicidal.—Used when the cells of a capsule open by dehiscence through the dorsal suture.

Lunate.—Crescent-shaped.

Lurid.—Of a dull dirty-brown color.

Lyrate.—Pinnatifid with the terminal lobe largest and rounded, the lower lobes small.

Marcesent.—Withering and persistent.

Marginate, Margined.——Furnished with a border peculiar in structure or appearance.

Maritime.—Belonging to the sea-coast.

Mealy.—Covered with a whitish mealy powder.

Membranous, Membranaceous. — Thin and rather soft and translucent, like membrane.

Mid-rib, or *Mid-nerve.*—The central and principal nerve of a leaf.

Monadelphous.—Having the stamens all united by their filaments into a column or tube.

Monandrous.—Having a single stamen.

Moniliform.—Resembling a chaplet or string of beads.

Monocephalous.—Bearing a single head of flowers.

Monocotyledon.—A plant whose embryo has a single cotyledon.

Monœcious.—With stamens and pistils in separate flowers upon the same plant.

Mucro, Mucronation.—A short and small abrupt rigid tip.

Multicipital.—Many-headed, applied to a much-branched rootstock.

Multifid.—Cleft into many lobes or segments.

Muricate.—Rough, with short hard points.

Muriculate.—Finely muricate.

Nectariferous.—Secreting nectar.

Nectary.—The manifest gland of a petal within.

Nerve.—A simple vein; a rib.

Nodding.—Curving downward; somewhat inclined from the perpendicular.

Node.—A knot or swelling; a place upon a stem where a leaf, or a pair, or a whorl of leaves, is borne.

Nodose.—Having knots or swelling joints.

Nut.—A hard indehiscent one-seeded fruit, usually resulting from a compound ovary.

Nutlet.—A small nut; also applied to the hard seed-like divisions of the fruit of the Labiatæ, Verbena, etc.

Obcompressed.—Flattened contrary to the direction of the sides, dorsally instead of laterally.

Obconical.—Resembling an inverted cone.

Obcordate.—Inverted cordate, the lobes at the upper end.

Oblanceolate.—Inverted lanceolate, with the broadest part toward the apex.

Oblate.—Flattened at top and bottom.

Oblique.—Turned to one side; unequally sided.

Oblong.—Considerably longer than broad and with nearly parallel sides.

Obovate.—Inverted ovate, the broader part toward the apex.

Obovoid.—Inverted egg-shaped, the broader part above.

Obtuse.—Blunt or rounded at the end.

Obversely.—In a reverse manner.

Ochroleucous.—Yellowish-white.

Opaque.—Dull, not shining.

Operculum.—A lid, separating by a transverse line of dehiscence.

Opposite.—Standing against or facing each other, as a stamen against a petal, or two leaves at the same node.

Orbicular.—Circular or nearly so.

Order.—A principal group next above the genus in rank, and including related genera more or less distinguished from others by certain common characters.

Ordinal.—Relating to orders.

Osseous.—Bony.

Oval.—Broadly elliptical.

Ovary.—The dilated portion of the pistil, bearing and containing the ovules.

Ovate.—Shaped like the longitudinal outline of an egg, the broader portion toward the base; also egg-shaped as applied to fruits, etc.

Ovoid.—Nearly egg-shaped.

Ovule.—A rudimentary organ which after impregnation becomes a seed.

Palate.—A protrusion at or near the throat of a bilabiate corolla.

Palea.—A chaff or chaffy bract; in grasses, the two inner bracts of the flower.

Paleaceous—Chaffy or furnished with chaff.

Palmate.—Of leaves, compound with the leaflets radiating from the summit of the petiole.

Palmately.—In a palmate manner.

Palmatifid.—Palmately cleft or divided.

Panicle.—A loose irregularly branched inflorescence.

Panicled, Paniculate.—After the manner of a panicle; bearing a panicle.

Papilionaceous.—Butterfly-like; applied to the peculiar irregular flower common in the Leguminosæ.

Papillose, Papillate.—Bearing minute thick nipple-shaped or somewhat elongated projections.

Pappus.—In Compositæ, the hairs, bristles or scales crowning the achene and taking the place of a calyx.

Papyraceous.—Having the texture of paper.

Parasitic.—Growing upon and deriving nourishment from another plant.

Parietal—Relating to or situate upon the walls of a cavity.

Parted.—Cleft nearly to the base.

Partition.—An inner wall or dissepiment.

Pectinate.—Comb-like: cleft into narrow closely-set segments.

Pedate.—Palmately divided or parted with the lateral divisions again 2-cleft.

Pedicel.—The footstalk or support of a flower.

Pedicellate.—Borne on a pedicel.

Peduncle.—A general or primary flower-stalk.

Pedunculate.—Furnished with a peduncle.

Peltate.—Shield-shaped; flat and attached to its support by its lower surface.

Pendulous.—Hanging nearly inverted from its support; of ovules, more or less drooping, as distinct from suspended.

Penicillate.—Resembling a brush of fine hairs.

Perennial.—Persistent a series of years.

Perfect.—Of a flower, having both stamens and pistil.

Perfoliate.—Of leaves, connate about the stem.

Perianth.—The calyx and corolla, when much alike and seeming like one floral circle.

Pericarp.—The seed-vessel or ripened ovary.

Persistent.—Not falling off; of leaves, continuing through the winter.

Personate.—Used of a labiate corolla with prominent palates closing the throat.

Petal.—One of the parts of a choripetalous corolla.

Petaline.—Relating to the inner segments of a perianth.

Petaloid.—Colored and resembling a petal.

Petiole.—The footstalk of a leaf.

Petioled, Petiolate.—Having a petiole.

Petiolule.—The footstalk of a leaflet.

Pilose.—Hairy, usually with soft rather remote hairs.

Pinnate.—Having its parts arranged in pairs along a common rachis.

Pinnately.—In a pinnate manner.

Pinnatifid.—Pinnately cleft into opposite nearly equal segments.

Pinnatisect.—Pinnately divided down to the midrib.

Pistil.—The female organ of a phanerogam, consisting of the ovary with its styles and stigmas.

Pistillate.—Having a pistil and no stamens, as distinct from perfect or staminate.

Pitted.—Marked with small depressions or pits.

Placenta.—That part of the ovary or fruit which bears the ovules and seeds.

Plane.—Having a flat surface.

Plicate.—Folded into plaits, like a fan.

Plumose.—Plume-like; having fine hairs on each side like a feather.

Pollen.—The powdery or sometimes waxy contents of the anther.

Polliniferous.—Bearing pollen. Used of deformed or reduced anthers which nevertheless yield pollen.

Polymorphous.—Of many forms; variable in form.

Pome.—A fleshy fruit, like the apple, enclosing several parchment-like or bony carpels.

Posterior.—In an axillary flower, the side toward the axis and away from the bract.

Prickle.—A small spine, an outgrowth of the bark or cuticle.

Prismatic.—Elongated, truncate at both ends, with angular circumscription.

Process.—Any projecting appendage; in mosses, the inner teeth or cilia of the peristome.

Procumbent.—Lying upon the ground.

Produced.—Extended or prolonged,

Proliferous.—Producing offshoots.

Prostrate.—Lying flat on the ground.

Pruinose.—Covered with a minute bloom or powder.

Puberulent.—Very minutely pubescent.

Pubescence.—A short soft hairiness, or, more generally, any kind of hairy or woolly indument.

Pubescent.—Covered with hairs, usually short and soft.

Pulverulent.—Dusty, as if covered with a minute powder.

Punctate.—Dotted with minute depressions, or with translucent internal glands or colored dots.

Puncticulate.—Very minutely punctate.

Pungent.—Terminating in a rigid and stout sharp point or prickle.

Putamen.—The bony or crustaceous shell inclosing the seed of a drupe.

Pyramidal.—Shaped like a pyramid; narrowing to an apex from an angular base.

Pyriform.—Pear-shaped.

Raceme.—A form of inflorescence with pedicellate flowers upon a simple prolonged axis, the flowers developing from below upward.

Racemose.—In racemes, or resembling a raceme.

Radiate.—Diverging from a common center; bearing ray flowers.

Radical.—Belonging to or proceeding from the root, or from the base of the stem.

Radicle.—That part of the embryo below the cotyledons, its stem-portion and the primal internode, developing the root from its lower extremity.

Ray.—One of the radiating branches of an umbel ; the marginal flowers, as distinct from those of the disk, in Compositæ, Umbelliferæ, etc.

Receptacle.—A more or less expanded or produced surface forming a common support for a cluster of organs (in a flower) or a cluster of flowers (in a head), etc.

Reclinate.—Reclining. With an erect or ascending base, the upper part recurved and trailing.

Rectangular.—Of an oblong right-angled figure.

Recurved.—Curved backward or downward.

Regular.—Symmetrical in form ; uniform in shape or structure.

Reniform.—Kidney-shaped: deeply cordate with the breadth exceeding the length.

Repand.—With the margin slightly sinuate or wavy.

Reticulated.—With markings or veinings resembling network.

Retrorse.—Turned backward or downward.

Retuse.—With a shallow or obscure notch at the rounded apex.

Revolute.—With the margins or apex rolled backward.

Rhachis.—The axis of a spike or of a compound leaf or frond.

Rhizomatous.—Producing rhizomes or of the character of a rhizome.

Rhizome, or *Rootstock.*—A somewhat horizontal underground rooting stem, producing a stem, leaves or flower-stalk at its apex or nodes.

Rhombic.—Obliquely four-sided.

Rhomboidal—Somewhat rhombic in outline.

Rib.—A principal and prominent nerve of a leaf.

Ribbed.—Furnished with prominent nerves.

Ringent.—Gaping; applied to a labiate corolla with widely separated lips and open throat..

Rootstock.—See Rhizome.

Rostellate. — Diminutive of rostrate; having a small beak.

Rostrate.—Beaked; bearing a slender terminal process.

Rosulate.—Collected in a rosette.

Rotate.—Wheel-shaped; of a corolla, spreading abruptly from near the base and nearly flat.

Rufous.—Reddish or brownish red.

Rugose.—Wrinkled; ridged.

Rugulose.—Finely or minutely wrinkled.

Runcinate.—Deeply toothed or incisely lobed, with the segments directed backward.

Runner.—A very slender prostrate branch (stolon), rooting and developing a new plant at the nodes or tip, as in the strawberry.

Saccate.—Sac-shaped; furnished with a sac or pouch-like cavity.

Sagittate.—Shaped like an arrow-head; triangular with basal lobes prolonged downward.

Salverform.—Narrowly tubular, with limb abruptly or flatly expanded.

Samara.—An indehiscent membranous winged one-seeded fruit, as in the Ash and Maple.

Sarcocarp.—The succulent part of a fleshy fruit.

Sarmentose—Producing long runners.

Scabrous.—Rough to the touch with minute rigid points.

Scales.—Usually variously modified bracts or leaves, thin and scarious, or coriaceous, fleshy, foliaceous, or woody, often imbricated.

Scape.—A naked peduncle rising from the ground.

Scarious.—Thin, dry and membranaceous, not green.

Scorpioid.—Incurved like the tail of a scorpion, applied to a unilateral circinately coiled inflorescence, unrolling as the flowers expand.

Scrobiculate.—Marked by minute depressions.

Scurf.—Small bran-like scales on the epidermis.

Secund.—Turned in one direction, as the leaves or flowers upon a stem.

Seed.—The ripened ovule, consisting of the embryo with its proper envelopes.

Segment.—One of the parts of a leaf or other organ that is cut or divided; more general than lobe.

Sepal.—A leaf or division of a calyx.

Sepaline.—Relating to the outer segments of a perianth.

Septicidal.—Dehiscing through the dissepiments and between the cells, or through the lines of junction of the carpels.

Septifragal.—Breaking away from the partitions on dehiscence; this and the last are terms applied to the valves of a loculicidal capsule.

Sericeous.—Silky; covered with soft straight appressed hairs.

Serotinous.—Produced late in the season.

Serrate.—Having teeth directed forward. like the teeth of a saw.

Serratures.—Teeth like those of a saw.

Serrulate.—Finely serrate.

Sessile.—Attached immediately to the point of support without footstalk.

Setaceous.—Bristle-like.

Setose.—Beset with bristles.

Sheathing.—Enfolding like a sheath.

Shrub.—A plant woody throughout, of less size than a tree.

Shrubby.—Having the character of a shrub.

Silicle.—A short cruciferous pod, not many times longer than wide.

Silique.—The usually elongated pod in Cruciferæ, having two valves separating from two parietal placentæ.

Silky.—See Sericeous.

Simple.—Of one piece; not compound.

Sinistrorse.—Turned to the left, as seen from the outside; but often used in the opposite sense.

Sinuate.—With a strongly wavy margin.

Sinuous.—Curving back and forth.

Sinus.—The open interval between lobes or segments.

Smooth.—Not rough ; the surface even.

Spadix.—A spike with usually a thickened fleshy rhachis and subtended by a spathe.

Span.—The distance between the extremities of the thumb and little finger when extended ; about nine inches.

Spathaceous.—Bearing or resembling a spathe.

Spathe.—One or more clasping and often sheathing bracts enclosing a flower cluster or inflorescence and mostly colored.

Spatulate.—Narrowed downward from an abruptly rounded summit.

Species.—A group of things of the same kind, having essentially the same characters.

Specific.—That which relates to a species.

Spicate.—In spikes or resembling a spike.

Spike.—A simple elongated inflorescence, with the flowers sessile or very nearly so.

Spine.—A sharp woody or rigid outgrowth from the stem, a modification of a branch, leaf or stipule.

Spinescent.—Ending in a spine or rigid point.

Spinose, Spiny. — Furnished with or resembling spines.

Spinulose.—Having diminutive spines.

Spur.—A usually slender tubular process from some part of a flower. often nectariferous.

Squarrose.—Roughened and jagged with projections spreading every way, as by the divaricately spreading ends of crowded leaves or bracts.

Stamen.—The pollen-bearing organ of the flower, consisting of an anther usually supported upon a stalk or filament.

Stamineal.—Relating to or consisting of the stamens.

Staminodium.—A sterile stamen or something taking the place of a stamen.

Stellate.—Star-shaped; radiating in fine lines from a centre, like the rays of an asterisk.

Stem.—The main axis of a plant.

Stemless.—Without manifest stem above ground.

Sterile.—Barren; not capable of producing seed; a sterile stamen is one not producing pollen.

Stigma.—That portion of the pistil without epidermis through which the pollen-tubes effect entrance to the ovules, very variable in shape and position.

Stigmatic.—Belonging or relating to the stigma.

Stipe.—The footstalk of a pistil raising it above the receptacle: in ferns, the naked stalk of the frond.

Stipitate.—Borne upon a stipe.

Stipular.—Belonging to the stipules.

Stipulate.—Possessing stipules.

Stipule.—An appendage to the base of a petiole, very various in form and character.

Stolon.—A horizontal prostrate offshoot from the base of a plant.

Stoloniferous.—Bearing or propagating by stolons.

Stone.—The hard endocarp or putamen of a drupe.

Striate.—Marked with fine longitudinal lines.

Strict.—Upright and very straight.

Strigillose.—Minutely strigose.

Strigose.—Beset with short straight stiff and appressed sharp-pointed hairs.

Strophiole.—An appendage at the point of attachment of some seeds.

Style.—That portion of the pistil between the ovary proper and the stigma, usually attenuated, often wanting.

Stylopodium.—A cushion-like expansion at the base of the style in Umbelliferæ.

Subtended.—Supported or surrounded, as a pedicel by a bract, or a flower cluster by an involucre ; fulcrate.

Subulate.—Awl-shaped.

Succulent.—Fleshy and juicy.

Sucker.—A shoot from the underground base of a stem, or from underground roots or rhizomes.

Suffrutescent. — Somewhat or slightly shrubby; woody at base.

Suffruticose.—Low and shrubby.

Sulcate.—Grooved or furrowed.

Superior.—Growing above; a superior ovary is one wholly above and free from the calyx; in a lateral flower, nearest to the axis.

Surculose.—Producing suckers.

Suture.—A line of union or of dehiscence.

Symmetrical.—Regular in shape if of a plant or tree as a whole; in the number of its parts, if spoken of a flower.

Sympetalous.—Having the petals united.

Synsepalous.—Having the sepals united.

Teeth. — Small marginal or terminal lobes of any kind.

Tendril.—A thread-like production from an axil, the extremity of a leaf or elsewhere, capable of coiling and used for climbing.

Terete.—Cylindrical or nearly so; not angled nor channelled.

Ternate.—In threes; in three divisions.

Tessellated.—Chequered; like mosaic or chequerwork.

Testa.—The outer seed-coat.

Tetradynamous.—With four long and two shorter stamens; applied to the Cruciferæ.

Tetragonal.—Four-angled.

Tetramerous.—Of a flower having its parts in fours.

Tetrandrous.—With four stamens.

Thorn.—See Spine.

Throat.—The orifice of a sympetalous corolla or calyx; the portion of the corolla immediately below the limb or between the limb and the tube.

Thyrse.—A contracted or close panicle.

Tomentose. — Pubescent with matted wool.

Tomentum.—Dense matted woolly pubescence.

Toothed.—Provided with teeth.

Top-shaped.—Inverted broad-conical.

Torose.—Swollen at intervals.

Tortuous,—Bending about irregularly.

Torulose.—Slightly torose.

Torus.—The receptacle of a flower; the apex of the flower-stalk, more or less modified to support the parts of the flower.

Transverse.—Across, from side to side.

Tree.—A woody branching plant, with erect trunk, ten feet high or more.

Triandrous.—With three stamens.

Triangular.—Three-angled.

Trichotomous.—Branching by threes.

Trifid.—Three-cleft.

Trifoliate.—Three-leaved.

Trifoliolate.—Having three leaflets.

Trimerous.—Having its parts in threes.

Tripinnate.—Three times pinnate.

Triquetrous.—Of a stem, etc., triangular with the sides somewhat concave or channelled.

Triternate.—Three times ternate.

Truncate.—Ending abruptly as if cut off transversely.

Trunk.—A main stem.

Tube.—Any elongated hollow body or part of an organ.

Tuber.—A thickened rhizome, with scattered buds or eyes.

Tubercle.—A small projection or pimple: a small tuber or a tuberous root.

Tuberculate. — Covered with small rounded prominences or knobs.

Tunicated.—Denoting bulbs made up of concentric seamless tunic-like coats, as in the onion.

Turbinate.—Top-shaped.

Turgid.—Puffed out; distended.

Twining.—Ascending by winding about a support.

Type.—The ideal pattern or form.

Typical.—That which corresponds to or represents the type. A typical species is one upon which the generic character was founded, or one which conforms most closely to the general characters of the genus, deviations from which form the basis for subgenera, etc. So the typical form of a species is that upon which the specific character is based, as distinguished from all varieties, sports, etc.

Umbel. — An umbrella-shaped inflorescence, the pedicels radiating from the summit of the common peduncle.

Umbellate. — Bearing or growing in umbels.

Umbelliferous. — Bearing umbels.

Umbellulate. — Bearing umbellets.

Unarmed. — Without prickles, spines, or the like.

Uncinate. — Hooked at the extremity.

Undulate. — Wavy, alternately raised above and depressed below the general plane.

Undershrub. — A very low shrub.

Unguiculate. — Of a petal, narrowed below into a claw or petiole-like base.

Unifoliolate. — A compound leaf-type reduced to a single leaflet.

Unilateral. — One-sided.

Unilocular. — One-celled.

Uniovulate. — Having a single ovule

Uniserial. — In one row or series.

Unisexual. — Of one sex ; of flowers having stamens only or pistils only.

Urceolate. — Cylindrical or ovoid, but contracted at or below the open orifice, like an urn or pitcher.

Utricle. — A small bladdery usually one-seeded pericarp indehiscent or bursting irregularly or circumscissile; any small bladder-like organ, or some times applied to forms of tissue-cell

Utricular. — Consisting of or belongin to utricles.

Valvate. — Opening by valves, as a capsule ; meeting by the edges, without overlapping, as sepals, etc., in æstivation.

Valve. — The several parts of a dehiscent pericarp ; the door-like lid by which some authers open.

Variegated. — Irregularly colored.

Variety. — The principal subdivisions of a species, differing from the type in certain constant characters of subordinate value.

Veined. — Furnished with veins.

Venation. — The mode of veining.

Ventral. — Belonging to the anterior or inner face of a carpel, etc., the opposite of dorsal.

Ventricose. — Swelling unequally or inflated on one side.

Venulose. — Abounding with veinlets.

Vermicular. — Worm-shaped.

Vernal. — Appearing in spring.

Vernation. — The folding of leaves in the leaf-bud.

Vernicose. — Appearing as if varnished.

Verrucose. — Covered with wart-like elevations

Versatile. — Swinging; turning freely on its support.

Vertical. — Upright; perpendicular to the plane of the horizon; longitudinal.

Verticil. — A whorl.

Verticillaster. — The pair of dense cymes at each node of some Labiatæ, simulating a verticil or whorl.

Verticillate. — Arranged in whorls.

Vesicle. — A small bladdery body.

Vesicular. — Composed of vesicles.

Villous. — Bearing long and soft straight or straightish hairs.

Virgate. — Like a wand or rod, slender straight and erect.

Viscid. — Glutinous, sticky.

Whorl. — An arrangement of leaves, flowers, etc., in a circle about the stem or axie

Wing. — Any membranous or thin expansion or appendage ; each lateral petal of a papilionaceous flower.

Woolly. — Clothed with long and twisted or matted hairs.